Straightforward Statistics With Excel®

Second Edition

To my parents and the four most amazing guys in my life: Bill, Ted, Tim, and Kingsley.

Straightforward Statistics With Excel®

Second Edition

Chieh-Chen Bowen
Cleveland State University

SAGE

Los Angeles | London | New Delhi
Singapore | Washington DC | Melbourne

SAGE

FOR INFORMATION:

SAGE Publications, Inc.
2455 Teller Road
Thousand Oaks, California 91320
E-mail: order@sagepub.com

SAGE Publications Ltd.
1 Oliver's Yard
55 City Road
London EC1Y 1SP
United Kingdom

SAGE Publications India Pvt. Ltd.
B 1/I 1 Mohan Cooperative Industrial Area
Mathura Road, New Delhi 110 044
India

SAGE Publications Asia-Pacific Pte. Ltd.
18 Cross Street #10-10/11/12
China Square Central
Singapore 048423

Sponsoring Editor: Leah Fargotstein
Product Associate: Ivey Mellem
Production Editor: Astha Jaiswal
Copy Editor: Quads Prepress (P) Ltd.
Typesetter: C&M Digitals (P) Ltd.
Indexer: Integra
Cover Designer: Candice Harman
Marketing Manager: Victoria Velasquez

Printed in Canada

ISBN: 9781544361963

FSC www.fsc.org MIX Paper from responsible sources FSC® C103567

This book is printed on acid-free paper.

21 22 23 24 25 10 9 8 7 6 5 4 3 2 1

BRIEF CONTENTS

DETAILED CONTENTS

PREFACE

The simple nature of statistics seems intuitive and logical after a while. Once you learn a few basic rules, it is as easy as riding a bicycle. Yes, it requires study and practice to do it at first. You will find the more practice you have, the easier it gets. Statistics have become an integral part of our daily lives. Its applications can be found in marketing, finance, human resources management, data mining, clinical research, political polling, and the list goes on and on. There is no way to avoid it, so you might as well join it and master it.

In the second edition, I decided to update the book to reflect the improved functions in Excel. Excel is a very versatile and constantly improving spreadsheet tool and is available on every PC or Mac. Excel offers the ultimate flexibility and customizability. It is only limited by the users' knowledge, skills, and ability to use it. Once you learn how to program a statistical formula into Excel, it will take care of the tedious number crunching. Learning to break down the statistical formulas according to the order of operations is a great way to understand the calculation process involved in statistics without worrying about making computational mistakes. On top of that, Excel 19 of Microsoft 365 for PC or Mac includes new functions to find critical values for Z, t, F, and χ^2 and p values associated with them. These functions are nice supplements or even replacements for most of the probability-based critical value tables at the end of statistics books. Instructors can choose to use either Excel functions or printed critical value tables to fit their purposes. The tables with critical values still serve as visual aides to explain the connection between the critical values and probabilities.

Here are the highlights of the expanded Excel sections in each chapter. The Excel section in Chapter 1 provides step-by-step instructions for using Excel. It starts with opening Excel, explaining the spreadsheet structure and how to program statistical operation in it. You also learn how to set up an Excel file and how to break down statistical formulas according to order of operations. In Chapter 2, you learn how to calculate answers when the data are reported in a frequency table and how to construct a bar graph and a histogram. In Chapter 3, you learn how to calculate mean, SS, variance, and standard deviation in a frequency table. You can take a breather in Chapter 4 because there are no new Excel functions added. In Chapter 5, the **NORM.S.DIST(z,TRUE)** function is introduced and used to find the percentile rank of a z value or the probability to the left of a *z* value (i.e., $P(Z < z)$. And the inverse function **NORM.S.INV(probability)** is introduced to find the specific *z* value given the probability. In Chapter 6, Excel is used to demonstrate the Rule of Large Numbers and the **NORM.S.DIST(z,TRUE)** function continues to be used. Chapter 7 remains focused on the hypothesis test procedure, so no new Excel functions added. In Chapter 8, the **T.INV(probability,df)** function is introduced to find the critical values of *t* given a probability and an associated *df* (i.e., $P(t < t_c)$. You learn to use the built-in functions to calculate $\overline{X}$ and *s* in Excel. You also learn to use the **Data Analysis** function to conduct "**t-Test Paired Two Sample for Means.**" In Chapter 9, you learn to use the **Data Analysis** function to conduct independent-samples *t* tests by using two separate procedures. These involve, first, conducting an equality of variance test by choosing "**F-Test Two-Sample for Variances**" and, second, based on the results of the *F* test, either conducting a "**t-Test: Two-Sample Assuming Equal Variances**" or a "**t-Test: Two-Sample Assuming Unequal Variances.**" In addition, the **T.DIST.2T(x,df)** function is introduced to find the *p* value associated with the calculated two-tailed *t* value, and the **T.DIST.RT(x,df)** function is introduced to find the *p* value associated with the calculated right-tailed *t* value. Furthermore, the **F.INV.RT(**α**,df$_1$,df$_2$)** is introduced to find the critical values of the folded *F* test given the α level, df_1, and df_2, and the **F.DIST.RT(x, df$_1$,df$_2$)** is used to find the *p* value associated with the calculated *F*. In Chapter 10, you learn how to construct a scatterplot to judge whether a linear relationship exists between two variables. You can either program the Pearson's *r* formula into Excel or use the built-in function **PEARSON(A1:An,B1:Bn)** to calculate the correlation between two variables with *n* cases. In Chapter 11, you learn to use the **Data Analysis** function to conduct **Regression** with one predictor in the equation. In Chapter 12, you learn to program one-way ANOVA formulas in Excel. Also, the **T.DIST.2T(x,df)** function is used to identify *p* values for calculated two-tailed *t* values. This function requires $x > 0$, so the absolute value of the calculated *t* should be used. In Chapter 13, the **CHISQ.DIST.RT(x,df)** function is introduced to identify *p* values for the calculated χ^2. The Excel functions highlighted here can serve as a quick guide to look up different Excel functions in relevant chapters.

Simply expanding content on Excel on the second edition would make the size of the second edition unmanageable. Therefore, I have removed the content related to SPSS. There are other step-by-step SPSS user's guides that focus on teaching students how to use SPSS to run statistical procedures. The second edition of *Straightforward Statistics* keeps its focus on introductory statistics topics and fully deploys Excel as a teaching tool to enhance students' understanding of the computation process. It greatly helps comprehension to be able to break down the statistical formulas according to the order of operations. You should reserve your cognitive resources for the logic of statistical thinking and understanding the purpose of each statistical procedure instead of being overwhelmed by tedious number crunching.

The improvements of the second edition are largely due to reviewers and colleagues who took time to email me their suggestions, comments, and corrections. Thank you! I am grateful for your time and effort.

INSTRUCTOR RESOURCES

Instructors, visit **edge.sagepub.com/bowen2e** for a test bank, PowerPoint slides, additional practice problems, and more.

PREFACE TO THE FIRST EDITION

In my 26 years of teaching statistics, I have encountered many students who are anxious, confused, and frustrated in their efforts, trying to learn and understand statistics—this was such a contrast to my own experience learning statistics. I loved learning statistics, and I still love using them to extract useful information out of data. Every statistical procedure has a clear structure and a definite purpose. It all seems intuitive and logical to me. I wished I could share my love of statistics and my views on the clear structure and purpose of every statistical procedure with as many people as possible, so I decided to write a book that helps students understand introductory statistics.

I believe that learning statistics is similar to building a Lego project. Lego pieces have a lot in common, allowing them to be connected together. There are special pieces to make particularly interesting shapes, but there are always basic pieces to connect different parts together. Once you learn how to use the basic pieces and how the basic pieces connect together, you can build complicated projects by connecting different parts together. Learning statistics is similar to that. It involves clear understanding of basic terms, theories, formulas, and purposes of statistical procedures. Complicated statistical concepts are simply modifications, extensions, and combinations of basic statistical concepts. In making the journey to learn and understand statistics, the process hopefully gets easier instead of more difficult. By the time you get to more complicated statistical concepts, you should already be well versed in all the basic components. Learning how to use the basic components allows you to connect the different parts together.

Failing to recognize the common components and to make connections from one statistical topic to the next will likely contribute to the anxiety, confusion, and frustration often experienced by those who are trying to learn and understand statistics. I carefully move from one topic to the next, pointing out how the topics are similar, what you should already know, and in what crucial ways the topics are different. Such explicit connection should equip you with the necessary knowledge and assurance to feel confident in tackling a new topic. My intention is to reinforce the accumulative nature of learning statistics. As you work through chapters and gain more knowledge, moving on to the more complicated topics in later chapters should be relatively easy, and hopefully, a piece of cake.

I believe good statisticians can explain statistics in sophisticated terms, but the best statisticians can explain statistics in simple terms. Of course, sophisticated language has its time and place in statistics because it requires very precise language to express definite meanings. However, the language does not need to be sophisticated all the time. Language should be a facilitator for learning statistics instead of a hurdle. My simple and straightforward approach to learning statistics is thus to identify the common components of statistics, and how different topics are connected together, so that it provides a holistic learning experience. To help you with this learning experience, you can test your knowledge with the frequent pop quizzes you will find in all chapters. Also, at the end of the chapters, there is a trove of resources to help you further understand and succeed in mastering the material.

My goal has been to write an introductory statistics book that covers all the important topics in a plain, clear, and straightforward manner, and I hope that with your success, I achieve that goal. This book is designed to help social sciences students who are learning college-level statistics for the first time. With that in mind, the book is presented with a balance between mathematical operations and conceptual explanations. Most important, both the mathematical operations and conceptual explanations are technically accurate.

You need to be an active learner to be successful in a statistics course. Starting on the first day of the class, read the book chapter that is going to be covered in advance, pay attention to the professor's lecture, complete homework assignments on time, form study groups, and, most critically, don't skip any classes. What you get out of a statistics course depends on what you put into it. Learning statistics is a joint effort between the students and the professors. Students need to be prepared to learn, and professors need to provide a straightforward path to help the students reach their goal. This book provides a clear road map for your joint journey.

Please Note: The Odd Answers to the Learning Assessment section contains the answers to odd-numbered questions from the Learning Assessments. The answers to the even-numbered questions can be found on the Instructor Resource site here https://edge.sagepub.com/bowen2e.

ACKNOWLEDGMENTS

I would like to thank Bill Bowen for his persistent support. Thanks are also due to Leah Fargotstein for believing in this book, providing boundless support, and suggesting me to focus on Excel in this edition.

I am grateful to the many reviewers for their valuable critiques on earlier versions of this edition of book to make it better:

Gabriel Aquino, Westfield State University

Cindy A. Boyles, University of Tennessee–Martin

David Bugg, State University of New York–Potsdam

Qingwen Dong, University of the Pacific

Michael Duggan, Emerson College

Paige Gordier, Lake Superior State University

William J. Hagerty, Olivet College

John F. Hoye, Kellogg Community College

Dan Ispas, Illinois State University

Tim McDaniel, Buena Vista University

Christopher J. McLouth, University of Kentucky

Ingrid A. Nelson, Bowdoin College

Sara Peters, Newberry College

Steven S. Prevette, Southern Illinois University–Carbondale

Michele M. Wood, California State University–Fullerton

ABOUT THE AUTHOR

Chieh-Chen Bowen is an associate professor in the Department of Psychology at Cleveland State University. She received her PhD in industrial and organizational psychology and a doctoral minor in statistics from Penn State. She teaches undergraduate and graduate-level statistics, personnel psychology, job analysis, and performance management. Her research interests are examining job applicants' reactions to cybervetting, comparing cross-cultural management practices, investigating women's issues in the workplace, and validating personality questionnaires for employee selection. She also provides expert witness consulting services to attorneys who are involved in employment discrimination cases. On her days off, she works on figuring out the probability of a black-and-white cat selectively shedding only white fur on a navy blue carpet while shedding black fur on white ceramic tiles.

1

INTRODUCTION TO STATISTICS

Learning Objectives

After reading and studying this chapter, you should be able to do the following:

- Define samples and populations
- Define descriptive statistics and inferential statistics
- Define scales of measurement
- Provide two or three examples of each scale of measurement
- Identify numeric characteristics of each scale of measurement

Why is statistics important? What do you get out of it? Before you spend a semester suffering through a statistics course, the least I can do is to shed some light on these two urgent questions. Statistics is an essential process behind how people make scientific advances and sound management decisions based on data. Statistics can be applied to many fields such as research, medicine, engineering, social sciences, education, management, as well as our daily life. Having knowledge in statistics provides you with necessary tools and conceptual framework in quantitative thinking to extract useful information out of raw data. The job market for people with statistics skills is growing. Many of my former students told me that statistics knowledge and skills set them apart from their coworkers and is the reason for their steady paychecks.

Alright! Consider yourself sufficiently buttered up. Now, please endure some basic but important stuff. Statistics is a science that deals with collecting, organizing, summarizing, analyzing, and interpreting numerical data. It is especially useful when using numerical data from a small group (i.e., a sample) to make inferences about a large group (i.e., a population). Learning statistics is similar to learning a new language, and understanding the vocabulary, meaning, and structure is critical from the very beginning. Therefore, definitions, numerical characteristics, and symbols for both samples and populations are clearly presented and explained in this book. Scales of measurement are introduced to show how each scale measures and describes different numerical attributes. After reading this chapter, you should have learned how to differentiate samples from populations as well as understand specific numerical attributes of scales of measurement.

WHAT IS STATISTICS?

Statistics is a branch of mathematics dealing with the collection, analysis, interpretation, and presentation of masses of numerical data. **Data** are defined as factual information used as a basis for reasoning, discussion, or calculation. The use of statistics originated in collecting data about states or communities as administrators studied their social, political, and economic conditions in the mid- to late 17th century. Statistics are used to describe the numerical characteristics of large groups of people. Understanding statistics is useful in answering questions such as the following:

- What kind of data need to be collected?
- How many data are needed to provide sufficient guidance?
- How should we organize the data?
- How can we analyze the data and draw conclusions?
- How can we assess the strength of the conclusions?

Population Versus Sample

Learning statistics is very similar to learning a new language. Before you can master a new language, there are basic vocabularies you need to memorize and understand. The basic vocabulary of statistics starts with terms describing the numerical characteristics of a **population** and numerical characteristics of a **sample**. A population is defined as the entire collection of everyone or everything that researchers are interested in measuring or studying. A sample is defined as a subset of the population, from which measures are actually obtained. For example, a Midwest urban university wants to gauge students' interest in the possibility of establishing a football team on campus. The population, in this example, is the number of students in the entire student body, $N = 17{,}894$. N, uppercase, refers to the size of the population; a lowercase n is used for sample size. It would be extremely time-consuming and very expensive to ask every student in the university his or her interest in having a university football team. No matter how hard you tried, it was highly likely that some people would simply refuse to answer survey questions.

To obtain information efficiently, the university administration assigned the task to the Student Government Association. This association put a survey on the university website and collected responses from 274 students. This group of 274 students was the sample from where measures were actually obtained. The answers from these 274 students formed the data. Data are required to run statistical analyses. Collecting and sorting data usually happen before calculating statistics. Both a population and a sample contain numerical attributes that researchers want to investigate. The numerical attributes of a population are called **parameters**, and they are usually denoted by Greek letters. Numerical attributes of a sample are called **sample statistics**, and they are usually denoted by ordinary English letters. Table 1.1 shows the basic vocabulary to describe populations and samples. It is important to master the basic vocabulary, so as to make effective communication and clear understanding of statistical concepts achievable. In this chapter, we cover the definitions of some basic population parameters and sample statistics. The corresponding mathematical formulas will be covered in the following chapters.

Descriptive Statistics and Inferential Statistics

Statistical procedures can be generally classified into two categories: (1) *descriptive statistics* and (2) *inferential statistics*. **Descriptive statistics** are statistical procedures used to describe, summarize, organize, and simplify relevant characteristics of data. They are simply ways to describe and understand the data, but they do not go beyond the data. Mean, standard deviation, and variance are examples of descriptive statistics.

TABLE 1.1 ● The Basic Vocabularies to Describe Numerical Characteristics of a Population and a Sample

Meaning of the Symbol	Population Parameters	Sample Statistics
Mean	μ (pronounced as 'mew')	$\bar{X}$
Standard deviation	σ (pronounced as 'sigma')	s
Variance	σ^2	s^2
Size, number of observations	N	n

Inferential statistics are statistical procedures that use sample statistics to generalize to or make inferences about a population. As a rule, inferential statistics are more complicated than descriptive statistics, and they will be introduced in Chapter 7 of this book. Such statistical procedures usually require samples to be unbiased representatives of the population. This can be achieved if samples are randomly selected from the population. As there are different ways to slice a pie, there are different ways to sample a population.

Sampling a Population

Let's start this section with examples of a population. For instance, all the cars built by an automobile manufacturer, the entire student body in a university, and every citizen in a particular country are all examples of a population. As you can imagine, the number of members in a population can get enormous and become impossible to measure or study. Sometimes, it is simply not feasible to use the population for research purposes. You understand why an automobile manufacturer does not use every car they make to conduct the crash safety test. There will only be crashed cars left! Therefore, scientists have to conduct their research with a carefully selected sample. For example, a small number of cars used to conduct a crash safety test, an online survey administrated to evaluate students' preference of living on campus versus off campus, public polling on a politician's popularity, and a survey on the most memorable advertisement during a Super Bowl: All of these are examples of conducting research using samples. The reason to use a sample to conduct research instead of using the population is that it is neither possible nor feasible to gain access to every member in the population and have everyone consent to participate in the research. The most obvious difference between a population and a sample is its size. Many samples can be selected from a population.

Two basic ways of sampling a population are nonprobability and probability sampling—examples are *convenience samples* and *random samples*, respectively. A **convenience sample** is one in which researchers use anyone who is willing to participate in the study. A convenience sample is created based on easy accessibility. For example, a student completes an assignment on measuring people's perceived job satisfaction by begging her Facebook friends to fill out a job satisfaction questionnaire. Various talent shows (e.g., *American Idol, The Voice, Dancing With the Stars, America's Got Talent*, etc.) on television ask the audience to vote for their favorite contestants, so they can move on to the next level of the competition. A convenience sample is a nonprobability sample because not all members in the population have an equal chance of being included in the sample. Well, you know that not everyone uses Facebook and not everyone watches talent shows.

A **random sample** is an ideal way to select participants for scientific research. A random sample is defined as being one in which every member in the population has an equal chance of being selected. Because of the random nature of the selection process, it creates an unbiased representation of the population. However, this ideal is easier said than done.

Can you think of an example of a random sample? This is a question I pose to my students in Introductory Statistics classes. An overwhelming majority of students' answers actually fall in the category of convenience samples, such as exit polling during an election, spot surveying in a shopping mall/library/cafeteria, using students in a class, and so on.

Random sampling does not just happen. It actually requires thoughtful planning and careful execution. How do we select a random sample? I'm glad you asked!

POP QUIZ

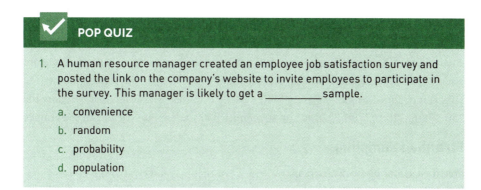

1. A human resource manager created an employee job satisfaction survey and posted the link on the company's website to invite employees to participate in the survey. This manager is likely to get a _________ sample.
   a. convenience
   b. random
   c. probability
   d. population

RANDOM SAMPLING METHODS

There are four commonly used methods to select a random sample: (1) simple random sampling, (2) systematic sampling, (3) stratified sampling, and (4) cluster sampling. The definition of each method and a practical example of each method will be provided in the following subsections.

Simple Random Sampling

A **simple random sample** of a sample size n is created in a way that all samples with the same sample size have the same chance of being selected. A simple random sample has a stronger requirement than a random sample. Each individual is chosen randomly, entirely by chance, and each subset of n individuals has the same probability of being chosen as any other subset of n individuals. For instance, a university was considering switching courses from four credit hours to three credit hours and the university administration sought students' opinions before making such changes. The targeted population is the entire student body in the university (N = 17,500). The administration would like to select 2% of the students to participate in a survey (n = 350). A simple random sample can be obtained by listing all students' names and giving each a unique identification number ranging from 1 to 17,500. A computer can then be used to randomly generate 350 numbers between 1 and 17,500. This creates a sample with n = 350. Such a procedure can be repeated many times to create many samples. Any one of these samples with $n = 350$ has an equal chance of being selected to participate in the survey.

Systematic Sampling

A **systematic sample** is obtained by selecting a sample from a population using a random starting point and a fixed interval. Typically, every "*k*th" member is selected from the population to be included in the sample. Systematic sampling is still thought of as being random, as long as the interval is determined beforehand and the starting point is selected at random. To choose a 2% systematic sample in our previous example, the university needs to randomly select a number between 1 and 50 as its starting point and then choose every 50th number to be included in the sample. For instance, to create a random sample with 350 students, Number 28 was randomly chosen as the starting point. The sample is obtained by picking every 50th student after that—meaning that we chose the 28th, 78th, 128th, 178th, 228th, and so on until 350 students are selected in the sample.

Stratified Sampling

Stratified sampling works particularly well when there are large variations in the population characteristics. Stratification is the process of grouping members of the population into relatively homogeneous subgroups before sampling. These homogeneous subgroups are called "strata." The strata need to be mutually exclusive: Every member in the population must be assigned to only one stratum. The strata should also be collectively exhaustive—meaning that no member in the population is excluded. A random sample from each stratum is independently taken in the same proportion as the stratum's size relative to the population. These subsets of the strata are then pooled to form a random sample. Therefore, the distribution of the key characteristics is the same as that in the population. For example, there are 60% female students and 40% male students in the university; to create a sample size of 350, a stratified sampling will randomly choose $60\% \times 350 = 210$ female students and $40\% \times 350 = 140$ male students.

Cluster Sampling

Cluster sampling works best when "natural" grouping (clustering) occurs in the population. Random sampling is conducted to select particular clusters to include in the sample. Once a cluster is selected in the sample, all individuals in the cluster are included in the sample. In the previous university example, there are, say, 900 courses offered in any given semester. A list of all 900 classes could be obtained from the registrar's office. A random sampling procedure is conducted to select 2% of classes to be included in the sample ($n = 18$ classes). Once the 18 classes are selected, every student in each of these classes is included in the sample. In this particular example, students need to be reminded not to answer the survey multiple times if more than one of the classes in which they are enrolled get selected in the sample.

Although all four random sampling methods are designed to obtain random samples in the university's effort to seek opinions from students, it is important to recognize that not everyone who is selected in the sample completes the survey. A low response rate ruins the nature of probability sampling, especially when there are systematic differences between students who take time to answer the survey versus students who either neglect or refuse to answer the survey.

POP QUIZ

2. When a large school district randomly selected three classes to conduct a learning environment study, every student in these three classes participated in a face-to-face interview with the school psychologist. This is an example of __________
   a. simple random sampling.
   b. systematic sampling.
   c. stratified sampling.
   d. cluster sampling.

SCALES OF MEASUREMENT

Researchers study physical or psychological characteristics by measuring them or asking questions about the attributes. In statistics, a **variable** refers to a measurable attribute. These measures have different values from one person to another, or the values change over time. Different values of a variable provide information for researchers. If there is no variation in the measures, the variable does not contain any information. For example, when I studied the variables that might be related to students' performance in a statistics course, I collected answers to the following questions: student's grade point average (GPA), number of mathematical courses taken at the college level, number of classes missed in the statistics course, number of tutoring sessions attended for the statistics course, and gender. The answers to these questions reflected the attributes of the students, and they usually varied from one individual to the next. I did not have to ask, "Are you currently enrolled in a statistics course?" Enrolling in a statistics course is an unnecessary question because only students who enrolled in a statistics course were selected for this study. A question producing identical answers from everyone provides no information at all. Only when answers to questions change from one person to the next do they provide information that we do not have before the questions are answered. Variables have different numerical values that can be analyzed, and hopefully, meaningful information can be extracted from them.

Scales of measurement provide a way to think systematically about the numerical characteristics of variables. **Scales of measurement** specifically describe how variables are defined and measured. Each scale of measurement has certain mathematical properties that determine appropriate applications of statistical procedures. There are four scales of measurement: (1) *nominal*, (2) *ordinal*, (3) *interval*, and (4) *ratio*. I will discuss them in the order from the simplest to the most complex. Remember that a higher level scale of measurement contains all the mathematical properties from a lower level scale of measurement plus something more.

Nominal Scale

In **nominal scales**, measurements are used strictly as identifiers, such as your student identification number, phone number, or social security number. The numbers on athletes' jerseys, for example, are simply used as identifiers. Among other things, jersey numbers allow referees to identify which player just committed a personal foul so a penalty can be properly assessed. In social sciences and behavioral sciences, nominal variables such as gender, race, religion, socioeconomic status, marital status, occupation, sexual orientation, and so on are often included in research. Nominal scaled variables allow us to figure out whether measurements are the same or different. The mathematical property of the nominal scale is simply $A = B$ or $A \neq B$.

Ordinal Scale

In **ordinal scales**, measurements not only are used as identifiers but also carry information about ordering in a particular sequence. The numbers are ranked or sorted in an orderly manner such as from the lowest to the highest or from the highest to the lowest. For example, the first time you are invited to a friend's house for dinner, at the outset, you need to find the house. Luckily, you have the address. You notice that houses on one side of the street have odd numbers, and houses on the other side have even numbers. The house numbers either increase or decrease as you walk down the street. Yes! House numbers are arranged in order, so they are ordinal. Therefore, you know 2550 is located in between 2500 and 2600. Can you imagine how confusing it would be trying to find a house for the first time if house numbers were nominal and displayed in random order? Another example is that at a swimming meet, gold, silver, and bronze medals are awarded to swimmers who finish in first, second, and third place. Although we don't know the time difference between the gold medalist and the silver medalist or the time difference between the silver medalist and the bronze medalist, we are sure that the gold medalist is faster than the silver medalist, and the silver medalist is faster than the bronze medalist.

Likert scales are often used to measure people's opinions, attitudes, or preferences. Likert scales measure attributes along a continuum of choices such as 1 = *strongly disagree*, 2 = *somewhat disagree*, 3 = *neutral*, 4 = *somewhat agree*, or 5 = *strongly agree* with each statement. The ratings from Likert scales belong in the category of ordinal scales. We know the rankings of the responses, but we are not sure whether the difference between 1 and 2 is the same as the difference between 2 and 3. For example, Melissa rates her satisfaction with an online purchase she made a week ago: 1 stands for *very unhappy*, and 5 stands for *very happy*. Melissa's answers are reflected by the bolded numbers in the table.

Characteristic	Very Unhappy	Somewhat Unhappy	Neutral	Somewhat Happy	Very Happy
Quality of product	1	2	3	4	**5**
Delivery time	1	2	3	**4**	5
Competitive price	1	2	**3**	4	5
Customer service	1	2	3	**4**	5

In this particular case, we learn that Melissa is *very happy* with the product quality, *somewhat happy* with the delivery time and customer service, and *neutral* on the pricing. But we don't know how much happier Melissa is with the product quality than with its price. It is not possible to quantify the difference between two numbers from an ordinal scale.

In summary, order contains information. Order gives us the direction of the rankings between two values. The mathematical properties of ordinal scales include everything a nominal scale has (i.e., identifiers) and something more: the direction of the rankings. The mathematical property of the ordinal scale is expressed as if $A > B$ and $B > C$, $A > C$.

Interval Scale

As measurements evolve to be more sophisticated, scientists go beyond merely figuring out the direction of the rankings. Being able to calculate the amount of the difference becomes the primary objective. The interval scale is designed to fulfill that particular objective. **Interval scales** can be used as identifiers, showing the direction of the rankings and something more: equal units. Interval scales not only arrange observations according to their magnitudes but also distinguish the ordered arrangement in equal units. When measurements come with equal units, they allow us to calculate the amount of difference or the distance between two measurements. Fahrenheit and Celsius temperature scales are two of the most commonly mentioned examples of interval scales, for which equal units exist and zero is an arbitrarily assigned measurement. We know that 0 degrees Celsius equals 32 degrees Fahrenheit. Zero in an interval scale simply represents a measurement.

Equal units have very important implications in statistics. Calculation of means and standard deviations, which are introduced in Chapter 3, require the variables to have equal units. Many measures in social science are interval scales, such as age, test scores, or standardized intelligence quotient (IQ) scores.

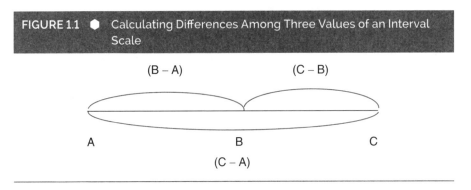

FIGURE 1.1 ● Calculating Differences Among Three Values of an Interval Scale

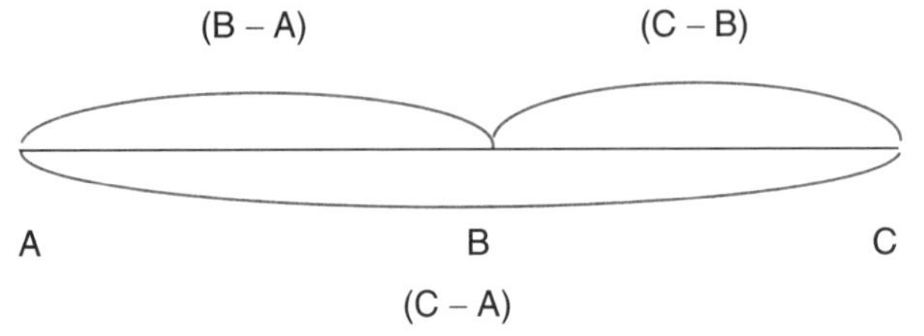

Interval scales allow us to do mathematical operations on the amount of difference between two values. The mathematical property of the interval scale is expressed as when A < B < C, the difference between A and C is the sum of the difference between B and A plus the difference between C and B, (C – A) = (B – A) + (C – B), as shown in Figure 1.1.

For example, age is an interval variable. The ages of three siblings in a family are 8, 13, and 17. The age difference between the youngest and the oldest is 17 – 8 = 9, which equals the sum of the differences between the oldest and the middle one, 17 – 13 = 4, plus the difference between the middle one and the youngest, 13 – 8 = 5.

Ratio Scale

The ratio scale is the most sophisticated scale of measurement. The **ratio scale** contains everything that lower level scales of measurement have, such as identifiers, direction of rankings, and equal units, but it also has something more: an **absolute zero**. Absolute zero means a complete absence of the attribute that you are measuring. Zero is not an arbitrarily assigned number. Most physical measurements, such as height and weight, belong in the ratio scales. Income is a good example of a ratio scale in the social sciences. Zero income means a complete absence of income.

Absolute zero makes it possible to measure the absolute amount of the variable, and it allows us to compare measures in terms of ratios. This unique mathematical property of ratio scales is expressed as if A = 2B and B = 2C, then, A = 4C.

After discussing scales of measurement, sometimes students have questions about why exam scores and IQ scores are classified as interval scales instead of ratio scales. Let's discuss the reasons why exam scores and IQ scores are classified as interval scales. The purpose of an exam is to assess your knowledge in a specific topic. The purpose of an IQ test is to measure and quantify intelligence. When an exam or IQ score turns out to be zero, it does not mean that the person completely lacks knowledge or intelligence. There might be other reasons to explain the zero, such as the test is in a language foreign to the test taker, the test taker is provided a wrong scantron for the test, or the test taker marks the answers on the wrong side of the scantron. The point is that exam or IQ scores do not have an absolute zero to show complete absence of the attribute; therefore, they belong in the category of interval scales.

It is useful to summarize all four scales of measurement with their mathematical properties in Table 1.2. This table also illustrates that the higher level scales of measurement have every attribute that the lower level scales have, plus something more. It is very common to label variables with nominal or ordinal scales of measurement as categorical variables and variables with interval or ratio scales of measurement as continuous or numerical variables.

TABLE 1.2 ● Mathematical Properties of Scales of Measurement

Scale of Measurement	Identifier	Direction of Ranking	Equal Units	Absolute Zero
Nominal	X			
Ordinal	X	X		
Interval	X	X	X	
Ratio	X	X	X	X

POP QUIZ

3. Travel speed is measured by the distance traveled divided by the time traveled. This is an example of a(n) __________ scale of measurement.
   a. nominal
   b. ordinal
   c. interval
   d. ratio
4. The answers to a marital status question are usually married, single, divorced, or cohabiting. This is an example of a(n) __________ scale of measurement.
   a. nominal
   b. ordinal
   c. interval
   d. ratio

VARIABLE CLASSIFICATIONS

There are other ways to classify variables. Some are related to the scales of measurement, but others are related to the research designs. We will cover the essential ones.

Discrete Versus Continuous Variables

Discrete variables are made of values that have clear separation from one number to the next. The answers for discrete variables can only be integers (i.e., whole numbers). For example, how many cars do you own? The answer can simply be counted as the number of cars that are registered under your name. The answers are likely to be 0, 1, 2, 3, and so on. You can't own 2.56 cars. The same goes with the answer to "How many pets do you have?" I hope, for your sake, you don't own 1.25 cats. Clearly, nominal and ordinal scales can also be classified as discrete variables.

Continuous variables are composed of values that do not have clear separation from one number to the next. There are numerous possible answers between two adjacent integers for continuous variables. Continuous variables are usually expressed with decimals or fractions. For example, the current world record in men's 50-meter freestyle was 20.16 seconds. Official swim time is measured at 100th of a second. Physical measures such as height and weight are similarly continuous variables. As a person who is "vertically challenged" (a politically correct way to say "short"), it is very important for me to list my height with as many numbers after the decimal point as possible given the limits of the preciseness of the ruler. Interval and ratio scales can also be classified as continuous variables.

Independent Variables Versus Dependent Variables

The distinction between independent variables and dependent variables is very important in experimental research. Generally speaking, there are three different types of empirical research: (1) *experimental*, (2) *quasi-experimental*, and (3) *nonexperimental*. **Experimental research** is usually conducted in a tightly controlled environment (i.e., research laboratories). Researchers deliberately choose variables to manipulate at different levels so as to investigate their effects on the variables that researchers are really interested in. The variables that are deliberately manipulated by researchers are called **independent variables**, predictor variables, or explanatory variables. **Dependent variables** are what researchers are really interested in studying. Dependent variables are also called criterion variables or response variables. Measuring and/or observing the changes in dependent variables as a result of the deliberate manipulation of the independent variables is the foundation of experimental research.

For example, researchers studied the effect of caloric consumption on the longevity of rats. Researchers selected a sample of 20 pairs of newborn rats in a lab. Each pair of rats was from the same mother. The researchers randomly assigned one in each pair to the control group and the other to the experimental group. The control group was fed a normal amount of calories every day. The experimental group was fed only 70% of the normal amount every day. At the end, researchers compared the life span of rats in the control group versus that of the experimental group. In this experiment, the rats' living environment was strictly controlled and kept constant by the researchers in terms of light, temperature, moisture, and sound. The independent variable was the amount of calories provided every day, which was deliberately manipulated by the researchers to set at 100% for the control group versus 70% for the experimental group. The dependent variable was the life span of the rats, which was the focus of this study. Researchers were really interested in finding out how variation in the amount of calories affected rats' life span by comparing rats in the control group with those in the experimental group.

In summary, experimental research is scientific research that has achieved three important features: (1) control to keep everything else constant, (2) manipulation of the independent variable, and (3) random assignment of participants to different conditions.

"Quasi" means "almost but not really." **Quasi-experimental research** has some but not all of the features of experimental research. More specifically, if one or more important features are not feasible but others remain intact, the research becomes quasi-experimental. For example, a school teacher investigated the effects of two instruction methods on sixth graders' science proficiency. One class was assigned to use inquiry through which students were supposed to find answers to their own questions while the instructor helped them design ways to do so. The other class was assigned to use standard lectures to teach science. Let's compare this study with the calories on rats' longevity study. The school study occurred in a school setting, not a research laboratory. These two groups of students were exposed to the particular instruction methods only during the science course instead of 24 hours a day. Both studies had manipulations on the independent variables. However, in the school study, it was not feasible to randomly assign every student to either the student-centered inquiry or the standard lecture group. This was done by assigning entire classes to either one of the conditions. Thus, it was clear that some of the control, manipulation, and random assignments were not achieved in the school study, but others remained intact; therefore, the school study was quasi-experimental. The independent variable in this research was the instruction methods, and the dependent variable was sixth graders' science proficiency scores.

Nonexperimental research is composed of studies in which none of the control, manipulation, or random assignment is attempted. Nonexperimental research happens in a natural environment where the research participants' behaviors, thoughts, opinions, interests, or preferences are observed and recorded. Surveys and public opinion polling belong in the category of nonexperimental research, and so do observational studies and cross-sectional studies. For example, an executive of a company was curious to find out whether its employees were happy with their work conditions. To satisfy this curiosity, an industrial and organizational psychologist was hired to design an employee job satisfaction survey to investigate how employees felt about their work conditions without any attempt to control, manipulate, or randomly assign employees to any particular condition. There was no real distinction between independent variables and dependent variables. Nonexperimental research is about observing and studying research participants in their natural settings without deliberately controlling the environment or manipulating their behavior or preferences.

There is no one single best way to conduct an empirical study. The selection of research design depends on the purpose of the research. If the identification of a causal relationship is desired, experimental research is more likely to achieve the purpose. The *extraneous variables* that exist with quasi-experimental and nonexperimental research designs will prevent a causal relationship being asserted as a result of the research. The **extraneous variables** are variables that are not included in the study but might have an impact on the relationship between variables in the study. Extraneous variables can be attributable to a lack of strict control of environmental variables or individual differences that research participants bring along with them. On the other hand, if participants' behavior in their natural habitat is the focus of the research, nonexperimental research is the right choice.

POP QUIZ

A company explored the effect of its compensation (pay) structure on job performance of salespeople. The company randomly chose three stores to conduct the study. Every salesperson in Store A was compensated with salary, every salesperson in Store B was compensated by sales commissions, and every salesperson in Store C was compensated by a low base salary plus an additional sales-based bonus. Six months later, job performance across all three stores was measured and reported.

5. What is the independent variable of this study?
6. What is the dependent variable of this study?
7. What type of research design is used in this study?

REQUIRED MATHEMATICAL SKILLS FOR THIS COURSE

What kinds of mathematical knowledge and skills are required for this statistics course? The answers to this question will help you go through this course smoothly and successfully. First, you need to master the order of operations. A statistical formula usually contains multiple parts of different math operations such as addition, subtraction, multiplication, division, exponentiation, and parentheses. Knowing which part of a mathematical operation to do first is the key to arrive at the correct answers. There is only one correct answer and an unlimited number of wrong answers for a statistical question. Knowing the order of operations provides a standard way to simplify and find the correct answer. There are several priority levels in order of operations shown in Table 1.3. The principle is to perform higher priority operations before moving on to lower priority ones. Many middle school math teachers use the phrase "please excuse my dear Aunt Sally" to help students memorize the order of operations: P-E-MD-AS. When dealing with many operations at the same level of priority, simplify the operations in the order they appear from left to right. Multiplication and division are at the same level of priority. Addition and subtraction are at the same level of priority. The strict rule of order of operations provides a standard way and the only correct way for everyone to interpret and solve statistical formulas containing complex operations. Therefore, when faced with the operations $8/2 \times 3 - 7$, you will be able to come up with the correct answer (i.e., 5). Order of operations is a fundamental math principle that needs to be strictly followed at all times. If you come up with wrong answers in your statistical exams, quizzes, or homework assignments, then chances are that you have messed up the order of operations somewhere. If I get a quarter every time any of my students makes a mistake on the order of operations, I will be a rich woman.

Second, learn to use your calculator correctly. When you don't use your calculator correctly, you will get wrong answers and lose points on homework, quizzes, and exams. Here is a simple check. Use your calculator to find the square of −3. Now, I will *pause* for your answer. Tell me the answer first before you move on to the next sentence.

If your answer is −9, you need figure out how to use your calculator correctly. If your answer is 9, you have passed the calculator test.

TABLE 1.3 ● Priority Levels in Order of Operations

Priority Level: Perform Higher Priority Operations First Before Moving on to Lower Priority Ones	Math Operations	Example
First priority	Parentheses: Simplify whatever is inside the parentheses before doing anything else	$(X-1)$, (X^2-1) or $(3-1) \times (4+2)$
Second priority	Exponentiation	X^2, $\sqrt{x}$ or 8^2
Third priority	Multiplication or division	XY, $\frac{X}{Y}$ or $2 \times 4/5$
Fourth priority	Addition or subtraction	$X_1 + X_2 + X_3$, $n-1$ or $5-7+8$

Third, refresh your knowledge of basic algebra. Sometimes you need to know the basic algebra to solve unknowns in the equations. Solving equations means figuring out a solution set for the equations. A very basic algebraic knowledge is that one linear equation can solve one unknown, two linear equations can solve two unknowns, and so on.

Statistical Notation

It is very common to add a set of values in statistics. To make it easier to communicate such a computation, a special notation is used to refer to the sum of a set of values. The uppercase Greek letter Σ (pronounced sigma) is used for summation. Here is our first summation operation:

$$\sum_{i=1}^{n} X_i$$

This summation operation means to calculate the sum of all values of X, starting at the first value, X_1, and ending with the last value, X_n.

In a summation operation, there are several components. They are clearly explained one by one in the following list.

1. The summation sign, Σ, means performing addition in this operation.
2. X is the variable being added.
3. The subscript i, which is placed next to X, indicates that there are i values of X.
4. There is a starting point and an ending point of the index i. This tells us how many values need to be added. Usually the starting point is $i = 1$, which is placed under the Σ, and the ending point is n, which is placed on top of the Σ.

$$\sum_{i=1}^{n} X_i = X_1 + X_2 + X_3 + \cdots + X_n$$

Sometimes, for the sake of simplicity, summation is expressed as ΣX.

$$\Sigma X = X_1 + X_2 + X_3 + \cdots + X_n$$

From now on this simple form of Σ will express summation. You have to do summation from the first value to the last value given to you in a problem statement. Let's go through several examples of the summation question to get you used to the simple form of Σ.

EXAMPLE 1.1

$X = 3, 5, 7,$ and 9. Find $\sum X$, $(\sum X)^2$, and $\sum X^2$.

There are four values in the X variable, and this problem has three sub-questions (1) $\sum X$, (2) $(\sum X)^2$, and (3) $\sum X^2$. You need to figure out the correct order of operations in each sub-question.

The first sub-question is $\sum X$, so there is only one operation involved.

$$\sum X = 3 + 5 + 7 + 9 = 24$$

The second sub-question $(\sum X)^2$ involves three operations: parentheses, addition, and exponentiation. Let's break it down. According to the order of operations, solving inside the parentheses is the first priority and there is an addition $(\sum X)$ inside the parentheses. Then you have to square the answer of $\sum X$. Therefore, you need to figure out $\sum X$ first.

$$(\sum X)^2 = (3 + 5 + 7 + 9)^2 = (24)^2 = 576$$

The third sub-question $\sum X^2$ involves two operations: exponentiation and addition. Order of operations mandates that the exponentiation has to be done before the addition. Therefore, you figure out the value of each X^2 and then add them up.

$$\sum X^2 = 3^2 + 5^2 + 7^2 + 9^2 = 9 + 25 + 49 + 81 = 164$$

EXAMPLE 1.2

$X = 2, 4, -6,$ and -9. Find $\sum X$, $(\sum X - 7)^2$, and $\sum X^2$.

There are four values in the X variable, and this problem has three sub-questions: (1) $\sum X$, (2) $(\sum X - 7)^2$, and (3) $\sum X^2$. The difference between Examples 1.1 and 1.2 is that some of the values are negative. Keep the negative sign consistent as stated in the problem.

The first sub-question is $\sum X$, so there is only one operation involved.

$$\sum X = 2 + 4 + (-6) + (-9) = -9$$

The second sub-question $(\sum X - 7)^2$ involves four different operations: parentheses with both addition and subtraction inside of it then exponentiation. Order of operations tells us to simplify whatever is inside the parentheses first. Therefore, you need to figure out $\sum X$ first then do the subtraction inside the parentheses before conducting the exponentiation.

$$(\sum X - 7)^2 = (-9 - 7)^2 = (-16)^2 = 256$$

The third sub-question $\sum X^2$ involves in two operations: exponentiation and addition. Order of operations mandates that the exponentiation has to be done before the addition. Therefore, you figure out the value of each X^2 and then add them up.

$$\sum X^2 = 2^2 + 4^2 + (-6)^2 + (-9)^2 = 4 + 16 + 36 + 81 = 137$$

You have just witnessed the order of operations and statistical notation of $\sum$ in action. It is obvious that the correct order of operations must be followed to arrive at the correct answer. There will be many statistical formulas involving different math operations, so make sure you have a solid understanding on solving these types of questions. There will be more practices like this in the Exercise Problems.

POP QUIZ

$X = 3, 6, 9, 12$, and 15.

8. Find $\sum(X - 9)^2$.
9. $\sum X^2 - 9$.

EXERCISE PROBLEMS

1. Researchers studied the effect of driving while texting on driving mistakes. College students were recruited to participate in the study. Due to risk of actual driving on the road, researchers conducted the research in a lab with a driving simulator. Students were randomly assigned to one of the two groups: (1) "driving without texting" or (2) "driving while texting." Driving mistakes were recorded by the simulator, which include driving more than 10 miles above or below the speed limit and failing to stay within the lane. What type of research was this study? What was the independent variable in this study? What was the dependent variable in this study?
2. Researchers investigated the gender gap in pay for physicians. An online salary survey was sent out to the members of the American Medical Association with an electronic link. What type of research was this study? What kind of sample was likely to be obtained by this online survey?
3. Where $X = 1, 2, 3$, and 4, and $Y = 2, 4, 6$, and 8, compute $\sum(X - 3)(Y - 5)$ and $\sum X^2Y^2$.

Solutions

1. The research is conducted in lab with a driving simulator. There are two conditions that are manipulated by the researchers. Research participants are randomly assigned to one of the two conditions. All three important features, namely (1) control, (2) manipulation, and (3) random assignment, are achieved in this study. Therefore, this study is an experimental research design. The independent variable measures the two driving conditions: (1) "driving without texting" and (2) "driving while texting." The dependent variable is the driving mistakes.

2. An online survey does not attempt to control, manipulate, or randomly assign participants to different conditions. Therefore, it is a nonexperimental design. Not all physicians belong to the American Medical Association. Therefore, sending a survey out to its members can't reach a random sample of physicians. The survey was voluntarily answered by people who received the link. Some physicians would simply ignore the email. Therefore, it was a convenience sample.

3. It is easy to create a table of X and Y to figure what need to be computed to solve the problems.

   $X = 1, 2, 3,$ and 4; $Y = 2, 4, 6,$ and 8; compute $\sum (X - 3)(Y - 5)$ and $\sum X^2Y^2$

You may solve this problem by using your calculator. I am going to show you how to use Excel to solve computation problems. Excel is a great tool to construct such a table to conduct mathematical operations. All instructions regarding Excel use Microsoft 365 Excel 2019. Most programming of mathematical operations is identical across different versions of Excel such as 2019, 2016, 2013, and 2010. Excel allows you to program mathematical operation formulas in each column. It is a great skill to have to be able to program math operations in Excel. Let's start by entering X values in column **A** and Y values in column **B**. Label **C1** as (X–3). Move the cursor to **C2** (column C, row 2), and type the formula that you want to create. In Excel, a function starts with a "=" which is a command to calculate. Excel is based on location of the variable. Therefore, to create the first operation of $(X - 3)$, grab the first value of X located in **A2** (column A, row 2) and subtract 3, in Excel language "**=A2–3**". The beauty of Excel is that once you create a formula for the first value of the variables involved, you may simply copy and paste the formula for the rest of the values in the sequence. Copy and paste can be done by moving the cursor to the lower right-hand corner of **C2** where the formula is successfully created, until a solid + shows up. Hold the left click on the mouse and drag to the last value of the variable, and then let go of the left click. You should see that the formula is copied with all cell references. Next, do the same procedure to create $(Y - 5)$ by labeling **D1** as (Y–5). Move the cursor to **D2**, and type "**=B2–5**", then hit **Enter**. Move the cursor to the lower right-hand corner of **D2** where the formula is successfully created, until a solid + shows up. Hold the left click on the mouse and drag to the last value of the Y, and then let go of the left click. According to the question, you need to find $\sum X - 3)(Y - 5)$, so you need to create $(X - 3)(Y - 5)$. To create $(X - 3)(Y - 5)$, label **E1** as (X–3)(Y–5), simply type "**=C2*D2**" in **E2**, and then hit **Enter**; the answer 6 appears in **E2** as shown in Figure 1.2.

FIGURE 1.2 ● Programming $(X-3)(Y-5)$ in Excel

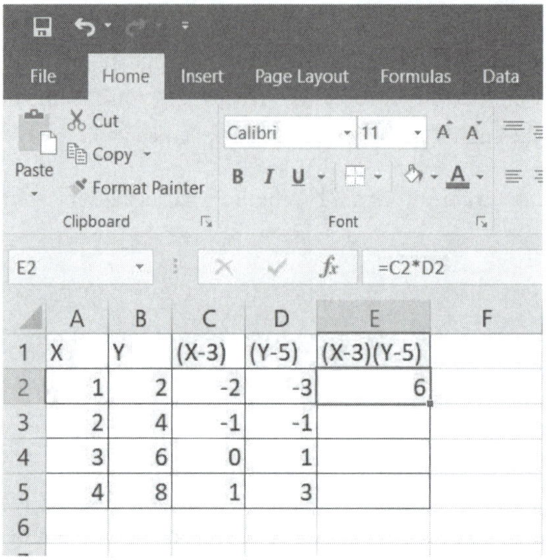

Perform the same copy and paste for the rest of values for $(X-3)(Y-5)$ to the last value in the sequence. Now move the cursor to **E6** where you want the answer of Σ(X–3)(Y–5) to show up typing "**=SUM(E2:E5)**", then hit **Enter**. The answer for Σ **(X–3)(Y–5)** appears in **E6** as shown in Figure 1.3.

FIGURE 1.3 ● Programming $\sum(X-3)(Y-5)$ in EXCEL

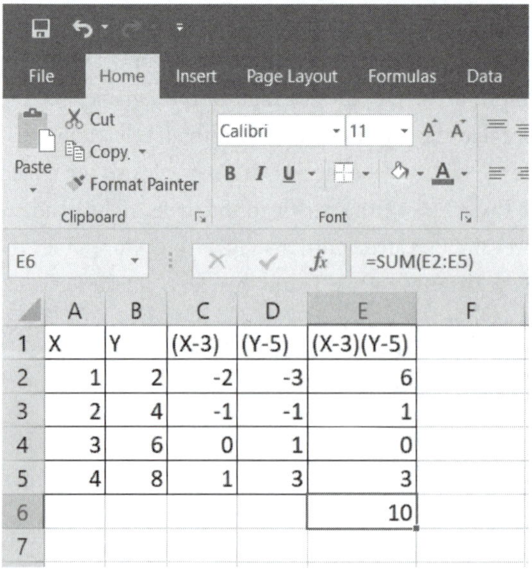

Next, you need to figure out how to perform necessary operations to solve $\sum X^2Y^2$ in Excel. Let's break it down. $\sum X^2Y^2$ involves exponentiation, multiplication, and addition. Order of operations dictates that exponentiation needs to be done first, next multiplication, and then addition. Each step is shown in a column of the table in Figures 1.4. In **F1**, label it as X^2. Excel calculates the square by using ^2. You may find the ^ symbol by holding shift and 6 simultaneously on your keyboard. In **F2**, type "**=A2^2**", which means grab the first value of X and square it. Hit **Enter** and the answer 1 appears in **F2**. Move the cursor to the lower right-hand corner of **F2** where the formula is successfully created, until a solid + shows up. Hold the left click on the mouse and drag to the last row of the variable, and then let go of the left click. You should see that the formula is copied with all cell references. The formula's operations are performed on the first row of the variable all the way to the last row of the variable. In **G1**, label it as Y^2. In **G2**, type "**=B2^2**", which means grab the first value of Y and square it. Copy and paste the formula to the last value of Y as shown in Figure 1.4.

FIGURE 1.4 ● Programming X^2 and Y^2 in Excel

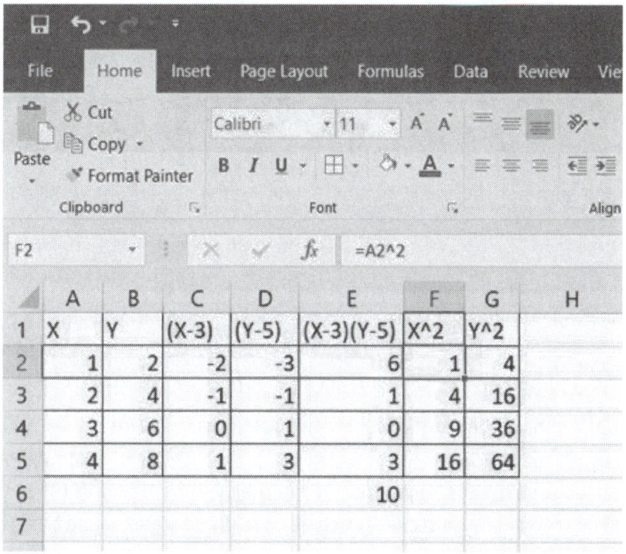

Now, the next step to solve this question is to create a formula for the multiplication of X^2Y^2. Label **H1** as X^2*Y^2. Move the cursor to **H2**. Type "**=F2*G2**", which means grab the first values of X^2 and Y^2 and multiply them. Copy and paste the function to the rest of the variables. Move the cursor to **F6** and conduct a summation by typing in "**=SUM(H2:H5)**". Hit **Enter** and the answer 1,416 appears as seen in Figure 1.5.

FIGURE 1.5 ● Programming $\sum X^2Y^2$ in Excel

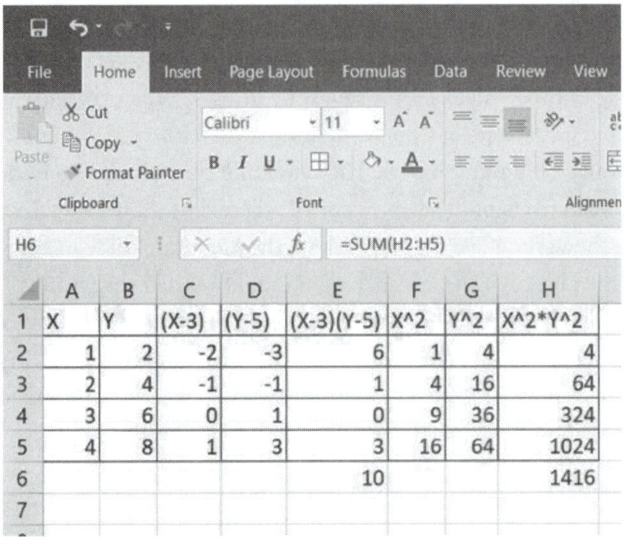

	A	B	C	D	E	F	G	H
1	X	Y	(X-3)	(Y-5)	(X-3)(Y-5)	X^2	Y^2	X^2*Y^2
2	1	2	-2	-3	6	1	4	4
3	2	4	-1	-1	1	4	16	64
4	3	6	0	1	0	9	36	324
5	4	8	1	3	3	16	64	1024
6					10			1416
7								

What You Learned

In this chapter, you have learned some of the basic vocabulary of statistics. Three major objectives of statistics are as follows:

1. To describe information contained in a sample
2. To design a sampling process, so that the selected sample is an unbiased representation of the population
3. To make inferences about a population from information contained in a sample

A brief definition of all the statistical terms mentioned in the chapter will be listed in the section "Key Words." You also need to sharpen some basic math skills such as order of operations and basic algebra to be able to follow formulas and solve equations. The correct order of operations is P-E-MD-AS.

Note: Each column of the Excel spreadsheet shows a particular step in the order of operations. Using Excel clearly shows the step-by-step procedures that lead to the correct answers. Excel is also widely available without incurring additional expenses for you. Excel is a great learning tool that solidifies the understanding order of operations in carrying out the calculation of statistical formulas.

Key Words

Absolute zero: Absolute zero means a complete absence of the attribute that you are measuring. It is not an arbitrarily assigned number. 10

Cluster sampling: Cluster sampling works best when "natural" grouping (clustering) occurs in the population. Random sampling is conducted to select which clusters are included in the sample. Once a cluster is selected in the sample, all individuals in the cluster are included in the sample. 6

Continuous variables: Continuous variables are values that do not have separation from one integer to the next. Continuous variables usually are expressed with decimals or fractions. 12

Convenience sample: A convenience sample is one in which researchers use anyone who is willing to participate in the study. A convenience sample is created based on easy accessibility. 4

Data: Data are defined as factual information used as a basis for reasoning, discussion, or calculation, so that meaningful conclusions can be drawn. 2

Dependent variable: A dependent variable is the variable that is the focus of researchers' interests and is affected by the different levels of an independent variable. 12

Descriptive statistics: Descriptive statistics are statistical procedures used to describe, summarize, organize, and simplify relevant characteristics of sample data. 3

Discrete variables: Discrete variables are values that have clear separation from one integer to the next. The answers for discrete variables can only be integers (i.e., whole numbers). 12

Experimental research: Experimental research is usually conducted in a tightly controlled environment (i.e., research laboratories). The three important features in experimental research are (1) control, (2) manipulation, and (3) random assignment. 12

Extraneous variable: The extraneous variables are variables that are not included in the study but might have an impact on the relationship between variables included in the study. 14

Independent variable: An independent variable is the variable that is deliberately manipulated by the researchers in a research study. 12

Inferential statistics: Inferential statistics are statistical procedures that use sample statistics to generalize to or make inferences about a population. 4

Interval scale: An interval scale not only arranges observations according to their magnitudes but also distinguishes the ordered arrangement in equal units. 9

Likert scales: Likert scales are often used to measure people's opinions, attitudes, or preferences. Likert scales measure attributes along a continuum of choices such as 1 = *strongly disagree*, 2 = *somewhat disagree*, 3 = *neutral*, 4 = *somewhat agree*, or 5 = *strongly agree* with each individual statement. 8

Nominal scale: In a nominal scale, measurements are used as identifiers, such as your student identification number, phone number, or social security number. 8

Nonexperimental research: Nonexperimental research is conducted to observe and study research participants in their natural settings without deliberately controlling the environment or manipulating their behaviors or preferences. 13

Ordinal scale: In an ordinal scale, measurements are used not only as identifiers but also to carry orders in a particular sequence. 8

Parameters: Parameters are defined as numerical characteristics of a population. 3

Population: A population is defined as an entire collection of everything or everyone that researchers are interested in studying or measuring. 2

Quasi-experimental research: Quasi-experimental research has some but not all of the features of experimental research. More specifically, if one or more of the control, manipulation, and random assignment features are not feasible but others remain intact, the research becomes quasi-experimental. 13

Random sample: A random sample is an ideal way to select participants for scientific research. A random sample occurs when every member in the population has an equal chance of being selected. 4

Ratio scale: A ratio scale contains every characteristic that lower level scales of measurement have, such as identifiers, direction of ranking, equal units, and something extra: an absolute zero. 10

Sample: A sample is defined as a subset of the population from which measures are actually obtained. 2

Sample statistics: Sample statistics are defined as numerical attributes of a sample. 3

Scales of measurement: Scales of measurement illustrate different ways that variables are defined and measured. Each scale of measurement has certain mathematical properties that determine the appropriate application of statistical procedures. 8

Simple random sample: A simple random sample is a subset of individuals (a sample) chosen from a larger set (a population). Each individual is chosen randomly and entirely by chance, and each subset of *k* individuals has the same probability of being chosen for the sample as any other subset of *k* individuals. 5

Statistics: Statistics is a science that deals with the collection, organization, analysis, and interpretation of numerical data. 2

Stratified sampling: Stratified sampling is the process of grouping members of the population into relatively homogeneous subgroups before sampling. A random sample from each stratum is independently taken in the same proportion as the stratum's size to the population. These subsets of the strata are then pooled to form a random sample. 6

Systematic sample: A systematic sample is achieved by selecting a sample from a population using a random starting point and a fixed interval. Typically, every "*k*th" member is selected from the total population for inclusion in the sample. 6

Variable: A variable refers to a measurable attribute. These measures have different values from one person to another, or the values change over time. 7

Learning Assessment

Multiple Choice: Circle the Best Answer in Every Question

1. A researcher was interested in the sleeping habits of college students. A group of 50 students were selected at random and interviewed. The researcher found that these students slept an average of 6.7 hours per day. For this study, the 50 students are an example of a __________.
   a. parameter
   b. statistic
   c. population
   d. sample

2. A quantity, usually an unknown numerical value that describes a population, is a __________.
   a. parameter
   b. statistic
   c. population
   d. sample

3. What additional characteristic is required on a ratio scale compared with an interval scale?
   a. Whether the measurements are the same or different
   b. The order of the magnitudes
   c. An absolute zero
   d. Scores with equal units

4. For $X = 0, 1, 6, 3$, what is $(\sum X)^2$?
   a. 20
   b. 46
   c. 64
   d. 100

5. Gender, religion, and ethnicity are measurements on a(n) __________ scale.
   a. nominal
   b. ordinal
   c. interval
   d. ratio

6. The measure of temperature in Fahrenheit is an example of a(n) __________ scale of measurement.
   a. nominal
   b. ordinal
   c. interval
   d. ratio

7. Which of the following pairs is usually unknown parameters of the population?
   a. $\bar{X}$ and μ
   b. s and σ
   c. s^2 and σ^2
   d. μ and σ

8. The measure of income is an example of a(n) __________ scale of measurement.
   a. nominal
   b. ordinal
   c. interval
   d. ratio

Free Response Questions

9. $X = 3, 4, 5$, and 7; compute $\sum X$, $(\sum X)^2$, and $\sum X^2$

10. $X = -3, 0, 1$, and 2; compute $\sum X$, $\sum (X - 1)^2$, and $\sum X^2 - 3$

11. $X = 3, 4, 5$, and 7; $Y = -1, 0, 1$, and 2; compute $\sum XY$ and $(\sum XY)^2$

12. $X = 3, 4, 5$, and 7; $Y = -1, 0, 1$, and 2; compute $\sum X^2Y^2$ and $\sum (X - 2)(Y - 3)$

13. $X = 4, 5, 6$, and 9; $Y = -1, -1, 1$, and 2; compute $(\sum XY)^2$ and $\sum (X - 5)(Y + 1)$

Answers to Pop Quiz Questions

1. a
2. d
3. d
4. a
5. Pay structure
6. Job performance of salespeople
7. Quasi-experimental research design with cluster sampling
8. $\sum (X - 9)^2 = 90$
9. $\sum X^2 - 9 = 486$

2

SUMMARIZING, ORGANIZING DATA, AND MEASURES OF CENTRAL TENDENCY

Learning Objectives

After reading and studying this chapter, you should be able to do the following:

- Construct a frequency distribution table with individual values to summarize data
- Create a frequency distribution table with equal intervals to summarize data
- Identify the differences between bar graphs and histograms
- Identify the common distribution shapes
- Define the positively skewed and negatively skewed distributions
- Differentiate among the three common central tendency measures: (1) mode, (2) median, and (3) mean
- Calculate mean from the formula
- Adjust the mean formula when using frequency tables

WHAT YOU KNOW AND WHAT IS NEW

You learned most of the basic vocabulary for statistics in Chapter 1. You learned how to distinguish between samples and populations and how to identify the different scales

FIGURE 2.1 ● Unsorted Data

of measurement. You learned the basics of summation notation and refreshed your mathematical skills. You also learned that different scales of measurement come with distinct mathematical attributes. All of these will be extremely useful and will continue to be relevant throughout this book.

One of the main purposes in statistics is to make sense out of data. It can be very confusing to encounter data without any particular pattern. There are some simple and commonly used methods to make data manageable. In this chapter, we will explore different ways to make sense out of nominal, ordinal, interval, and ratio variables by summarizing and organizing them and finding typical values to represent a group of observations.

With the help of three images, we illustrate why summarizing and organizing help us understand and extract useful information out of data. Data are facts presented in numerical form. Unfiltered or unprocessed data are disorienting and confusing. Data can simply look like an unruly mess such as the big pile of miscellaneous pieces shown in Figure 2.1.

FIGURE 2.2 ● Organized or Sorted Data

When data are organized or sorted according to a particular pattern, such as by type, we can quickly get a rough idea about this pile of miscellaneous pieces. It seems that there are two kinds of pasta: (1) macaroni and (2) shells, and three kinds of snacks: (1) almonds, (2) Wasabi peas, and (3) chocolate-covered espresso beans, as shown in Figure 2.2. Sorting is very useful and commonly used in tracking inventories in retail business.

When data are summarized in a way that provides easy-to-process information about the numeric attributes, we can make sense out of them immediately. The following graph provides the number of pieces in each type. A summary graph might not include all the numerical characteristics of the data such as weight and size of each piece, but it provides meaningful information on the number of pieces in every type as shown in Figure 2.3. The height of the bar represents the quantity of each item. The transformation from

FIGURE 2.3 ● Number of Pieces in Each Category

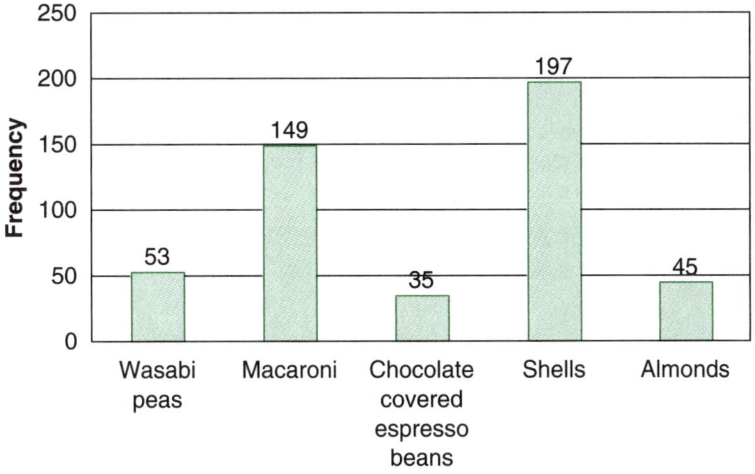

Figure 2.1 to Figures 2.2 and 2.3 illustrates the reason why it is helpful to organize and/or summarize data.

FREQUENCY DISTRIBUTION TABLE

Organizing and summarizing of data in a table form presents an easily accessible way to display the data at a glance. A **frequency distribution table** lists all values or categories along with a corresponding tally count for them. **Distribution** is defined as arrangement of values of a variable as they occur in a sample or a population. Both categorical variables (i.e., nominal or ordinal variables) and numerical variables (i.e., interval or ratio variables) can be organized in frequency distribution tables (often simply referred to as "frequency tables").

Organizing and Summarizing Categorical Variables

When dealing with categorical data (i.e., nominal or ordinal scales of measurement), the process of organizing and summarizing is fairly simple. The observations are organized or sorted into categories, and then the tally count of each category is reported as **frequency**, which is simply referred to as f in this book. When you add up the fs of each category, the sum of all fs equals the sample size, n. Example 2.1 is used to demonstrate such a relationship between f and n.

EXAMPLE 2.1

Suppose that I asked 20 people "what kind of job do you do?" The answers turned out to be "truck driver," "retail salesperson," "secretary," "personal assistant," "day care worker," "construction worker," "car wash worker," "server," "bartender," "librarian," "customer service representative," "call center worker," and so on. To organize these answers, I decided to group similar jobs into occupational categories such as professional, sales, service, clerical, and laborer. Then, I counted the number of answers in each occupation category. The results are shown in Table 2.1.

TABLE 2.1 ● Frequency Distribution Table of Occupations

X (Occupation)	*f* (Frequency)	Relative Frequency
Professional	3	.15
Sales	6	.30
Service	5	.25
Clerical	4	.20
Laborer	2	.10
Total	20	1.00

Based on the data in Table 2.1, when I add up all the frequencies, 3 + 6 + 5 + 4 + 2, the total frequency equals the sample size, $n = 20$. The last column of the frequency distribution table is called the relative frequency. **Relative frequency** is the same as proportion or percentage. It is defined and calculated as the frequency of a category (f) divided by the sample size (n). It provides the percentage of each category. Based on Table 2.1, the majority (i.e., 55%) of the 20 people worked in either sales or service jobs.

The sum of all relative frequencies should add to 1.00. It is a good way to check your math. If the sum does not add to 1.00, there are two possible explanations. First, the math is wrong. You need to double-check your calculations. Second, if the answer is .99, 1.01, or something very close to 1.00, it might be due to rounding.

$$\text{Relative frequency (proportion) for a category} = \frac{\text{Frequency of a category}}{\text{Total frequency}} = \frac{f}{n}$$

$$\text{Percentage for a category} = \frac{\text{Frequency of a category}}{\text{Total frequency}} \times 100\% = \frac{f}{n} \times 100\%$$

Organizing and Summarizing Numerical Variables

When dealing with numerical variables, frequency tables are constructed in a logical way to present the data in an orderly fashion with a reasonable number of categories. An example is used to explain the concepts of "orderly fashion" and "reasonable number of categories."

EXAMPLE 2.2

Technology has become an integral part of our daily lives. This is especially true in higher education. Computer networks keep us connected electronically. Just like any technology, sometimes problems happen. Our university kept a record of the number of computer network interruptions per day on campus. The following numbers showed the number of computer network interruptions in the month of September.

1 3 1 1 0 1 0 1 1 0

2 2 0 0 0 1 2 1 2 0

0 1 6 4 3 3 1 2 4 0

Please organize the daily number of computer network interruptions into a frequency table and calculate the total number of interruptions for the month.

Let's rearrange the numbers and put them in a frequency distribution table. The table needs to have a reasonable number of categories, so the table is not too long or too short. When the table is too long, the information is too detailed to be absorbed quickly. When the table is too short, there is not enough distinction among the categories. What is a reasonable number of categories? The answer to that question is based on the capacity of human short-term memory: 7 ± 2. A frequency table with categories between 5 and 9 is reasonable. The short-term memory capacity is defined as the capacity of holding a small amount of information in the brain in an active, immediately available fashion so that we are able to receive, process, and remember the information. According to Miller (1956), the capacity for short-term memory across many different types of stimuli such as words, notes, sounds, or pitches in various laboratory experiments all came to the same magical number 7, plus or minus 2.

Next, let's talk about the concept of "orderly fashion." There are many different kinds of "orders." One is ascending order, in which numbers are arranged from the lowest to the highest. Another is descending order in which numbers are arranged from the highest to the lowest. I prefer constructing a frequency table in an ascending order, from the lowest to the highest. To do this in the example of network interruptions, first, you need to identify the minimal value and maximal value. The minimal value is 0, and the maximal value is 6. Using the number of network interruptions as individual categories, thus, you have seven total categories. Tally the count for each category and report it as the frequency for that category. Therefore, the values of the network interruptions are arranged in an ascending order, lowest to highest, and the frequency of each value is counted, thus, you have constructed a frequency distribution table as shown in Table 2.2.

TABLE 2.2 ● Frequency Distribution Table for Daily Network Interruptions on a University Campus in September

X (Number of Interruptions)	*f* (Frequency)	Relative Frequency
0	9	.300
1	10	.333
2	5	.167

(Continued)

(Continued)

X (Number of Interruptions)	*f* (Frequency)	Relative Frequency
3	3	.100
4	2	.067
5	0	.000
6	1	.033
Total	30	1.000

When adding up all frequencies, the total is 30. There were originally 30 numbers reported in the month of September, one for each day of the month. Relative frequency (or proportion) is calculated and reported in the third column of Table 2.2. The sum of all relative frequencies is 1.000. From this table, we learn that about 63% of the time, the network is functioning fairly well with one or no interruptions in a day. Some of you might have noticed that the value 5 is not in the reported daily network interruption data. It is also acceptable to construct the frequency distribution table without the value 5 as shown in Table 2.2a. Because the frequency for the value 5 is 0, it does not affect the frequency count or the calculation of relative frequency. Either Table 2.2 or Table 2.2a is an acceptable way to construct a frequency table for these data.

TABLE 2.2a ● Alternative Frequency Distribution Table for Daily Network Interruptions on a University Campus in September

X (Number of Interruptions)	*f* (Frequency)	Relative Frequency	*fX*
0	9	.300	0
1	10	.333	10
2	5	.167	10
3	3	.100	9
4	2	.067	8
6	1	.033	6
Total	30	1.000	43

The frequency table summarizes the number of daily network interruptions in ascending order with corresponding frequency. Such an organization makes it easy to calculate the total number of network interruptions in September. Instead of adding 30 numbers from the first number to the last number, you can use the modified formula $\sum X = \sum fX$. The frequency table shows every distinct value with its corresponding frequency. In this example, 0 occurs 9 times, 1 occurs 10 times, 2 occurs 5 times, 3 occurs 3 times, 4 occurs twice, and 6 occurs once. Therefore, you have to multiply the frequency by every distinct value and add them all up. $\sum X = \sum fX = 0 \times 9 + 1 \times 10 + 2 \times 5 + 3 \times 3 + 4 \times 2 + 6 \times 1 = 43$. There are multiplications and additions involved in $\sum fX$; the order of operations dictates that the multiplications have to be done before the additions. The total number of interruptions for the month was 43.

When dealing with data spanning over a large range, creating a frequency table with **equal intervals** to cover the entire range is a good solution. Equal intervals are created by including the same number of values in each interval in a frequency table. We will use a couple of examples to illustrate this process.

EXAMPLE 2.3

A first-grade teacher asked parents to estimate their children's average daily screen time outside of academic learning. The reported screen times for 25 first graders are shown below. The unit of measurement is a minute.

79, 45, 66, 89, 97, 55, 61, 86, 93, 81, 80, 73, 76,

84, 81, 67, 92, 75, 76, 69, 57, 88, 84, 59, 72

Please create a frequency table with a reasonable number of categories in an orderly fashion to organize the reported screen time.

It is difficult to get a sense of the distribution of the times by looking at 25 numbers in no particular order. The minimal screen time was 45 and the maximal value was 97. The range is maximal value minus the minimal value, 97 – 45 = 52. It is not advisable to create a frequency table using individual values with such a large range. Equal intervals are usually created by the intervals of 5 or 10 for convenience, if feasible. To create equal intervals to cover this range of 52, an interval of 10 is appropriate. The lowest interval is between 40 and 49, so it covers the minimal value, and the highest interval is between 90 and 99, so it covers the maximal value. There should not be any overlap between two adjacent intervals. A frequency table with equal intervals is constructed in Table 2.3 for these 25 first graders.

TABLE 2.3 ⬢ Frequency Distribution Table for 25 First Graders' Screen Times

Screen Time Interval	*f* (Frequency)	Relative Frequency
40–49	1	.04
50–59	3	.12
60–69	4	.16
70–79	6	.24

(Continued)

(Continued)

Screen Time Interval	*f* (Frequency)	Relative Frequency
80–89	8	.32
90–99	3	.12
Total	25	1.00

Table 2.3 summarizes and organizes the 25 first graders' screen times with useful insights. It showed that 16% of the first graders' screen times were below 60 minutes, 56% of the first graders' screen times were between 70 and 89 minutes, and 12% of the first graders' screen times were higher than 90 minutes.

The same principles that were used in constructing a frequency table for discrete variables can also be applied to continuous variables. Continuous variables mean that the observations may take any value between two integers in the forms of fraction or decimal point. It is difficult to obtain precise measures for continuous variables. Therefore, the concept of **real limits** is designed to cover a range of possible values that may be reflected by a continuous measure. The lower limit is the value minus ½ of the unit and the upper limit is the value plus ½ of the unit. Physical measurements, such as height and weight, are continuous variables. For example, when 1 pound is used as a unit to measure a person's weight and John's weight is 180 pounds, the lower limit of John's weight is 179.5 pounds and the upper limit is 180.5 pounds.

Let's apply real limits to online purchase measures in Example 2.4.

EXAMPLE 2.4

Online shopping has become a routine part of our daily activities. We can shop without leaving the comfort of our home, and the merchandise is delivered to our door. There were 59 students in one of my statistics classes. Their answers to the amount of money they spent on online shopping last week were organized in Table 2.4. What was the estimated total online spending for these students last week?

TABLE 2.4 ● Frequency Distribution Table for 59 Students' Online Purchases Last Week

X (Online Purchases)	*f* (Frequency)	Relative Frequency
\$0–\$19	5	.08
\$20–\$39	7	.12

X (Online Purchases)	*f* (Frequency)	Relative Frequency
\$40–\$59	25	.42
\$60–\$79	12	.20
\$80–\$99	7	.12
\$100–\$119	2	.03
\$120–\$139	1	.02
Total	59	.99

When we made online purchases, the transactions were calculated to two places after the decimal point, such as \$34.67 or \$117.92. The dollar amount spent on the online purchase is a continuous variable so real limits apply here. For example, the real limits for the highest interval \$120–\$139 are \$119.5 as the lower limit and \$139.5 as the upper limit. Just to make it clear that each interval includes the lower limit but not the upper limit (i.e., lower limit $\leq X <$ upper limit), there is no overlap between two adjacent intervals. You might have already noticed that the total relative frequency adds to .99. This is simply due to rounding.

To obtain an estimated total dollar amount spent by 59 students' online purchases last week, the midpoint for each interval needs to be calculated. The formula for calculation of the midpoint for an interval is $X_{\text{midpoint}} = \dfrac{\text{lowend} + \text{highend}}{2}$.

The frequency table with midpoints is shown in Table 2.4a. In a frequency table consisting of equal intervals, the formula for $\sum X$ needs to be modified to $\sum fX_{\text{midpoint}}$. The estimated total amount is the sum of the midpoint of each interval times its corresponding frequency. Again, multiplications and additions are involved in $\sum fX_{\text{midpoint}}$. Multiplications have to be done before additions. The estimated total dollar amount for 59 students' online purchases was \$3,300.50.

TABLE 2.4a ● Frequency Distribution for 59 Students' Online Purchases Last Week With Midpoint of Each Interval

X (Online Purchases)	*f* (Frequency)	X_{midpoint}	fX_{midpoint}
\$0–\$19	5	\$9.5	\$47.50
\$20–\$39	7	\$29.5	\$206.50
\$40–\$59	25	\$49.5	\$1,237.50
\$60–\$79	12	\$69.5	\$834.00
\$80–\$99	7	\$89.5	\$626.50
\$100–\$119	2	\$109.5	\$219.00
\$120–\$139	1	\$129.5	\$129.50
Total	59		\$3300.50

Let's review the process of constructing a frequency table with equal intervals. The key points are summarized below:

1. Identify the minimal and maximal values of the variable. Calculate the range.
2. Choose an appropriate and convenient interval, usually in 5s or 10s if feasible, to construct a frequency table with 7 ± 2 equal intervals.
3. Create the equal intervals in an ascending order.
4. Make sure that there is no overlap between two adjacent intervals. For continuous variables, each interval includes the lower limit but not the upper limit.
5. The midpoint for each interval is calculated as

$$X_{\text{midpoint}} = \frac{\text{low end} + \text{high end}}{2}.$$

POP QUIZ

1. In a statistics class, students' quiz scores on a pop quiz with only four questions were reported in the frequency table below. Use the frequency table to calculate the total quiz score for the class.

X, Quiz	*f* (Frequency)
0	2
1	1
2	5
3	8
4	9

What is the total quiz score for the class?

a. 10
b. 25
c. 71
d. 100

2. When constructing a frequency table with equal intervals of a continuous variable X, each interval consists of same width. To make sure that there is no overlap between two adjacent intervals, each interval actually includes

a. lower limit $< X <$ upper limit.
b. lower limit $\le X <$ upper limit.
c. lower limit $< X \le$ upper limit.
d. lower limit $\le X \le$ upper limit.

GRAPHS

Creating a graph is another common way to organize and summarize data. There is a specialized field in statistics called data visualization. **Data visualization** is the graphic representation of data. This is especially important in the era of big data because graphs can better communicate information than millions of unsorted data points. A well-organized graph can deliver useful numerical information in a quick and easy-to-understand fashion. Data visualization tools are available in many statistics software packages. In this book, we cover basic chart functions in Excel. Two commonly used graphs are (1) bar graphs and (2) pie charts.

Bar Graphs and Histograms

When **bar graphs** are used to organize categorical data, the horizontal scale (x-axis) represents the values (or categories) of the variable and the vertical scale (y-axis) represents the frequency or relative frequency of each value. When using the data from Table 2.1 to construct a bar graph, the result is shown in Figure 2.4a with the x-axis representing occupations and the y-axis representing the corresponding frequency of each occupation.

It is also possible to create a bar graph with multiple sets of bars to represent multiple categorical variables. For example, male and female students' jobs can be reported separately as shown in Figure 2.4b where males' occupations are represented in lighter blue bars and females' occupations are represented in darker blue bars. There is no darker blue bar in the last occupational category, laborer, because there are no females in this category.

FIGURE 2.4a ● Bar Graph of Occupational Categories

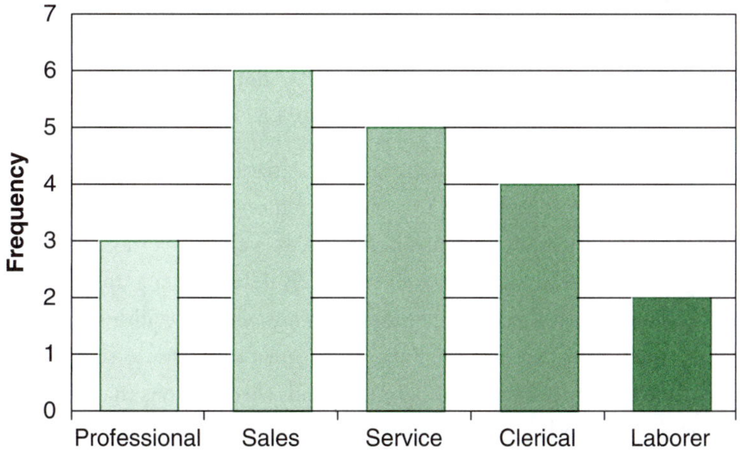

FIGURE 2.4b ● Bar Graph of Males' and Females' Occupations

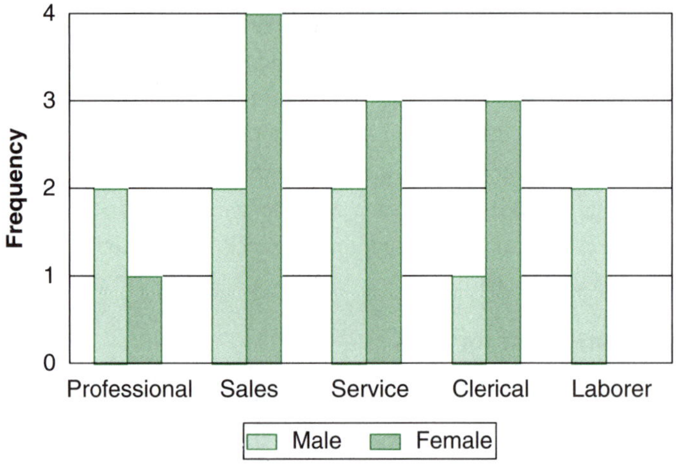

Bar charts can be applied to discrete data. When using data from Table 2.2 to construct a bar graph, the *x*-axis represents the number of reported daily network interruptions, and they are lined up from the smallest to the largest value from left to right. The *y*-axis represents the frequency, and the bar height represents frequency of each value. Frequency can only be integers. There are no valid answers between two integers: (1) k and (2) $k + 1$, because the number of daily network interruptions has to be a whole number. It is not possible to have 2.15 network interruptions in any particular day. Therefore, the bars are separate from one another, showing a gap in between two adjacent values. In bar graphs, bars do not touch each other as shown in Figure 2.5. Looking at Figure 2.5, you got an idea that most days (i.e., 19 out of 30 days), the university computer networks operated smoothly with no interruptions or only one interruption. According to Table 2.2 data, none of the days had five interruptions so there was not a bar that represented 5 interruptions. The last bar represented six daily interruptions happened once in September.

When using a graph to summarize a continuous variable, we need to consider the real limits of the values. Due to the fact that there are numerous possible answers between two adjacent values, there are no gaps between two values. We need to create a new type of graph called the **histogram**. A histogram is defined as a graphical presentation of a continuous variable. The bars of a histogram are of equal width to represent equal intervals of values, and they are touching each other to illustrate that the values are continuous. The difference between a bar graph and a histogram is that the two adjacent bars are touching each other in a histogram, but there are gaps between the bars in a bar

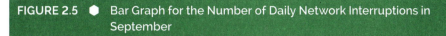

FIGURE 2.5 ● Bar Graph for the Number of Daily Network Interruptions in September

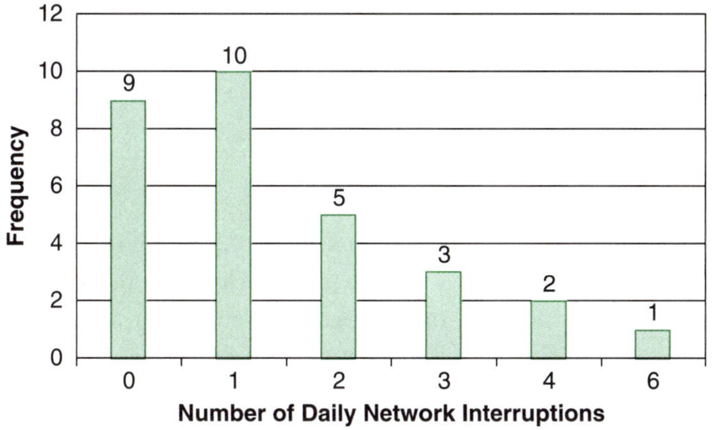

graph. Bar graphs are used for discrete variables, and histograms are used for continuous variables. The height of the bar represents the corresponding frequency for each interval. Using online purchase data in Table 2.4, we can construct a histogram with $20 intervals as shown in Figure 2.6.

The highest amount interval $120–$139 is actually bounded by a lower limit of $119.5 and an upper limit of $139.5. The lower limit is included but the upper limit is not

FIGURE 2.6 ● Histogram of 59 Students' Online Purchase Amount Last Week

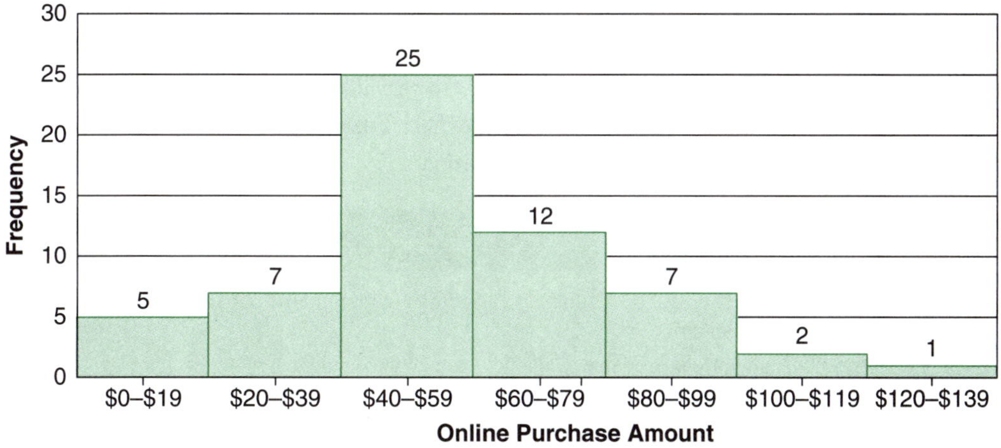

included in each interval; therefore, the real limits of the interval \$120–\$139 refer to \$119.5 ≤ X < 139.5, and the real limits of the second highest interval \$100–\$119 refer to \$99.5 ≤ X < \$119.5. Such a rule is to make sure that there are no overlaps between intervals for continuous variables. The tallest bar represents \$40–\$59, which means 25 students spent such an amount for online purchases last week. More students belong in this category than any other category. A step-by-step instruction on how to create this histogram in Excel is provided at the end of this chapter.

FIGURE 2.7 ● Pie Chart of the Occupational Categories

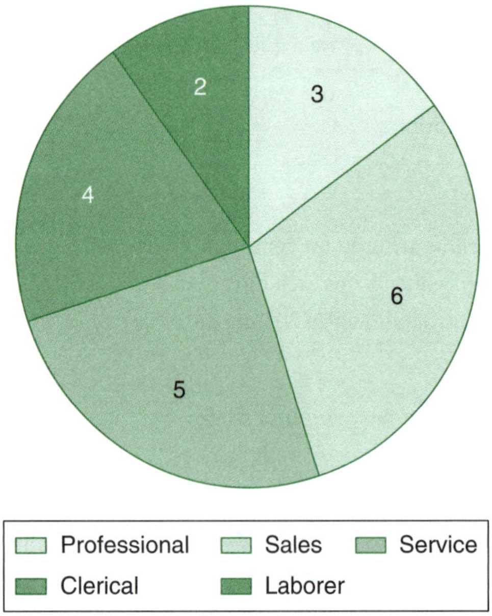

Pie Charts

A **pie chart** usually depicts categorical data in a circle where the size of each slice of the pie corresponds to the frequency or relative frequency of each category. Pie charts are visually intuitive because the entire circle is 360 degrees. All categories add up to 100%, which constitute the entire 360 degrees. Using the occupational data in Table 2.1, a pie chart is constructed in Figure 2.7. It presents a clear image that sales and service job categories cover more than half of the pie, which means more than 50% of respondents hold jobs in these two categories.

Pie charts are extremely popular in presenting reports on budgets, government spending, or TV polling. When used appropriately, pie charts can be a powerful visual aid to present information. The right conditions to use pie charts include (a) the total adds to a meaningful sum, (b) there are no overlaps between categories, and (c) there is a reasonable number of categories. As you can imagine, if the number of categories is more than 10, each slice becomes very small. There will be too much information crowded into the pie chart. The disadvantage of a pie chart is that it doesn't present any statistical information other than frequency or proportion.

POP QUIZ

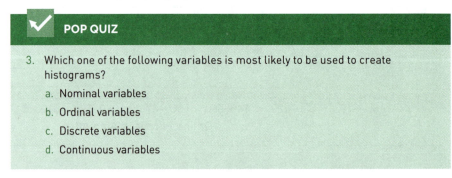

3. Which one of the following variables is most likely to be used to create histograms?
   a. Nominal variables
   b. Ordinal variables
   c. Discrete variables
   d. Continuous variables

COMMON DISTRIBUTION SHAPES

A distribution refers to all possible values and the number of times every value occurs in a sample or a population. In either a bar graph or a histogram where values are lined up from the smallest (the left side of the *x*-axis) to the largest (the right side of the *x*-axis), the height of the bar represents frequency or relative frequency (i.e., proportion) of each value; therefore, the shape of the distribution is in full display. Common shapes of data distributions include uniform distribution, normal distribution, and skewed distribution.

Uniform Distribution

In a **uniform distribution**, every value appears with the same frequency, proportion, or probability. Many things have uniform distribution such as dice, coins, and cards. For example, a die has six sides with 1, 2, 3, 4, 5, or 6 dots on each side. The relative frequency or probability of throwing a die and obtaining any of 1, 2, 3, 4, 5, or 6 dots is evenly distributed as 1/6, which is shown in Table 2.5 and Figure 2.8. The probability distribution of the number of dots on a die is called a uniform distribution. A fair die produces the same probability of landing on every side.

Normal Distribution

When you throw two dice at the same time and add the number of dots on the dice, the answers are 2, 3, 4, 5, 6, 7, 8, 9, 10, 11, or 12 dots. The probabilities of obtaining those dots and various combinations for creating the number of dots are listed in Table 2.6. The total probability adds to 1.001 due to rounding. The bar graph of the sum of dots from two dice is shown in Figure 2.9.

TABLE 2.5 ● Probability of Throwing a Die and Obtaining a Specific Number of Dots

X (Number of Dots)	*f* (Frequency)	Relative Frequency (Probability)
1	1	1/6 = .167
2	1	1/6 = .167
3	1	1/6 = .167
4	1	1/6 = .167
5	1	1/6 = .167
6	1	1/6 = .167
Total	6	1.002

FIGURE 2.8 ● Uniform Distribution of Number of Dots on a Die

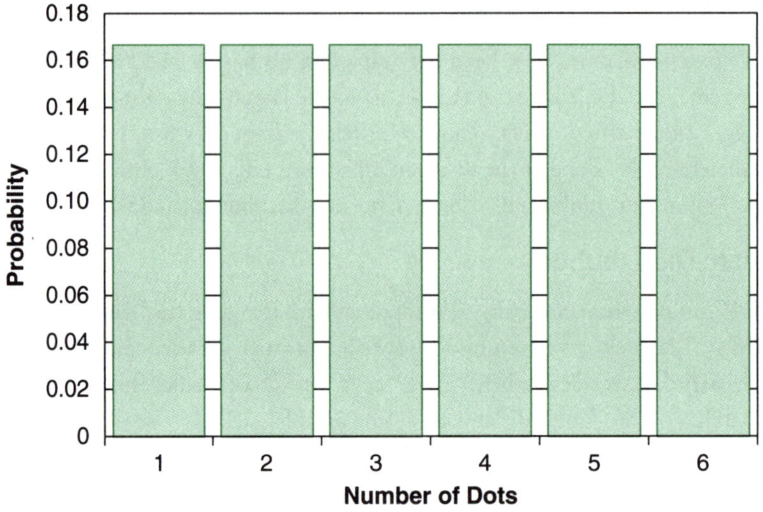

TABLE 2.6 ● Total Number of Dots From Two Dice and the Probabilities

X (Number of Dots)	*f* (Frequency)	Probability
2 (1, 1)	1	1/36 = .028
3 (1, 2); (2, 1)	2	2/36 = .056
4 (1, 3); (2, 2); (3, 1)	3	3/36 = .083
5 (1, 4); (2, 3); (3, 2); (4, 1)	4	4/36 = .111
6 (1, 5); (2, 4); (3, 3); (4, 2); (5, 1)	5	5/36 = .139
7 (1, 6); (2, 5); (3, 4); (4, 3); (5, 2); (6, 1)	6	6/36 = .167
8 (2, 6); (3, 5); (4, 4); (5, 3); (6, 2)	5	5/36 = .139
9 (3, 6); (4, 5); (5, 4); (6, 3)	4	4/36 = .111
10 (4, 6); (5, 5); (6, 4)	3	3/36 = .083
11 (5, 6); (6, 5)	2	2/36 = .056
12 (6, 6)	1	1/36 = .028
Total	36	1.001

FIGURE 2.9 ● Probability of Number of Dots From Two Dice

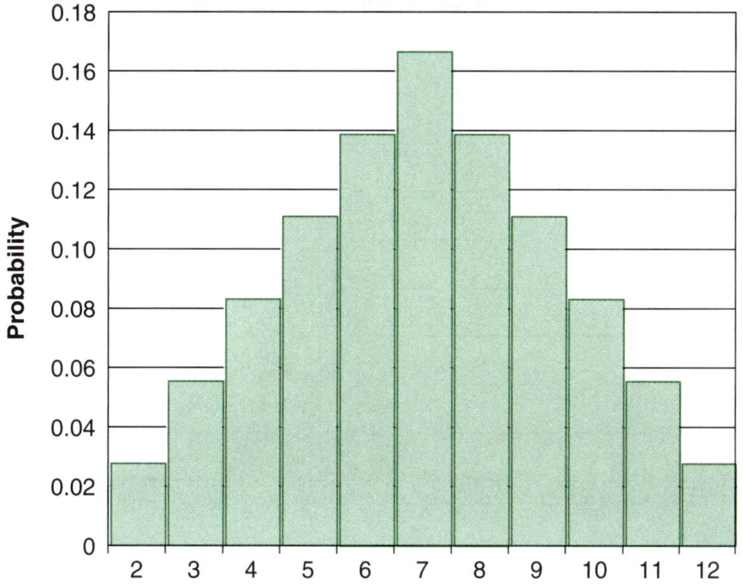

The bar graph shown in Figure 2.9 has the typical characteristics of a **normal distribution**: (a) the distribution peaks at the center, (b) the distribution symmetrically tapers off on both sides and the left side is a mirror image of the right side, and (c) 50% of the distribution is above the mean and the other 50% is below the mean.

Many statistical procedures require sample data with a distribution that is approximately normally distributed. A normal distribution is a very important data distribution pattern in statistics. When one examines a distribution from a large sample size, the edges of the histograms are likely to be smoothed out and almost become a curve as seen in Figure 2.10, showing the bar graph of the probability of getting a certain number of heads from 100 coin tosses.

When you trace the outline of the bar graph in Figure 2.10, you create a **line graph**. A line graph is defined as a graph that displays quantitative information with a line or curve that connects a series of adjacent data points. This particular line graph shows a normal distribution curve. A normal distribution curve peaks at the center of the data (i.e., the mean) and then symmetrically tapers off on both sides with 50% of the distribution above the mean and 50% of the distribution below the mean. Draw a line straight through the mean, and the left side and the right side are mirror images of each other as shown in the line graph in Figure 2.11.

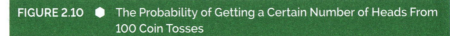

FIGURE 2.10 ● The Probability of Getting a Certain Number of Heads From 100 Coin Tosses

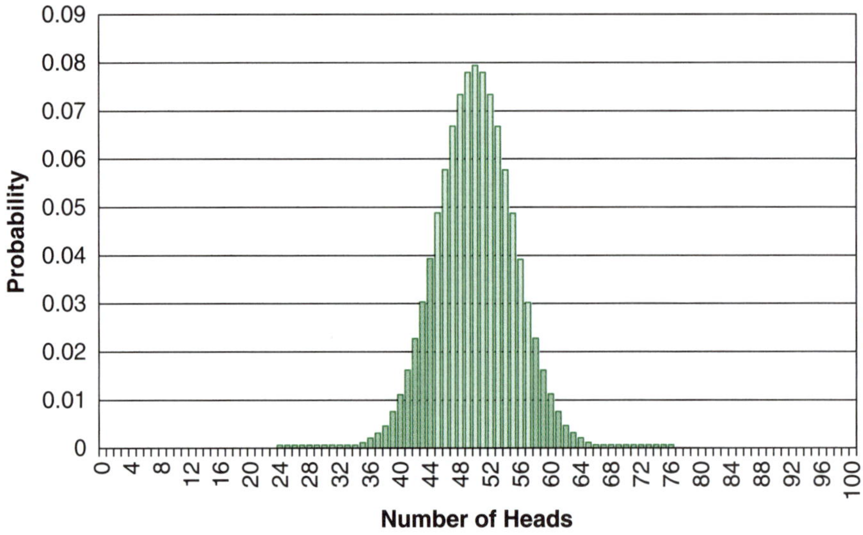

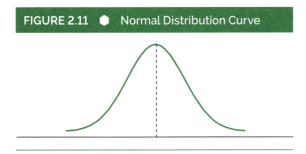

FIGURE 2.11 ● Normal Distribution Curve

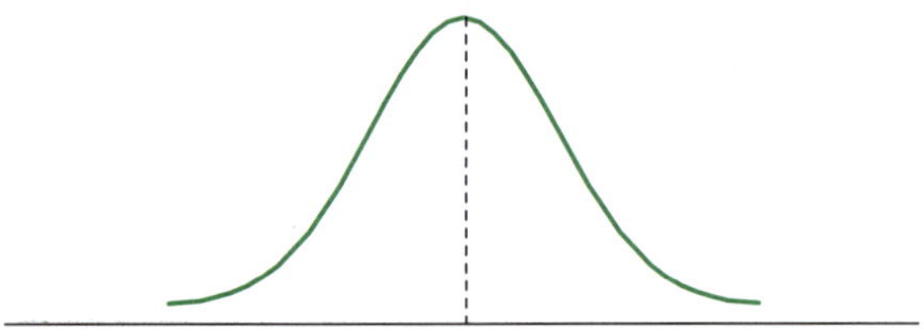

Skewed Distribution

Not all data are normally distributed. A **skewed distribution** happens when values are not symmetrical, and they concentrate more on one side than on the other. When the distribution mostly concentrates on the left side with a long tail to the right side, it is labeled as **positively skewed distribution** or right skewed as shown in Figure 2.12. For example, income distribution in the United States is positively skewed with a very long tail to the right side. It means that most people earn a modest income while the top 1% is making hundreds times more than regular working-class salary.

When the distribution mostly concentrates on the right side with a long tail to the left side, it is labeled as **negatively skewed distribution** or left skewed as shown in Figure 2.13. For example, when students are asked what grades they expect to get on a statistics exam, the majority of the students answer 80s or 90s with very few low scores. The labeling of the skewness depends on the direction of the tail. When the tail points to the high end of the distribution, the data distribution is positively skewed (or right skewed) and when the tail points to the low end of the distribution, the data distribution is negatively skewed (or left skewed).

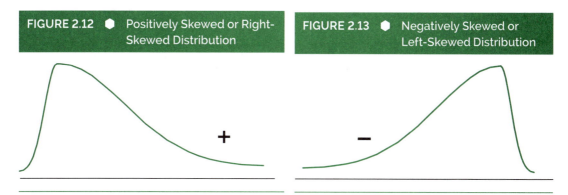

FIGURE 2.12 ● Positively Skewed or Right-Skewed Distribution

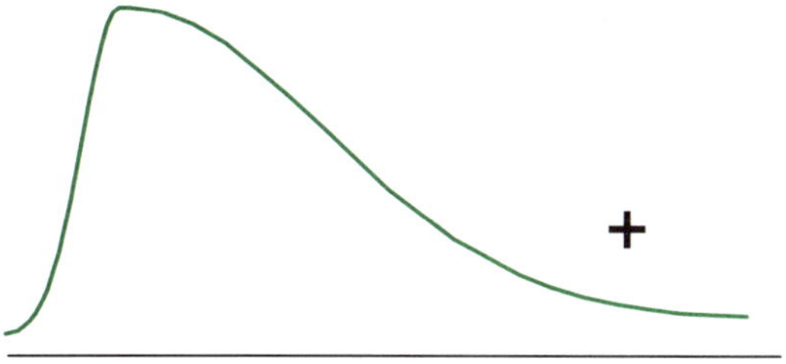

FIGURE 2.13 ● Negatively Skewed or Left-Skewed Distribution

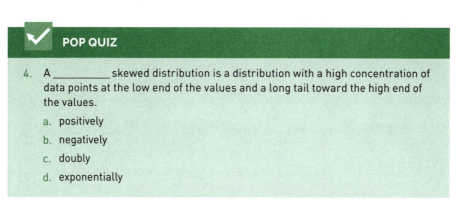

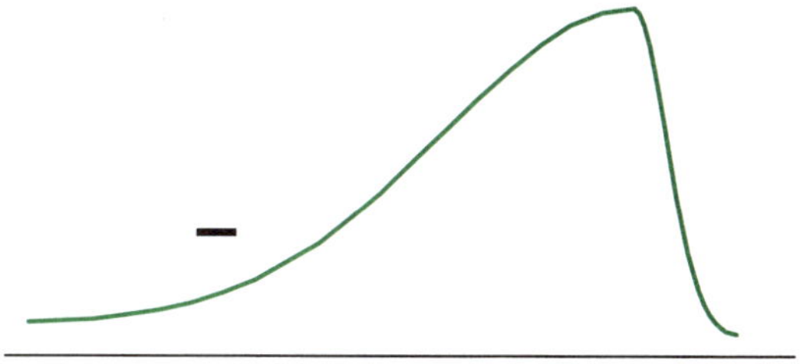

POP QUIZ

4. A ________ skewed distribution is a distribution with a high concentration of data points at the low end of the values and a long tail toward the high end of the values.
   a. positively
   b. negatively
   c. doubly
   d. exponentially

MEASURES OF CENTRAL TENDENCY

Besides tables and charts, we also use numeric numbers to describe attributes of a sample. **Central tendency** is defined as a single number to describe the center of a distribution. The center of a distribution is usually the most typical or representative value of all the values. There are three measures that are commonly used to represent central tendency: *Mean*, *median*, and *mode*. These three common measures of central tendency will be presented in order of mathematical complexity from the simplest to the most sophisticated. Therefore, we will discuss mode first, median second, and mean last.

Mode

The **mode** is defined as the value or category with the highest frequency in a distribution. It means the value or category that occurs most often. To identify the mode in a distribution, you may do a simple tally count of individual values or look into a frequency table and identify the *value* with the highest frequency without performing any other mathematical operation. Frequency counts can be applied to all four scales of measurement: (1) nominal, (2) ordinal, (3) interval, and (4) ratio. Let's look at an example.

EXAMPLE 2.5

According to T. Rowe Price's most recent annual Parents, Kids, Money Survey, one of the questions for the parents was "Which of the following best describe how you feel about savings for your kids' college education?" The frequency and percentage of the answer to this question is shown in Table 2.7. Based on Table 2.7, what is the mode of how parents feel about saving for their kids' college education?

TABLE 2.7 ● Frequency and Percentage of How Parents Feel About Saving for Kids' College Education

Which of the Following Best Describe How You Feel About Saving for Your Kids' College Education?	Frequency	Percentage (Relative Frequency)
Was not able to pay ANY of the cost	164	18.5%
Was able to pay SOME of the cost	397	44.8%
Was able to pay MOST of the cost	219	24.7%
Was able to pay ALL of the cost	107	12.1%
Total	887	100.1%

Note: Data from T. Rowe Price, "11th Annual Parents, Kids & Money Survey—College Savings Results" (2019) via Slideshare.com.

The savings for college education question was answered on a nominal scale of measurement. Mode can be applied to a nominal variable. The category with the highest frequency was "able to pay SOME of the cost." Notice that 44.8% of parents belonged in this category. Only 12.1% of the parents felt that they were able to pay all of the cost for their kids' college education, 24.7% of the parents were able to pay most of the cost, and still 18.5% of the parents were not able to pay any of the cost. Therefore, the vast majority of the students would still have to be responsible for at least part of the cost for their college education. The total percentage added up to 100.1% due to rounding.

Mode can be applied to all four different scales of measurement. In the next example, we will demonstrate how mode can be applied to an ordinal scale of measurement.

EXAMPLE 2.6

A middle school conducted an anonymous survey on the occurrences, severity, and types of bullying incidents at school. The answers to the question "How often have you been bullied at school in the past couple of months?" were tallied and reported in Table 2.8. What is the mode of this distribution?

TABLE 2.8 ● Occurrences of Bullying Incidents at a Middle School

Occurrences of Bullying Incidents	Frequency	Relative Frequency
I have not been bullied at school	320	65%
It has only happened once or twice	121	24%
2 or 3 times a month	25	5%
About once a week	17	3%
Several times a week	13	3%
Total	496	100%

The frequency table reported the occurrences of bullying incidents in five categories. The answers to "How often have you been bullied at school in the past couple of months?" are classified as an ordinal scale of measurement. The answers indicate an increasing level of occurrences of bullying incidents from "not being bullied" to "several times a week." The mode is the category with the highest frequency. The highest frequency is 320. The mode is the value or category with the highest frequency, so the answer is "I have not been bullied at school in the past couple of months." As is indicated by the relative frequency column, 65% of children reported that they were not bullied at school. It was a good news that a majority of the students did not have the personal experience of being bullied. The results also indicated that there were 30 students who had encountered regular weekly bullying. The school needs to take action to protect these students so they don't continue to suffer such mistreatment or abuse at school.

What you need to know about mode is that

1. it can be applied to nominal, ordinal, interval, or ratio data, and
2. a data set may have one mode, multiple modes, or no mode. For example, if two values or categories both have the same highest frequency, then the distribution has two modes, which is called a bimodal distribution. When none of the values is repeated in a distribution, or when each value is repeated the same number of times, there is no mode. Usually, it is common to have one mode in a distribution.

Median

The **median** is defined as the value right in the middle of the distribution when all values are sorted from the lowest to the highest. Based on the definition, the data need to be sorted. Therefore, they need to have at least ordinal attributes. It is important to note that median can be meaningfully applied to only ordinal, interval, or ratio variables.

The median is the middle position of sorted observed values. Let's clarify the term *middle position*. When the sample size n is an odd number, there is only one middle position in the distribution. When the sample size n is an even number, there are two middle positions in the distribution. Finding the median involves finding the middle location(s) of the distribution. This process is illustrated by Examples 2.7 and 2.8.

EXAMPLE 2.7

The following numbers are the fuel efficiency, in miles per gallon, reported on five randomly selected cars of the same model by the car manufacturer: 31, 28, 46, 39, and 41. What is the mode and median fuel efficiency of these five cars?

These five values all occur once in the sample. There is no mode in this case.

The sample size, $n = 5$, is an odd number. There is one middle position when sample size is an odd number. However, the values were reported in an arbitrary order so they need to be sorted first. The sorted values from the lowest to the highest are 28, 31, 39, 41, and 46. When n is an odd number, the middle location is determined by $L = \frac{n}{2}$ then round it up to the next integer. In this example, $L = \frac{5}{2} = 2.5$ then round up to 3. The third position of the sorted data is the median. The value of the third position is 39. Therefore, the median fuel efficiency for these five cars is 39 miles per gallon.

EXAMPLE 2.8

Let's use the frequency table of internet network interruption example in Table 2.2 to illustrate how to figure out the mode and median by introducing a new concept, cumulative relative frequency. **Cumulative relative frequency** is defined as the accumulation of the relative frequency for a particular value together with all the relative frequencies for the values lower than it. The cumulative relative frequency only applies to variables with orders (i.e., ordinal, interval, or ratio variables). Median can be identified in a frequency table as the first value with the cumulative relative frequency over .50. Let's use the network interruptions example with cumulative relative frequency calculated in Table 2.9.

TABLE 2.9 ● Frequency Table for Daily Network Interruptions With Cumulative Relative Frequency

X (Number of Interruptions)	*f* (Frequency)	Relative Frequency	Cumulative Relative Frequency
0	9	.300	.300
1	10	.333	.633
2	5	.167	.800

X (Number of Interruptions)	*f* (Frequency)	Relative Frequency	Cumulative Relative Frequency
3	3	.100	.900
4	2	.067	.967
5	0	.000	.967
6	1	.033	1.000
Total	30	1.000	

Median can be identified as the first value with cumulative relative frequency over .50. In this case, the first cumulative relative frequency over .50 is .633, and its corresponding value is 1. If you use the same method as Example 2.7 by identifying two middle locations in a sample size $n = 30$, the values of 15th and 16th positions are both 1; therefore, they are averaged to obtained the median = 1. You get the same results using different methods to obtain the median.

Here is a summary of the steps to identify the median in a distribution. Median can only be obtained in ordinal, interval, or ratio variables.

1. SORT the data from the smallest value to the largest value.
2. If n is an odd number, the location for the middle position is $L = n/2$, then round up to the next integer. Identify median as the value of the Lth location in the distribution.
3. If n is an even number, there are two middle positions in the data. They are located in Lth and $(L + 1)$st positions. Median is the average of the values at those two middle locations.
4. In a frequency table, the median is the first value with cumulative relative frequency over .50.

What you need to know about the median is that

1. its calculation does not involve every value in the variable, and
2. it is not affected much by extreme values of the variable. It remains relatively stable even with extreme values in the variable.

Mean

The **mean** (also referred to as the average, or arithmetic mean) is a measure of central tendency that is determined by adding all the values and dividing by the number of observations. The formula for the sample mean is $\bar{X} = \frac{\Sigma X}{n}$ and the formula for the population mean is $\mu = \frac{\Sigma X}{N}$.

The common element of the formulas for $\bar{X}$ and μ is ΣX. As mentioned in Chapter 1 in the section on Statistical Notations,

$$\Sigma X = X_1 + X_2 + X_3 + \cdots + X_n$$

The summation of *X*, ΣX, denotes adding all values of *X*, from the first value to the last value. The mean is calculated by the sum of all values divided by the number of observations, $\bar{X} = \frac{\Sigma X}{n}$ or $\mu = \frac{\Sigma X}{N}$.

When applying the mean formula using frequency tables where individual values occur different numbers of times, the formula needs to be modified to $\bar{X} = \frac{\Sigma X}{n} = \frac{\Sigma fX}{n}$. Such a modification allows the impact of frequencies to be fully incorporated. The same modification also applies to calculation of the population mean. We will use an example to illustrate this process.

EXAMPLE 2.9

The numbers of times a group of college students used Grubhub to have food delivered to them during last week are reported in Table 2.10. What are the mode, median, and mean of these students' number of Grubhub usage last week?

TABLE 2.10 ● Students' Grubhub Usage

X (Grubhub Usage)	*f* (Frequency)	Relative Frequency	Cumulative Relative Frequency
0	1	.02	.02
1	2	.04	.06
2	2	.04	.10
3	14	.28	.38
4	15	.30	.68
5	16	.32	1.00
Total	50		

The highest frequency is 16 in Table 2.10 and its corresponding value is 5. The mode is 5.

The median is easy to find with the cumulative relative frequency column. It can be readily seen that 4 is the first value with cumulative relative frequency over .50. The median is 4.

The formula for mean is adjusted to $\bar{X} = \frac{\Sigma fX}{n}$ because data are presented in a frequency table. According to the formula, the numerator and the denominator need to be figured out before the division. For the numerator, the order of operations dictates the multiplication of *fX* be conducted first. Therefore, the column of *fX* is added to Table 2.10a. Then at the end, the summation $\sum fX$ is calculated and the answer is 188. For the denominator, the sample size is 50. The mean is $\bar{X} = \frac{\Sigma X}{n} = \frac{\Sigma fX}{n} = \frac{188}{50} = 3.76$. The rule of rounding in reporting median or mean is to report the answer with one more decimal place than the original data. Since the original quiz scores are reported as integers, the answer for the mean should be rounded one place after the decimal point. Mean is 3.8. That is, the students' mean number of usage of Grubhub is 3.8.

TABLE 2.10a ● Students' Grubhub Usage With *fX* Column

X (Grubhub Usage)	*f* (Frequency)	*fX*
0	1	0
1	2	2
2	2	4
3	14	42
4	15	60
5	16	80
Total	50	188

When dealing with data spanning over a large range, creating a frequency table with equal intervals is a good solution. Identifying measures of central tendency involved in frequency tables with equal intervals needs to be explicitly stated here. Let's use Example 2.10 to illustrate this process.

EXAMPLE 2.10

Depression can affect anyone. Table 2.11 shows the ages for a random sample of $n = 35$ adult patients who are diagnosed with depression. What are the measures of central tendency for the patients' ages?

(Continued)

(Continued)

TABLE 2.11 ● Frequency Table of 35 Adult Depression Patients' Ages

Patient's Age	*f* (Frequency)
20–29	7
30–39	9
40–49	8
50–59	7
60–69	4
Total	35

The mode of patients' ages is 30–39 and the median 40–49. When you are asked to estimate the mean of the patients' ages from the frequency table, there is no way to find where the ages are located within the interval. The best single value to estimate each interval is the midpoint of the interval, which is defined by the average of the endpoints in the interval, $X_{midpoint} = \frac{\text{low end} + \text{high end}}{2}$. When the frequency table consists of equal intervals, the formula for the mean $\bar{X} = \frac{\Sigma X}{n} = \frac{\Sigma fX}{n}$ needs to be modified to $\bar{X} = \frac{\Sigma X}{n} = \left(\Sigma fX_{\text{midpoint}} / n.\right)$ The midpoint for the first interval 20–29 is $(20 + 29)/2 = 24.5$. The midpoints are shown in Table 2.11a. The estimated total score is the sum of the midpoint of each interval times its corresponding frequency. Again, multiplication and addition are involved in $\Sigma fX_{\text{midpoint}}$. Multiplication operations have to be done before addition operations. The estimated total age for 35 patients is 1477.5. The mean is $1477.5/35 = 41.2$. Clearly, this estimated mean is slightly different from the actual mean if you have all the individual patient's ages and can add them all up. However, when you are presented with a frequency table without all individual values, this estimated mean provides a quick answer that is very close to the actual mean due to the fact that some numbers in the interval are higher than the midpoint and others are lower than the midpoint. The positive differences and the negative differences might cancel each other out to some extent.

TABLE 2.11a ● Frequency Table of 35 Adult Depression Patients' Ages With ΣfX midpoint

Patient's Age	*f* (Frequency)	X_{midpoint}	fX_{midpoint}
20–29	7	24.5	171.5
30–39	9	34.5	310.5
40–49	8	44.5	356
50–59	7	54.5	381.5
60–69	4	64.5	258
Total	35		1477.5

TABLE 2.12 ● Summary Table on What Central Tendency Measures Apply to Which Scales of Measurement

Central Tendency	Scales of Measurement			
Mode	Nominal	Ordinal	Interval	Ratio
Median		Ordinal	Interval	Ratio
Mean			Interval	Ratio

It is important to know which statistical procedures can be appropriately applied to what scales of measurement. In Table 2.12, I summarize what measures of central tendency can be appropriately applied to which scales of measurement.

In the next sections, we will discuss calculating mean in different situations that involve slight modification of the mean formula.

Weighted Mean

The **weighted mean** is defined as calculating the mean when data values are assigned different weights, w. The formula for weighted mean is $\bar{X} = \frac{\Sigma wX}{\Sigma w}$. Weighted mean is appropriate when every value in the data is not treated equally. One of the most relevant examples for calculating a weighted mean for students is to show how GPA (grade point average) is calculated. Here is a statement I've often heard from students, "I got an A in one course and a C from another course, my average for this semester was a B." Was this statement correct? Not necessarily, as it depends on whether these two courses had the same number of credit hours. If John got an A in a 1-credit-hour American Sign Language course and a C in a 4-credit-hour Behavioral Science Statistics course, then his semester average would be closer to C than to B. When calculating GPA, courses with a higher number of credit hours are considered more heavily than courses with a lower number of credit hours. The credit hours of the courses are the weights that need to be considered. In a four-point system, the grades are worth A = 4, B = 3, C = 2, D = 1, and F = 0 points. Full-time undergraduate students usually take four to six different courses in a semester. Let's demonstrate how to calculate a student's GPA by using an example.

EXAMPLE 2.11

Chris received a B in Abnormal Psychology (3 credit hours), an A in Behavioral Science Statistics (4 credit hours), an A in Rock and Blues (1 credit hour), a C in Psychology of Women (3 credit hours), and a B in Biology (5 credit hours) last semester. What is Chris's GPA for last semester?

(Continued)

(Continued)

Let's start by organizing all the courses in a table including the courses, grades, and credit hours as shown in Table 2.13.

TABLE 2.13 ● Chris's Courses, Grades, and Credit Hours

Course	X (Grade)	w (Credit Hours)
Abnormal Psychology	B = 3	3
Behavioral Science Statistics	A = 4	4
Rock and Blues	A = 4	1
Psychology of Women	C = 2	3
Biology	B = 3	5

According to the formula for weighted mean, $\bar{X} = \frac{\Sigma wX}{\Sigma w}$, the multiplication wX needs to be conducted first. Therefore, the column of wX is created in Table 2.13a.

TABLE 2.13a ● Chris's Courses, Grades, and Credit Hours With *wX* Column

Course	X (Grade)	w (Credit Hours)	wX
Abnormal Psychology	B = 3	3	9
Behavioral Science Statistics	A = 4	4	16
Rock and Blues	A = 4	1	4
Psychology of Women	C = 2	3	6
Biology	B = 3	5	15
Total		16	50

Once the column of wX is created, the $\Sigma wX = 50$ is calculated at the end of the column, which provides the total weighted points earned. Then the value of the total weighted points is divided by the total number of credit hours (the total weight). This division gives the weighted mean = 50/16 = 3.125. Again, the rounding off rule states the answer of a calculation is one more decimal place than the original data. The original grades are reported as integers. The weighted mean should be reported at one place after the decimal point, Chris's GPA for last semester is 3.1.

What you need to know about mean is that

1. it applies to interval or ratio measures which come with equal units,
2. its calculation involves every value in the variable, therefore,
3. it is sensitive to the impact of *outliers*.

Outliers refer to extreme values in a distribution. Outliers usually stand far away from the rest of the data points. They are not representative of the sample or population. In some statistical procedures, outliers can have heavy influences on the results. In central tendency measures, mean is the only one that is heavily influenced by outliers. Mode and median are not likely to be influenced by outliers.

Locations of Mean, Median, and Mode in Different Shapes of Distribution

In a normal distribution, the highest frequency is in the middle, then it symmetrically tapers off on both sides. The right side is a mirror image of the left side; thus, mean, median, and mode are located in the same spot, as shown in Figure 2.14.

The mode is the value with the highest frequency, which is at the peak exactly in the middle. The median is the midpoint of the distribution, so it is also precisely in the middle. Because of the symmetrical tapering off on both sides, the values above the mean on the right side compensate for the values below the mean on the left side perfectly. Therefore, the mean stays exactly in the middle as well. In a normal distribution, the mean, median, and mode are located at the same spot, exactly in the middle.

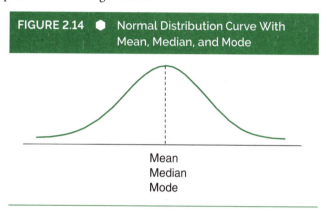

FIGURE 2.14 ● Normal Distribution Curve With Mean, Median, and Mode

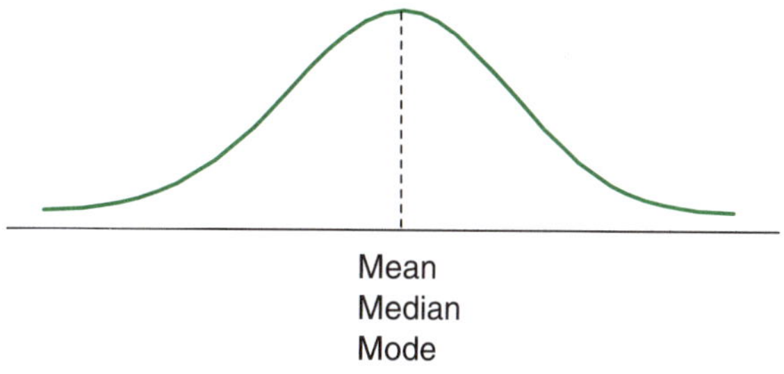

The locations of the mean, median, and mode in skewed distributions are quite different. In a positively skewed (right skewed) distribution, the mode is the value with the highest frequency. This occurs at the peak, which is located to the left side as marked in Figure 2.15. The mean is the arithmetic average of all values and is sensitive to the impacts of outliers. Therefore, the long tail to the right side is pulling the mean toward the tail. The median is in between the mode and the mean.

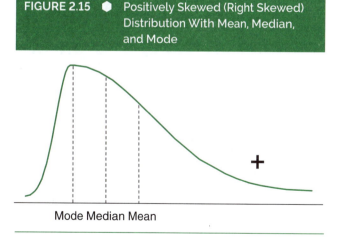

FIGURE 2.15 ● Positively Skewed (Right Skewed) Distribution With Mean, Median, and Mode

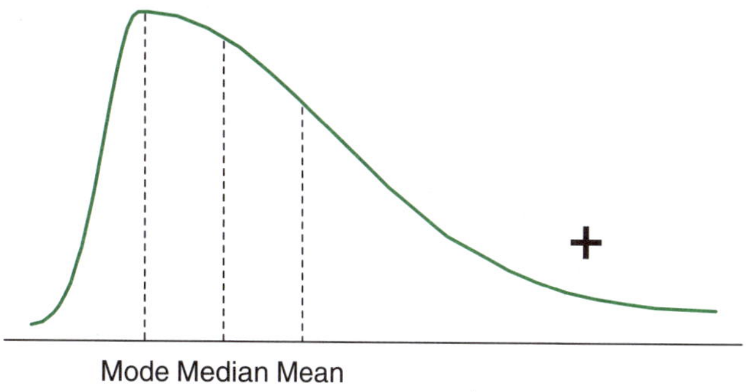

In a negatively skewed distribution, the peak is located to the right side as marked in Figure 2.16. The peak is the mode which is the value with the highest frequency. The mean is the arithmetic average of all values and is sensitive to the impacts of outliers.

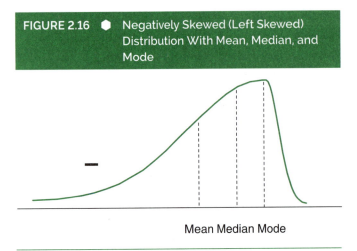

FIGURE 2.16 ● Negatively Skewed (Left Skewed) Distribution With Mean, Median, and Mode

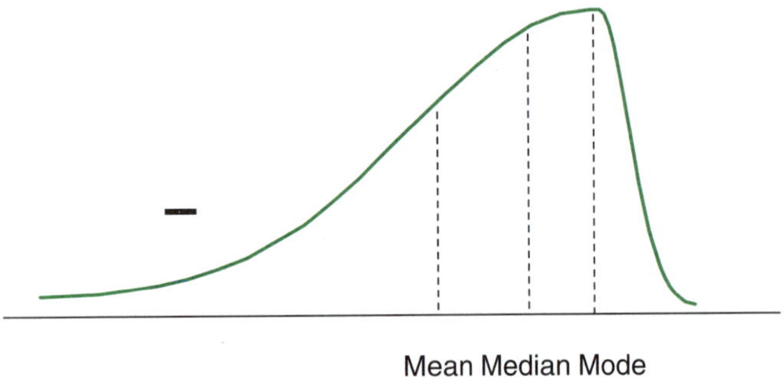

Therefore, the long tail to the left side is pulling the mean toward the tail. The median is in between the mode and the mean.

There are three measures for central tendency. Which measure of central tendency is the best? The answer to that question is not simple. It depends on the shape of the distribution and the purpose for the central tendency. For example, let's use annual income in the United States as an example. Income is known to have a positive skewed distribution. Few outliers with extremely high annual salary would have inflated the mean salary; therefore, the mean is less accurate in describing ordinary people's annual salary. The mean is higher than the median in a positively skewed distribution, because every number is used to calculate the mean, the outliers pull the mean toward the tail (high end). To avoid outliers unduly influencing the central tendency measures, the median is a more appropriate measure to represent ordinary people's annual salary than is the mean.

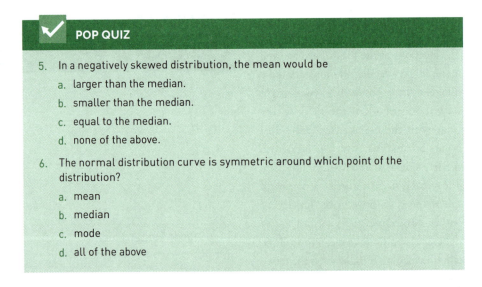

POP QUIZ

5. In a negatively skewed distribution, the mean would be
   a. larger than the median.
   b. smaller than the median.
   c. equal to the median.
   d. none of the above.
6. The normal distribution curve is symmetric around which point of the distribution?
   a. mean
   b. median
   c. mode
   d. all of the above

EXCEL STEP-BY-STEP INSTRUCTION FOR CALCULATING $\sum X$ FROM A FREQUENCY TABLE AND CONSTRUCTING A BAR GRAPH AND A HISTOGRAM

To calculate $\sum X$ using a frequency table in Excel, we use the network interruption example. First, enter the data showing daily network interruptions in September in Excel as shown in columns **A** and **B** in Figure 2.17.

FIGURE 2.17 ● Programming *fX* in Excel

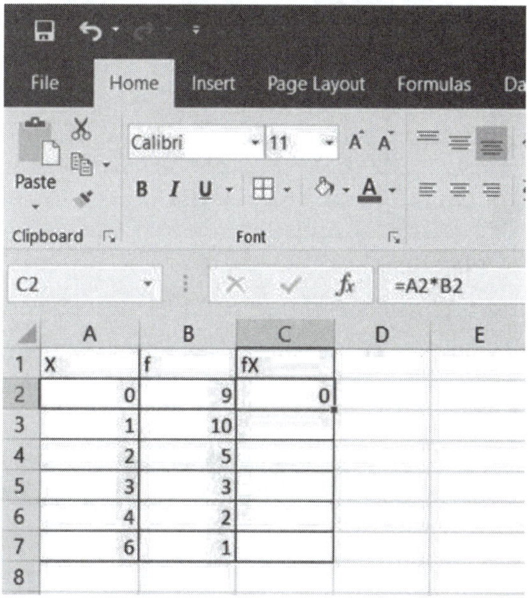

The $\sum X$ formula needs to be adjusted to $\sum fX$ using the frequency table. You need to calculate the multiplicative product of the frequency (f) times the value X by moving the cursor to **C1** and label it as fX, then move to **C2** and type "**=A2*B2**" for the multiplication. All formulas start with "=," **A2** is the first value of X, and **B2** is the corresponding frequency. Then hit **enter.** Notice **C2** shows 0, which is the result of **A2*B2** = 0 × 9 as shown in Figure 2.17.

Then move the cursor to the lower right corner of **C2** until + shows up; then left click and hold the mouse and drag it to **C7**, and then release the mouse. You will see that the rest of the multiplicative products of *fX* automatically show up in **C3** to **C7**. Now you need to sum up all the *fX* values from **C2** to **C7**. Move the cursor to **C8** and type "**=SUM(C2:C7)**". Then hit **enter.** The answer for the total number of network interruptions is 43 as shown in Figure 2.18. If you have trouble following the description of steps here, I have good news for you. I created a 3-minute YouTube video for you. You may find it by using this URL https://www.youtube.com/watch?v=14G91DzRMvs.

The following graphing instructions apply to Microsoft Excel 2019. Graphing functions are likely to vary across different versions of Excel. However, Excel 2013, 2016, and 2019 all have the same functionality for creating a bar graph. To create a bar graph of daily network interruptions, first highlight the data range including the X values, their

FIGURE 2.18 ● Programming $\sum fX$ in Excel

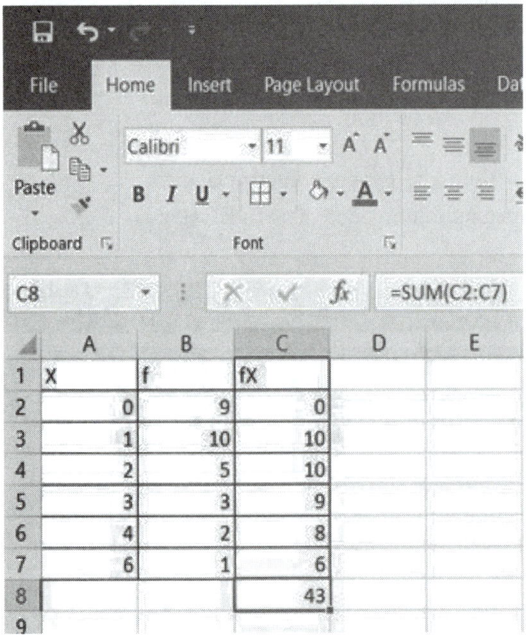

corresponding frequencies and also include the variable names; **A1:B7**. Once these two columns are highlighted, click the **Insert** tab on the top of the tool bar; then click on **Recommended Charts** as shown in Figure 2.19. A new pop-up window appears to display the recommended charts, choose the second chart, **Clustered Column**, on the left panel as shown in Figure 2.20, then click **OK.** Congratulations, you just created a basic bar chart inside Excel. If you want to refine the chart, you can create a chart title and axis title and place frequency on top of each bar in the chart, so your refined chart looks like the one shown in Figure 2.21. Learning by doing is the best way to get familiar with software functions. Click on different parts of the graphs to see options available. Try different options to see if you like the results. If you make mistakes (we all do), there is always the undo button on the upper left corner to restore the graph. Have fun creating your own graphs.

FIGURE 2.19 ● Finding the Bar Graph Function in Excel

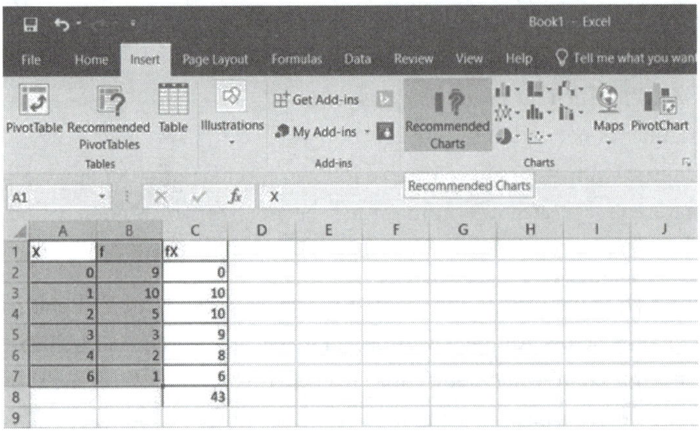

FIGURE 2.20 ● A Pop-Up Window Displaying Recommended Charts

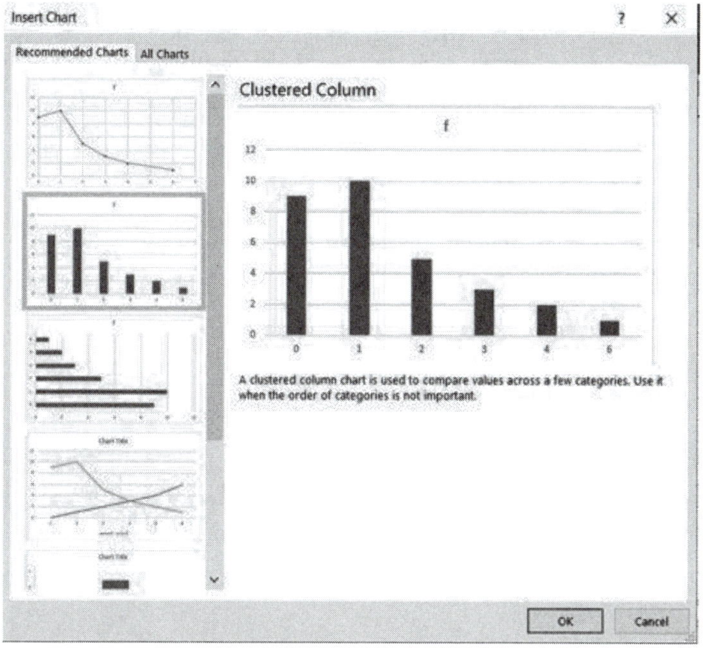

FIGURE 2.21 ● Refined Bar Chart With Axis Titles and Frequency on Top of Each Bar

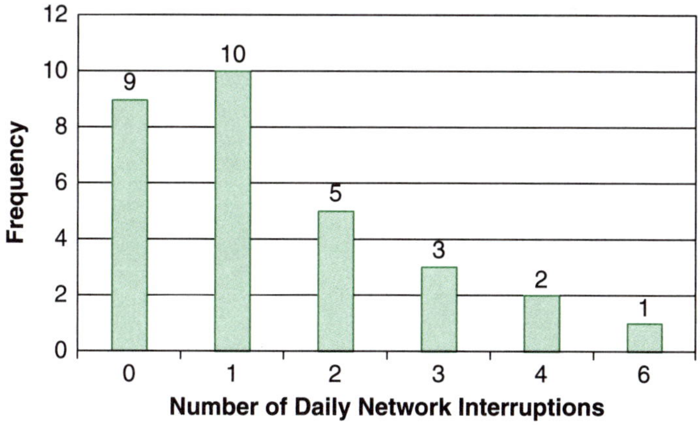

EXERCISE PROBLEMS

1. Workouts are necessary to keep you healthy. Although we all know the importance of keeping an active lifestyle, school and work demands usually get in the way. Here is a list of 20 college students' number of workouts during last week.

   5, 3, 4, 4, 3, 3, 5, 2, 3, 1,

   1, 2, 4, 3, 0, 5, 4, 3, 5, 5

   a. Organize them into a frequency table of the number of workouts for the 20 college students.

   b. Construct a bar graph based on the frequency table.

   c. Identify the measures of central tendency for the number of workouts for these 20 college students.

2. Student loans are at the historical record high. Many college students are incurring large amount of debts while pursuing their degrees. Here is a frequency table of student loans from a random sample of $n = 35$ students in Table 2.14.

   a. Create a histogram based on the frequency table and comment on the shape of the distribution.

   b. What is the estimated mean for the student loans based on the frequency table?

TABLE 2.14 ● Frequency Table of Student Loans

Student Loan (in Dollars)	*f* (Frequency)
0–9,999	1
10,000–19,999	1
20,000–29,999	2
30,000–39,999	3
40,000–49,999	4
50,000–59,999	6
60,000–69,999	8
70,000–79,999	10

Solutions

1. The minimal value of the number of workouts is 0 and the maximal value is 5. The frequency table can be created using the individual values.

a. The frequency table of 20 students' number of workouts is shown in Table 2.15.

TABLE 2.15 ● Frequency Table of Number of Workouts Among 20 Students Last Week

Number of Workouts	*f* (Frequency)
0	1
1	2
2	2
3	6
4	4
5	5
Total	20

b. The bar graph of the number of workouts among 20 college students is shown in Figure 2.22.

FIGURE 2.22 ● Bar Graph of 20 Students' Number of Workouts Last Week

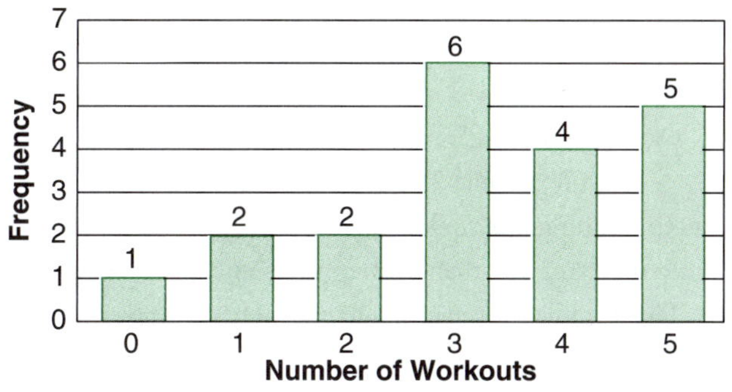

c. The measures of central tendency are mode, median, and mean. The mode is 3, median is 3, and the mean is $\bar{X} = \frac{\sum X}{n} = \frac{\sum fX}{n} = \frac{65}{20} = 3.25.$

2.

a. The histogram was created based on the frequency table of a random sample of 32 students' loans in Table 2.14. The distribution of student loans has a long tail to the left and peaks at the high end of the distribution. The distribution of student loans has the characteristics of a negatively skewed distribution. The histogram is shown in Figure 2.23. The difference between creating a bar graph versus a histogram in **Excel** is in the **Chart Tools**. The default setting automatically creates bar graphs where bars are not touching one another. Click on **Format** tab, and click **Format Selection** on the upper left side. Then a pop-up window showing **Format Data Series** appears. The second sliding ruler shows **Gap Width**. Move the ruler all the way to the left, 0% showing no gap between bars. Your bar graph turns into a histogram.

FIGURE 2.23 ● Histogram of Student Loans

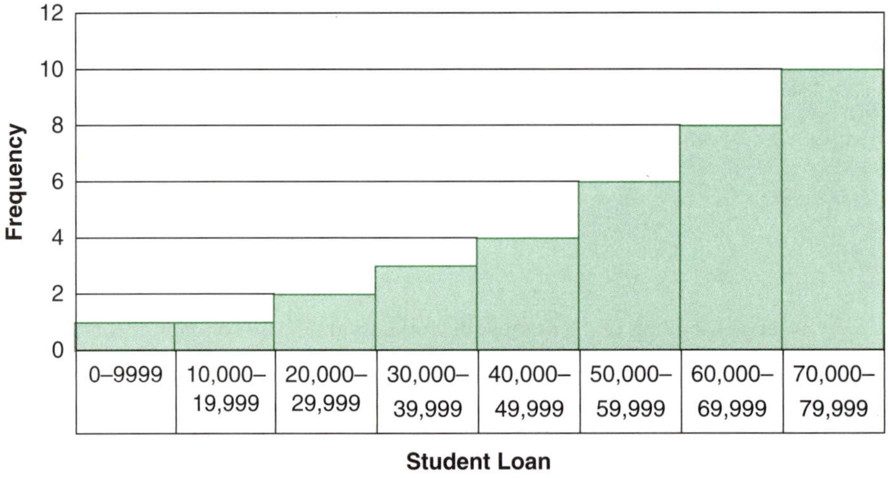

b. To calculate the estimated mean of the student loan based on the frequency table, you need to calculate the midpoint in each interval as shown in the $X_{midpoint}$ column in Table 2.15a. Consider the impact of the frequency, you need to do the multiplication of $f \times X_{midpoint}$ as shown in the last column of Table 2.15a then sum up the numbers at the bottom of the table. The formula for the estimated mean needs to be modified to $\bar{X} = \frac{\sum fX_{midpoint}}{n} = \frac{1954983}{35} = 55856.6$ The estimated mean for the 35 students' loan is $55,856.6. Many students have expressed that the student loan feels like a weight tied around their neck forcing them to assume jobs with the highest paychecks instead of pursuing their passion.

TABLE 2.15a ● Frequency Table of Student Loans With Midpoint in Each Interval

Student Loan (in Dollars)	f (Frequency)	$X_{midpoint}$	$fX_{midpoint}$
0–9,999	1	4,999.5	4,999.5
10,000–19,999	1	14,999.5	14,999.5

Student Loan (in Dollars)	f (Frequency)	$X_{midpoint}$	$fX_{midpoint}$
20,000–29,999	2	24,999.5	49,999
30,000–39,999	3	34,999.5	104998.5
40,000–49,9991	4	44,999.5	1,79,998
50,000–59,999	6	54,999.5	3,29,997
60,000–69,999	8	64,999.5	5,19,996
70,000–79,999	10	74,999.5	7,49,995
Total	35		19,54,983

What You Learned

In this chapter, you have learned to construct frequency tables and graphs to present a quick glance and easy-to-understand data summary. When constructing a frequency table, here are the keys to success:

1. Identify the minimal and maximal values of the variable. Calculate the range.
2. Construct a frequency table with 7 ± 2 categories. When the range is too large to construct a table with individual values, choose an appropriate and convenient interval, usually in 5s or 10s, if feasible.
3. Create the categories or equal intervals in an ascending order.
4. Make sure that there are no overlaps between any two adjacent intervals. For continuous variables, each interval includes the lower limit but not the upper limit.

In bar graphs and histograms, the graphs consist of bars of equal width. The bar width presents the value or interval. The bar height represents the frequency or relative frequency of each value or interval. Bar graphs are used to organize discrete variables, and histograms are used to organize continuous variables. In pie charts, the graphs consist of different slices. The size of the slice is proportional to the frequency or relative frequency of each category.

Central tendency is defined as using one single value to represent the center of the data. There are three common measures: mode, median, and mean.

1. Mode is the value with the highest frequency in the distribution.
2. Median is the midpoint of a sorted distribution. Median is not affected much by extreme outliers.

3. Mean is the arithmetical average of all values. The formula for the population mean is $\mu = \frac{\Sigma X}{N}$ and the formula for the sample mean is $\overline{X} = \frac{\Sigma X}{n}$. When using a frequency table to calculate mean, the formula needs to be modified to $\overline{X} = \frac{\Sigma fX}{n}$ to incorporate the impact of different frequencies on individual values in the table. Same modification also applies to $\mu = \frac{\Sigma fX}{N}$. Mean is highly sensitive to the impact of outliers.
4. When estimated mean is calculated from a frequency table with equal intervals, the formula needs to be modified to $\overline{X} = \frac{\Sigma fX_{\text{midpoint}}}{n}$.

Common distribution shapes include uniform distribution, normal distribution, and skewed distribution.

Key Words

Bar graph: A bar graph uses bars of equal width to show frequency or relative frequency of discrete categorical data (i.e., nominal or ordinal data). Adjacent bars are not touching each other. 37

Central tendency: Central tendency is defined as utilizing a single value to represent the center of a distribution. There are three commonly used measures for central tendency: mean, median, and mode. 45

Cumulative relative frequency: Cumulative relative frequency is defined as the accumulation of the relative frequency for a particular value and all the relative frequencies of lower values. 48

Data visualization: Data visualization is the graphic representation of data. 37

Distribution: Distribution is the arrangement of values of a variable as it occurs in a sample or a population. 29

Equal intervals: Equal intervals are created in a frequency table by including the same number of values in each interval. 33

Frequency: Frequency is a simple count of a particular value or category occurring in a sample or a population. 29

Frequency distribution table: A frequency distribution table lists all distinct values or categories arranged in an orderly fashion in a table, along with tally counts for each value or category in a data set. 29

Histogram: A histogram is a graphical presentation of a continuous variable. The bars of a histogram are touching each other to illustrate that the values are continuous. 38

Line graph: A line graph is a graphical display of quantitative information with a line or curve that connects a series of adjacent data points. 43

Mean: the mean is defined as the arithmetic average of all values in a data distribution. 49

Median: the median is defined as the value in the middle position of a sorted variable arranged from the lowest value to the highest value. 47

Mode: The mode is defined as the value or category with the highest frequency in a data distribution. 45

Negatively skewed distribution: When the data points mostly concentrate on the high-end values with a long tail to the low-end values, the distribution is negatively skewed. 44

Normal distribution: A normal distribution curve peaks at the mean of the values and symmetrically tapers off on both sides with 50% of the data points above the mean and 50% of the data points below the mean. Draw a line straight through the mean, and the left side and the right side are mirror images of each other. 43

Outliers: Outliers refer to extreme values in a distribution. Outliers usually stand far away from the rest of the data points. 55

Pie chart: It is a circular graph that uses slices to show different categories. The size of each slice is proportional to the frequency or relative frequency of each category. 40

Positively skewed distribution: When the data points mostly concentrate on the low-end values with a long tail to the high-end values, the distribution is positively skewed. 44

Real limits: Real limits cover a range of possible values that may be reflected by a single continuous measure. The lower limit is the value minus ½ of the unit and the upper limit is the value plus ½ of the unit. 34

Relative frequency: Relative frequency is defined as the frequency of a particular value or category divided by the total frequency (or sample size). Relative frequency is also called proportion. 30

Skewed distribution: A skewed distribution happens when data points are not symmetrical, and they concentrate more on one side of the mean than on the other. 44

Uniform distribution: In a uniform distribution, every value appears with the same frequency, proportion, or probability. 41

Weighted mean: The weighted mean is defined as calculating the mean when data values are assigned different weights, *w*. The formula for weighted mean is $\bar{X} = \frac{\Sigma wX}{\Sigma w}$. 53

Learning Assessment

Multiple Choice Questions: Choose the Best Answer in the Following Questions.

1. A __________ skewness is a distribution with a high concentration of data points at the high end of the distribution and a long tail toward the low end of the distribution.

a. negative
b. positive
c. discrete
d. continuous

2. A proportion describes the frequency (f) of a category in relation to the total frequency or sample size (n), f/n It is also called a __________.
   a. frequency
   b. relative frequency
   c. cumulative relative frequency
   d. correlation

3. The appropriate number of categories in a frequency table is
   a. 5 ± 2
   b. 7 ± 2
   c. 9 ± 2
   d. 11 ± 2

4. What is the sample size according to the frequency table of students' homework grades?

X (Homework Grade)	f (Frequency)
0	2
6	4
7	4
8	12
9	14
10	17

   a. 40
   b. 41
   c. 51
   d. 53

5. Calculate mean quiz score of students based on the following frequency table.

X (Quiz Grade)	f (Frequency)
0	2
1	4

X (Quiz Grade)	*f* (Frequency)
2	4
3	12
4	14
5	17

a. 2.7
b. 3.1
c. 3.6
d. 4.8

6. What is wrong with the following frequency table?

X (Exam Grade)	*f* (Frequency)
40–50	2
50–60	4
60–70	4
70–80	12
80–90	4
90–100	7

a. The frequency table does not consist of equal intervals
b. The number of intervals is too large
c. The number of intervals is too small
d. There are overlaps between two adjacent intervals

7. In a sample where outliers are present, the worst statistical measure to represent the central tendency is __________.
a. the mean
b. the median
c. the mode
d. all of the above

8. In a positively skewed distribution, scores with the highest frequencies are __________.
a. on the high end (right side) of the distribution
b. on the low end (left side) of the distribution
c. in the middle of the distribution
d. represented at two distinct peaks

9. In a frequency distribution, the __________ is the score or category that has the highest frequency.
   a. mean
   b. median
   c. mode
   d. range

10. When there are 12 observations in a small sample, the median is calculated by
    a. the sixth position of an ascending sorted data
    b. the average value of the sixth position and the seventh position of an ascending sorted data
    c. the sixth position of the data as they were presented without sorting
    d. the average value of the sixth position and the seventh position of the data as they were presented without being sorted

11. Mean is an appropriate statistical measure for central tendency for __________ scales of measurement.
    a. nominal, ordinal, interval, and ratio
    b. ordinal, interval, and ratio
    c. both interval and ratio
    d. only ratio

12. In a positively skewed distribution, the mean would be
    a. larger than the median
    b. smaller than the median
    c. equal to the median
    d. none of the above

Free Response Questions

13. There is a small city in Ohio with a population of 179 people and 0.27 miles on a busy interstate highway. It is famous for its speed traps. A speed trap is defined as a section of a road where police, radar, or traffic cameras frequently check the speed of motorists and strictly enforce traffic regulations. A municipality can intentionally lower the speed limits to catch motorists cruising by without paying attention to the speed signs to boost its revenues. Table 2.16 shows a summary of citation amount intervals and frequency associated with each interval. You may answer the question by using your calculator and draw the histogram by hand or you may use the Excel step-by-step instructions to answer the following questions.
    a. What was the total dollar amount collected from the traffic citations in this small city in the past 12 months?

b. What is distribution shape of the traffic citations? Create a histogram and comment on the shape of the distribution of traffic citations.

c. What are the mode, median, and mean of the traffic citations?

TABLE 2.16 ● Frequency Table of Traffic Citation Amounts

X (Citation Amount in Dollars)	*f* (Frequency)
90–109	12
130–149	25
150–169	46
170–189	105
190–209	852
210–229	907
230–249	981

14. A summary of all employees' annual salary from a local company is shown in Table 2.17. You may answer the question by using your calculator and draw the histogram by hand or you may use the Excel step-by-step instructions to answer the following questions.

 a. What is the distribution shape of the employees' income? Create a histogram of employees' annual salary and comment on the shape of the distribution.

 b. What are the mode, median, and the estimated mean of employees' annual salary?

TABLE 2.17 ● Frequency Table of Employees' Annual Salary

X (Employee Annual Salary)	*f* (Frequency)
20,000–39,999	329
40,000–59,999	342
60,000–79,999	198
80,000–99,999	60
100,000–119,999	13
120,000–139,999	3
140,000–159,999	2
160,000–179,999	1

15. A brief survey of college students revealed their reported hand washing routine immediately after returning to their private living space from outside during COVID-19 pandemic. The results are shown in Table 2.18. Use the frequency numbers in the table to create a bar chart to describe the shape of the distribution of college students' hand washing behavior during the pandemic.

TABLE 2.18 ● Frequency Table of College Students' Hand Washing Behavior

X (Hand Washing Behavior)	*f* (Frequency)
Never	2
Seldom	4
Sometimes	23
Most of the time	39
Always	53

Answers to Pop Quiz Questions

1. c
2. b
3. d
4. a
5. b
6. d

3

MEASURES OF VARIABILITY

Learning Objectives

After reading and studying this chapter, you should be able to do the following:

- Describe and differentiate among the three common variability measures: (1) range, (2) variance, and (3) standard deviation
- Calculate variance and standard deviation using formulas
- Adjust formulas to calculate variance and standard deviations when using frequency tables
- Calculate variance and standard deviation using Excel
- Apply the correct formulas to calculate variance and standard deviation in populations or samples

WHAT YOU KNOW AND WHAT IS NEW

In Chapter 2, you learned to organize and summarize data by constructing frequency tables or graphs to provide useful insights at a quick glance. You also learned how to identify and/or calculate measures of central tendency.

In this chapter, you will learn new quantitative terms and formulas, such as *variability measures* to describe how individual values differ from one another. Psychology is all about studying individual differences. Why is variability important? Variability is an essential part of descriptive statistics, but it plays a huge role in inferential statistics as well. It provides useful quantitative information about how accurately we can use sample statistics to represent population parameters. There are several ways to quantify the variation or dispersion of values of a variable. Using quantitative terms to describe variables is an efficient way to communicate numerical attributes of them. The definitions of these quantitative terms will be clearly explained and illustrated in this chapter.

MEASURES OF VARIABILITY

Variability describes the extent to which observed values from a variable are dispersed, spread out, or scattered. Basically, variability measures tell us how individual values differ from one another in a sample or a population. When variability is large, we have large differences among the values. When variability is small, the values are similar to one another. There are three common **variability measures**: (1) range, (2) variance, and (3) standard deviation. These measures of variability involve calculating differences that require an interval or ratio scale of measurement.

Range

Range is defined as the difference between the maximum and the minimum in the observed values of a variable. As stated in the definition, range only takes the two most extreme values in its calculation, so it is highly sensitive to the impact of outliers. It is a simple calculation but not particularly useful in describing differences among individual values other than the maximum and minimum. Let's use an example to illustrate this point.

$$\text{Range} = \text{Maximum} - \text{Minimum}$$

EXAMPLE 3.1

In a family of five, the ages of the five family members are 17, 22, 26, 48, and 57. What is the range of this population?

The reason these five numbers are referred to as a population is that there are five members in this family and every member's age is listed in the problem statement. The range is calculated by the formula, Range = Maximum − Minimum = 57 − 17 = 40.

The range provides a simple but unreliable measure of variability. Statisticians were not satisfied with the results. Therefore, better ways of measuring variability were invented to ensure that two considerations are met: (1) the difference is calculated between each individual value and the mean and (2) every value is included in the calculation. Variance and standard deviation are developed to provide more precise measures of variability. The formulas for calculating variance and standard deviation are different in a population versus a sample, so they are discussed in separate subsections.

Variance and Standard Deviation for a Population

Variance and standard deviation are designed to study individual differences. I plan to illustrate the process of evolving from deviation to **standard deviation**. To quantify the individual differences, we calculate the difference between individual values and their mean. In statistics, deviation means difference between two values. In a population, deviation from the mean is defined and calculated as $X - \mu$. Deviation from the mean can be calculated for every value. When we have many numbers, we have the tendency of adding them all up. Unfortunately, that's where the problem occurs. Let's illustrate this problem with an example.

EXAMPLE 3.2

In a family of five, calculate the deviation from mean age for all family members: 17, 22, 26, 48, and 57. What is the sum of deviations?

Just to clarify, population does not always contain large number of people. How big the population is depends on your research interest. My interest here is only on this particular family. The example lists every member's age, so it is a population. Let's put the calculation process in Table 3.1. $\mu = \Sigma X/N = 170/5 = 34$. Deviation from mean is $(X - \mu)$, which is calculated for every value as shown in the column. The sum of deviations is added up at the bottom of the column.

TABLE 3.1 ● Ages of Family Members and Deviation From the Mean

X (Age)	$X - \mu$
17	−17
22	−12
26	−8
48	14
57	23
$\Sigma X = 170$	$\Sigma(X - \mu) = 0$

The sum of all deviations adds up to 0. This is not a special case. It is a mathematical certainty for every variable. Let's prove this point by breaking up $\Sigma(X - \mu)$ into $(X_1 - \mu) + (X_2 - \mu) + (X_3 - \mu) + \cdots + (X_N - \mu)$. Let's rearrange this process and present it vertically.

(Continued)

(Continued)

$$
\begin{array}{rccc}
 & (X_1 & - & \mu) \\
 & (X_2 & - & \mu) \\
 & (X_3 & - & \mu) \\
 & \cdots & & \\
+) & (X_N & - & \mu) \\
\hline
 & (\Sigma X) & - & (N\mu) = \Sigma X - \Sigma X = 0
\end{array}
$$

The terms to the left of the minus sign become $X_1 + X_2 + X_3 + \cdots + X_N = \Sigma X$, and the terms to the right of the minus sign become μ times N, that is $N\mu$. Based on the mean formula $\mu = \Sigma X / N$; therefore, multiply N on both sides of the equation and you get $N\mu = \sum X$ As the last step of the calculation, you get $\sum X - \sum X = 0$.

Now, it is mathematically proven that in every situation the sum of deviations is always 0. This is what I referred to as a problem earlier in the subsection. Any mathematical operation that always produces an answer of 0 is not useful in providing any information. The positive deviations completely cancel out the negative deviations. Therefore, the sum of deviations concept needs to be adjusted before it can provide useful information. The useful information is the magnitude of deviations. To retain the magnitude but avoid positive deviations perfectly canceling out negative deviations, the negative signs need to be modified. Squaring is a universal mathematical operation to get rid of negative signs. Therefore, the **sum of squared deviations (*SS*)** was invented. In short, it is called Sum of Squares = $SS = \Sigma(X - \mu)^2$. The **mean squared deviation** is obtained by dividing the sum of squared deviations by the population size. The mean squared deviation is also called the **variance** = $\sigma^2 = SS/N$. Once the differences have been squared, they are no longer at the same scale as the original data. Therefore, a square root operation is needed to reverse it back to the original scale, and it is labeled as the standard deviation. The formula for standard deviation is $\sigma = \sqrt{\text{Variance}} = \sqrt{SS/N}.$ The standard deviation is the most commonly used measure of variability in statistics.

EXAMPLE 3.3

Let's continue to use the example of family members' ages in Table 3.1. What are the variance and standard deviation of their ages?

Based on the formula for $SS = \Sigma(X - \mu)^2$, the order of operations dictates that the column of $(X - \mu)$ be constructed first; then the column for $(X - \mu)^2$ needs to be created as the squared deviation from mean.

Then the sum adds up at the end of the column. A step-by-step calculation process is demonstrated in Table 3.1a.

TABLE 3.1a ● Family Members' Ages, Their Deviations, and Sum of Squared Deviations

X (Age)	$X - \mu$	$(X - \mu)^2$
17	−17	289
22	−12	144
26	−8	64
48	14	196
57	23	529
$\sum X = 170$	$\sum(X - \mu) = 0$	$\sum(X - \mu)^2 = 1{,}222$

Once the table is completed, all the numbers needed for the variance and standard deviation are available.

$$\text{Variance} = \sigma^2 = \frac{SS}{N} = \frac{\sum(X-\mu)^2}{N} = \frac{1222}{5} = 244.4$$

$$\text{Standard deviation} = \sigma = \sqrt{\text{Variance}} = \sqrt{244.4} = 15.6$$

Based on the rounding rule, the final values for the variability measures are reported at one more decimal place than in the original data. It is important to note that the rounding rule applies to the final answer, not the intermediate numbers in the calculation process; otherwise, rounding errors cumulate throughout the calculation steps. The unit for variance is the square of the unit for the original variable, and the unit for range and standard deviation is the same as the original variable. In the case of ages of family members, the unit for variance is year^2, and the unit for range and standard deviation is year. The formula for population variance and standard deviation is useful in theory and in explaining the concepts. But in reality, they are rarely known. This is because the population mean, μ, is usually unknown, and it is almost impossible to obtain measures from every member in the population when the population size is large.

Variance and Standard Deviation for a Sample

In most cases, not every member in the population is available and willing to provide information for research purposes. It is necessary for researchers to use sample statistics to make inferences about the population parameters.

Sample mean, $\bar{X}$, is an unbiased estimate of the population mean, μ, and sample variance, s^2, is an unbiased estimate of the population variance, σ^2. The calculation of sample variance follows the same basic principles used in calculating population variance

A Trip Down the Memory Lane

Frequency tables are useful to summarize data. In Chapter 2, we mentioned that considering the impact of frequency in mathematic operations when using frequency tables by modifying $\sum X$ to $\sum fX$. In Chapter 3, you learn more mathematic formulas such as SS, variance, and standard deviation. You need to consistently apply the same principle to consider the impact of frequency by multiplying the f with the original formula before you sum up the results, for example, modifying $SS = \sum(X - \bar{X})^2$ to $SS = \sum f(X - \bar{X})^2$

except for a few changes in notation that signify the difference between population parameters and sample statistics. Sample mean is $\bar{X}$, and sample size is n. Because the population mean, μ, is unknown, you must calculate the sample mean, $\bar{X}$, to estimate μ. Continue to calculate the deviation from the mean $(X - \bar{X})$ for every individual X value, and then square the deviation $(X - \bar{X})^2$, to avoid positive deviations completely cancel out the negative deviations. Then add up the squared deviations to form the sum of squares, $SS = \sum(X - \bar{X})^2$. Up to this point, the calculations are identical both for a population and for a sample.

$$\bar{X} = \frac{\sum X}{n}$$

$$SS = \sum(X - \bar{X})^2$$

Now, you have to learn the most important distinction between the population variance formula and the sample variance formula. It is called the degrees of freedom, which is calculated as $df = (n - 1)$ for sample variance. The **degrees of freedom** of a statistical procedure is defined as the number of values involved in the calculation that can be free to vary in the sample. The degrees of freedom are usually determined by these two factors: (1) the sample size (n) and (2) the number of parameters (k) that need to be estimated. When you have to estimate one parameter, you lose 1 df. Therefore, the general rule for determining the degrees of freedom is $(n - k)$.

EXAMPLE 3.4

TABLE 3.2 ● Height Measures of a Sample of Four Boys

X (Height)
67
71
72
$X_4 = ?$
$\bar{X} = 69$

In a sample of four boys, the average height was 69 inches. The heights of three boys were measured at 67, 71, and 72 inches (Table 3.2). What was the fourth boy's height?

This question illustrated the point that when the sample mean was known, only $(n - 1)$ values could vary freely. Once the sample mean and height measures of the three boys were given, the fourth boy's height was determined. It can't freely assume any other value.

Apply the formula $\bar{X} = \frac{\sum X}{n}$ and plug in the numbers into the formula $69 = \frac{\sum X}{4}$, therefore, $\sum X = 276$.

$$X_1 + X_2 + X_3 + X_4 = 276$$

$$67 + 71 + 72 + X_4 = 276$$

$$210 + X_4 = 276$$

$$X_4 = 276 - 210$$

$$X_4 = 66$$

This example was meant to illustrate that estimating the population mean by calculating the sample mean results in the loss of 1 *df*. Therefore, only (n – 1) values in the sample can vary freely. Such a restriction applies when calculating a sample variance and standard deviation.

The sample variance is the mean squared deviations, which is the sum of squared deviations divided by the degrees of freedom, $s^2 = \frac{SS}{df} = \frac{SS}{n-1}$, and the sample standard deviation, s, is the square root of the sample variance $s = \sqrt{\frac{SS}{df}} = \sqrt{\frac{SS}{n-1}}$.

You are going to learn many statistical formulas to analyze data. When there are large number of observations in a data set, using hand calculator to solve the problems is no longer efficient. We are going to learn to program formulas into Excel and let the computer do the calculation for you. Before we start, here is a list of basic math operations besides the regular +, –, *, / in Excel as shown in Table 3.3. This table will help you through the process.

TABLE 3.3 ● Basic Excel Math Formula Symbols and Descriptions

Symbol	Description
=	This is an equal sign and denotes the beginning of a formula or a math operation.
:	This is colon and is used in a formula to involve a range of cells (e.g., A2:A21).
$	This is a dollar sign and can be created by holding shift and 4 simultaneously on the keyboard. It is used to create an absolute cell reference. It allows you to always refer to a specific cell even when you copy the formula across a range of cells.
( )	These are round brackets and are used as parentheses in formulas.
^	This is a carat symbol and can be created by holding shift and 6 simultaneously on the keyboard. It is used to indicate an exponent or power in math operations.

In other words, in the process of calculating variance, sample mean, $\bar{X}$, is used to estimate the population mean. This restricts the calculation of the sample variance and, accordingly, 1 *df* is lost. Hence, the degrees of freedom for sample variance are (n – 1). Let's use an example to explain the degrees of freedom concept.

EXAMPLE 3.5

Statistics courses require students to study on their own outside of the class time to be able to learn well. I conducted a brief survey on how many hours students studied statistics on their own outside of class time last week. In a random sample of 20 students, their reported hours of study last week were summarized in Table 3.4. What are the variability measures for these 20 students' study hours?

a. Use adjusted statistical formulas when using a frequency table to answer this question.

b. Use Excel to calculate the variance and standard deviation of these 20 students' study hours.

TABLE 3.4 ● Frequency Table of 20 Students' Study Hours

X (Number of Study Hours)	*f* (Frequency)
0	3
3	7
5	7
6	2
8	1

Answers:

a. The measures of variability are range, variance, and standard deviation. To answer part a, you need to use the adjusted formulas because the data were presented in a frequency table.

We can conduct statistical calculations using frequency tables. The measures of variability include range, variance, and standard deviation.

Range is Maximum – Minimum = 8 – 0 = 8

Both variance and standard deviation start with $SS = \Sigma(X - \bar{X})^2$. In frequency tables, individual values occur different numbers of times, so the impact of frequency needs to be incorporated. Therefore, the original *SS* formula, $\Sigma(X - \bar{X})^2$, is modified to $SS = \Sigma f(X - \bar{X})^2$. Based on the order of operations in the *SS* formula, the first step is to create column $(X - \bar{X})$, and the second is to square it to create column $(X - \bar{X})^2$. Remember to consider the impact of frequency by creating $f(X - \bar{X})^2$. These computations are shown in Table 3.4a.

TABLE 3.4a ● Adjusted Formulas Using a Frequency Table of 20 Students' Study Hours

X (Study Hours)	*f* (Frequency)	*fX*	$X - \bar{X}$	$(X - \bar{X})^2$	$f(X - \bar{X})^2$
0	3	0	−3.8	14.44	43.32
3	7	21	−0.8	0.64	4.48
5	7	35	1.2	1.44	10.08
6	2	12	2.2	4.84	9.68

X (Study Hours)	f (Frequency)	fX	$X-\bar{X}$	$(X-\bar{X})^2$	$f(X-\bar{X})^2$
8	1	8	4.2	17.64	17.64
	20	76			85.2

The Mean, $\bar{X} = 76/20 = 3.8$

At the end of the column $f(X-\bar{X})^2$, we sum every number to get the $SS = \sum f(X-\bar{X})^2 = 85.2$.

Sample variance $s^2 = SS/df = SS/(n-1) = 85.2/19 = 4.48$.

Sample standard deviation $s = \sqrt{\text{Variance}} = \sqrt{4.48} = 2.12$.

The result showed that some students did not study outside of the class time while others might have studied 6 or 8 hours. The average number of study hours was 3.8 hours, the variance was 4.48 and the standard deviation was 2.12.

b. Table 3.4 presents a frequency table of 20 students' study hours. Such a frequency table offers a quick summary of students' study hours at a glance. In Excel, no such summary is needed. You simply keep record of 20 students' study hours individually. To set up an Excel data file, you need to treat each row as a person and each column as a variable. Therefore, you enter each student's number of study hours in a separate cell. For example, there were three students with 0 hours, so you need to enter 0 three times in three different cells. Continue entering the data the same way until all 20 students' study hours are recorded in Excel. Now, please take a moment to compare Table 3.4 and Table 3.4b because these two tables contain exactly the same information but in different format. Table 3.4 is a frequency table that provides summary at a quick glance. Table 3.4b recorded the number of study hours one student at a time in Excel. Excel can handle 20 or hundreds or even thousands of observations with ease, so it does not require the organizing or summarizing work.

TABLE 3.4b ● 20 Students' Study Hours in Excel

Number of Study Hours
0
0
0
3
3
3
3
3
3
3
5
5
5
5
5
5
5
6
6
8

There are two different ways to use Excel to get the mean, variance, and standard deviation of a sample of 20 students' study hours: (1) Use the statistics formulas and program the math operations into Excel to get answers and (2) use the statistical

(Continued)

(Continued)

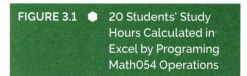

FIGURE 3.1 ● 20 Students' Study Hours Calculated in Excel by Programing Math054 Operations

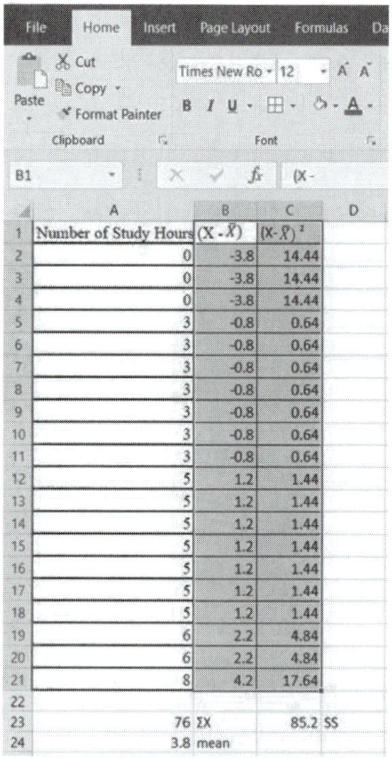

	A	B	C	D
1	Number of Study Hours	$(X-\bar{X})$	$(X-\bar{X})^2$	
2	0	-3.8	14.44	
3	0	-3.8	14.44	
4	0	-3.8	14.44	
5	3	-0.8	0.64	
6	3	-0.8	0.64	
7	3	-0.8	0.64	
8	3	-0.8	0.64	
9	3	-0.8	0.64	
10	3	-0.8	0.64	
11	3	-0.8	0.64	
12	5	1.2	1.44	
13	5	1.2	1.44	
14	5	1.2	1.44	
15	5	1.2	1.44	
16	5	1.2	1.44	
17	5	1.2	1.44	
18	5	1.2	1.44	
19	6	2.2	4.84	
20	6	2.2	4.84	
21	8	4.2	17.64	
22				
23	76	ΣX	85.2	SS
24	3.8	mean		

functions existing in Excel to get answers. I will show you both ways.

(1) Being able to follow statistical formulas to program math operations into Excel provides greatest flexibility for users to be able to do any analysis they need to do. To calculate variance and standard deviation, we need to calculate *SS* first. To calculate the *SS*, we need to find out the $\bar{X}$. Excel can handle tedious calculations quickly, regardless of the sample size. Just for the visual effect, I will show answers at the end of the data. The formula for the mean is $\bar{X} = \frac{\sum X}{n}$. Move the cursor to **A23**, type "**=SUM(A2:A21)**". Then hit **Enter**. The answer for the total number of study hours for 20 students is $\sum X = 76$. You probably noticed that Excel does not label the result, so make it a habit to label the result in a nearby cell such as **B23**, type in "**ΣX**" to label the answer of 76. In **A24**, type "**=A23/20**" to calculate Mean $\bar{X} = \frac{\sum X}{n} = \frac{76}{20} = 3.8$. Or alternatively, in **A24**, you may type "**=AVERAGE(A2:A21)**" to get the mean of 20 students' study hours. In a nearby cell such as **B24**, type "**mean**" to label the answer of 3.8 as shown in Figure 3.1. There are usually more than one way to reach the same goal in Excel.

The value of $\bar{X}$ allows us to calculate the *SS*. Its formula is $SS = \sum(X-\bar{X})^2$. Based on the order of operations, $(X-\bar{X})$ and $(X-\bar{X})^2$ need to be calculated next. In **B1**, label the column as $(X-\bar{X})$ I have to admit that it took a long time for me to create the label $(X-\bar{X})$ in Excel so I suggest that you use (X-mean) as a substitute label. In **B2**, type "**=A2-A$24**", which shows the operation of first $(X-\bar{X})$ The **$** in the Excel program allows us to create an absolute cell reference in **A24** to calculate each individual value of *X* and its difference from the $\bar{X}$ for the subsequent operations. Therefore, it is possible to copy the formula to produce the rest of $(X-\bar{X})$. Then move the cursor to the lower right corner of **B2** until + shows up, then left click and hold the mouse and drag it to **B21**, then let go. You will see the rest of $(X-\bar{X})$ automatically show up in **B3** to **B21**. In **C1**, label the column as $(X-\bar{X})^2$ or (X-mean)^2 as a substitute label In **C2**, type "**=B2^2**", which shows the operation of first $(X-\bar{X})^2$ Then move the cursor to the lower right corner of **C2** until + shows up then left click and hold the mouse and drag it to **C21**, then let go. You will see the rest of $(X-\bar{X})^2$ automatically show up in **C3** to **C21**. In **C23**, type "**=SUM(C2:C21)**" to obtain $SS = \sum(X-\bar{X})^2 = 85.2$, and in **D23** type "**SS**" to label the answer of 85.2 as shown in Figure 3.1. Once we know $SS = 85.2$, it is easy to figure out sample variance, $s^2 = \frac{SS}{n-1} = \frac{85.2}{19} = 4.48$ and standard deviation, $s = \sqrt{s^2} = \sqrt{4.48} = 2.12$.

The results calculated by Excel matched with the ones from frequency table.

(2) Next, I will show you how to solve the same problem by using the existing statistical functions in Excel. Let's calculate sample mean, variance, and standard deviation of 20 students' study hours. Use the same data setup as in (1), move the cursor to a blank cell, **A23**, where you want the answer of the sample mean to appear and type "=**AVERAGE(A2:A21)**", then hit **Enter**. This command means calculating the sample mean from the first student's study hours to the 20th student's study hours. Then 3.8 appeared in **A23**. You just got the answer of 3.8 hours as the mean for 20 students' study time without programming any math operation on your own as shown in Figure 3.2. Then move cursor to **B22** where you want the answer of the sample variance to appear and type "=**VAR.S(A2:A21)**", then hit **Enter**. This command means calculating sample variance from the first student to the 20th student's study hours. Thus, 4.484 appeared in **B22**. To calculate the sample standard deviation, move the cursor to **B23**, where you want the answer to appear and type "=**STDEV.S(A2:A21)**", then hit **Enter**. This command means calculating sample standard deviation from the first student to the 20th student's study hours. Thus, 2.118 appeared in **B23** as shown in Figure 3.2. The existing statistical functions in Excel incorporate first few letters of the math operations to form the commands. It is obvious that **AVERAGE** stands for mean, **STDEV.S** stands for sample standard deviation, and **VAR.S** stands for sample variance. Due to the fact that there are different formulas for population standard deviation and population variance, you could probably guess their corresponding commands in Excel to be **STDEV.P** and **VAR.P** if you ever have the chance to actually analyze population data.

FIGURE 3.2 ● 20 Students' Study Hours Calculated in Excel by Using Existing Statistical Functions

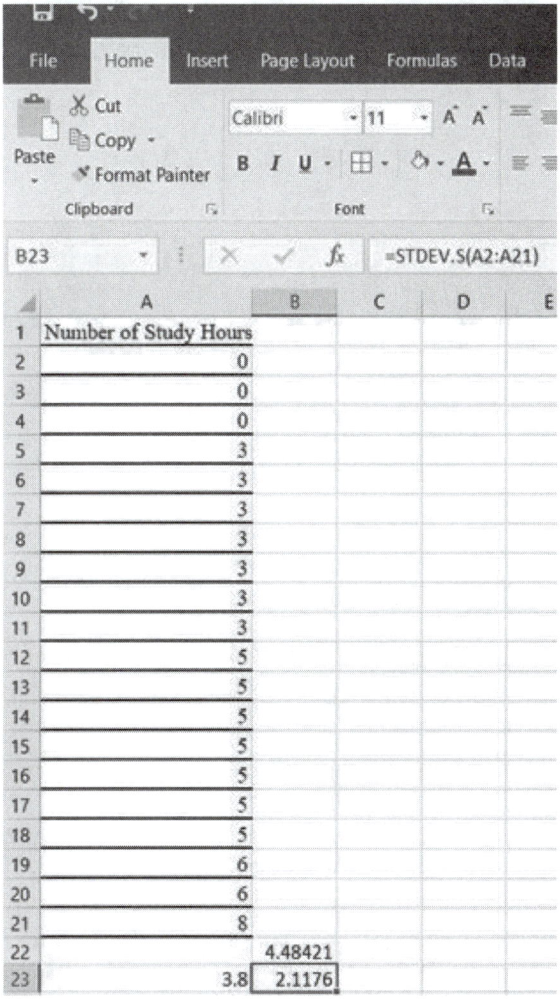

In the beginning of the chapter you are told that variability measures show how individual values differ from one another in a sample or a population. When variability is large, we have large differences among the values. When variability is small, the values are similar to one another. Let's use an example to illustrate this point. I will use the existing statistical functions in Excel to do the calculations.

EXAMPLE 3.6

TABLE 3.5 ● Male and Female Students' Homework (HW) Grades

Males' HW	Females' HW
1	3
1	3
4	4
5	4
5	4
6	5
9	5
9	5
	5
	6
	6
	6
	7
	7

A homework assignment was worth 10 points. The instructor randomly selected a sample of male students' homework and a sample of female students' homework grades. The grades are reported in two separate columns in Table 3.5. Use Excel to find out the mean, variance, and standard deviation for male and female students' HW grades. Discuss the results.

Here are the step-by-step instructions to use the existing Excel functions.

1. Enter the data in Table 3.5 into Excel. Pick a blank cell, **A17**, where you want the answer of the males' mean HW grades to show up, type "=**AVERAGE(A2:A9)**", then hit **Enter**. The answer, 5, shows up. Pick a blank cell, **B17**, where you want the answer of the females' mean HW grades to show up, type "=**AVERAGE(B2:B15)**", then hit **Enter**. The answer, 5, shows up as displayed in Figure 3.3.
2. Pick a blank cell, **A18**, where you want the variance of males' HW grades to show up, type "=**VAR.S(A2:A9)**", then hit **Enter**, the answer, 9.428571, shows up. Pick a blank cell, **B18**, where you want the variance of the females' HW grades to show up, type "=**VAR.S(B2:B15)**", then hit **Enter**. The answer, 1.69230769, shows up as displayed in Figure 3.3.
3. Pick a blank cell, **A19**, where you want the standard deviation of males' HW grades to show up, type "=**STDEV.S(A2:A9)**", then hit **Enter**, the answer, 3.070598, shows up. Pick a blank cell, **B19**, where you want the standard deviation of the females' HW grades to show up, type "=**STDEV.S(B2:B15)**", then hit **Enter**. The answer, 1.30088727, shows up as displayed in Figure 3.3.

4. Excel existing statistical functions do not label the answers, so I use nearby cells to label them by moving the cursor to **C17** and typing "**sample mean**", to **C18** and typing "**sample variance**", and to **C19** and typing "**sample standard deviation"** as shown in Figure 3.3.

Using the existing statistical functions to calculate sample mean, variance, and standard deviation is easy once you know the command and know how to specify the data range in them. Let's discuss the results. The males and females had the same mean but very different variances and standard deviations. Male students had a variance of 9.43 and a standard deviation of 3.07. Female students had a variance of 1.69 and a standard deviation of 1.30. Male students' variance and standard deviation were much larger than female students'. Male students had more extreme scores on both ends and female students had scores near the middle of the distribution. Variance and standard deviation reflect the magnitude of the spread among individual values. Large variance and standard deviation indicate that individual values vary wildly and small variance and standard deviation indicate that individual values are similar to one another.

FIGURE 3.3 ● The Mean, Variance, and Standard Deviation of Males' and Females' Homework Grades

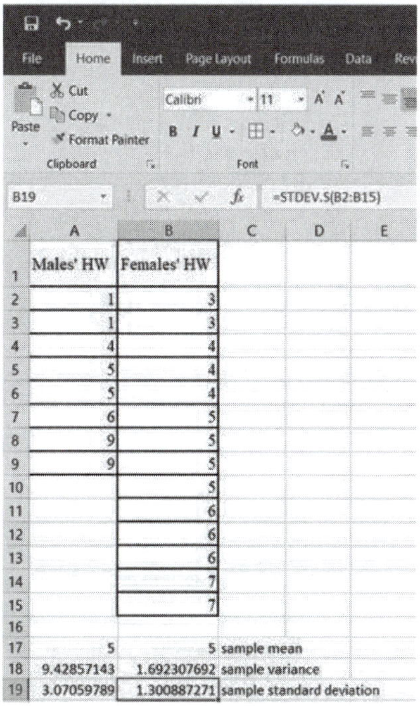

Excel provides prepopulated functions for you to use. Table 3.6 provides a list of basic statistical functions that we discussed in Examples 3.5 and 3.6.

To summarize all the formulas in mean, variance, and standard deviation and when formulas need to be modified when data are presented in frequency tables, all formulas in either populations or samples are listed in Table 3.7.

TABLE 3.6 ● Excel Basic Statistical Functions

Function	Formula	Example
To add up a total	=SUM(cell range)	=SUM(A2:A9)
To calculate a mean	=AVERAGE(cell range)	=AVERAGE(A2:A9)
To calculate a population variance	=VAR.P(cell range)	=VAR.P(A2:A9)
To calculate a sample variance	=VAR.S(cell range)	=VAR.S(A2:A9)
To calculate a population standard deviation	=STDEV.P(cell range)	=STDEV.P(A2:A9)
To calculate a sample standard deviation	=STDEV.S(cell range)	=STDEV.S(A2:A9)

TABLE 3.7 ● A Summary Table of Statistical Formulas in Populations or Samples

Statistical Function	Population	Sample
Mean	$\mu = \frac{\Sigma X}{N}$	$\bar{X} = \frac{\Sigma X}{n}$
Mean modified in frequency tables	$\mu = \frac{\Sigma fX}{N}$	$\bar{X} = \frac{\Sigma fX}{n}$
Sum of squared deviation (*SS*)	$SS = \Sigma(X - \mu)^2$	$SS = \Sigma(X - \bar{X})^2$
SS modified in frequency tables	$SS = \Sigma f(X - \mu)^2$	$SS = \Sigma f(X - \bar{X})^2$
Variance	$\sigma^2 = \frac{SS}{N}$	$s^2 = \frac{SS}{n-1}$
Standard deviation	$\sigma = \sqrt{\frac{SS}{N}}$	$s = \sqrt{\frac{SS}{n-1}}$

POP QUIZ

1. In the process of calculating the sample variance, the sample mean is used to estimate the population mean. This puts one restriction on the number of values in the sample that can vary freely. Therefore, the degrees of freedom for the sample variance are
   a. n.
   b. $(n - 1)$.
   c. $(n - 2)$.
   d. $(n - 3)$.

DISCOVER THE HIDDEN FUNCTIONS IN EXCEL: ANALYSIS TOOLPAK INSTALLATION INSTRUCTIONS

Excel is a widely used spreadsheet program that can be used to create text, numbers, and formulas. It also comes with prepopulated statistical functions to cover many topics such as different types of *t* tests, correlations, regressions, and ANOVAs (analyses of variance), which are discussed in this book. These prepopulated functions are add-ins in Excel that need to be installed by the users. The following link is provided by Microsoft Office Support, which include separate tabs for Windows and MacOS installation instructions (https://support.office.com/en-us/article/load-the-analysis-toolpak-in-excel-6a63e598-cd6d-42e3-9317-6b40ba1a66b4).

Here is the instruction for Microsoft 365 for Mac, Excel 2019, Excel 16, and Excel 2013. If you use different versions of Excel or you use MacOS, please refer to the previous link for instruction.

1. Open the **Excel** worksheet in which you want to use the **Analysis ToolPak**, or create a new worksheet. Click the **File** tab, click **Options**, and then click the **Add-Ins** category.
2. In the **Manage** box, select **Excel Add-ins**, and then click **Go**.
3. In the **Add-Ins** box, check the **Analysis ToolPak** check box, and then click **OK**.
   - If **Analysis ToolPak** is not listed in the **Add-Ins available** box, click **Browse** to locate it.

- If you are prompted that the Analysis ToolPak is not currently installed on your computer, click **Yes** to install it.

4. To verify the **ToolPak** has been successfully installed, click the **Data** tab on the worksheet, you will see a **Data Analysis** option shows up on the upper right side of the toolbar as shown in Figure 3.4.

FIGURE 3.4 ● The Screenshot Shows That You Have Successfully Installed the Analysis ToolPak

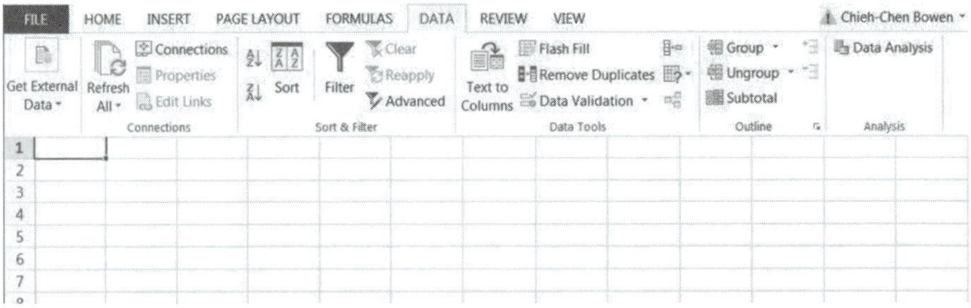

TABLE 3.8 ● Homework Grades for a Sample of 21 Students

X (Homework)	*f* (Frequency)
0	2
7	3
8	4
9	5
10	7

TABLE 3.9 ● Rollover Ratings on SUVs

X (Number of Stars)	*f* (Frequency)
1	3
2	7
3	14
4	16
5	10

EXERCISE PROBLEMS

1. In a sample of 21 students, homework grades are presented in Table 3.8. What are the range, variance, and standard deviation of students' homework grades in this sample?

2. Rollover accounts for 33% of all passenger vehicle fatalities. Taller and narrower vehicles are more likely to roll over. Rollover ratings are from 1 to 5 stars. The higher the number of stars, the less likely is the SUV (sports utility vehicle) to roll over. The results of rollover ratings on a sample of 50 used SUVs are presented in Table 3.9. What are the range, variance, and standard deviation of SUV's rollover ratings in this sample?

Solutions

1. The range of the distribution is the Maximum – the Minimum = 10 – 0 = 10.

I mentioned that there are usually more than one way to reach the goal in Excel, so I decided to show you a different way to solve problem by programming the modified mean and *SS* formulas for frequency tables as shown in Table 3.7 using Excel. Once you learn how to program the statistical formulas in Excel yourself, you can do any operations. To calculate the variance, we need to calculate the *SS* (sum of squares). To calculate *SS*, we need to know $\bar{X}$ first because *SS* is the sum of squared differences between individual value, *X*, and the sample mean, $\bar{X}$. In a frequency table, the mean formula is modified to $\bar{X} = \frac{\sum fX}{\sum f} = \frac{\sum f x}{n}$. Therefore, a column of *fX* needs to be created in the frequency table. Excel can take care of the calculation process regardless of how many values in the sample. Enter the homework grades in column **A** and the frequency in column **B**. You need to calculate the multiplication of the frequency (*f*) times the grade, *X*, by moving the cursor to **C1** and label it as "**fX**". Move the cursor to **C2** and type "**=A2*B2**" for the multiplication. All formulas start with "=" as a command to calculate. **A2** is the first value of *X*, and **B2** is the corresponding frequency. Then hit **Enter**. Notice **C2** shows 0, which is the result of **A2*B2** = 0 × 2 = 0 as shown in Figure 3.5.

FIGURE 3.5 ● Programming *fX* in Excel

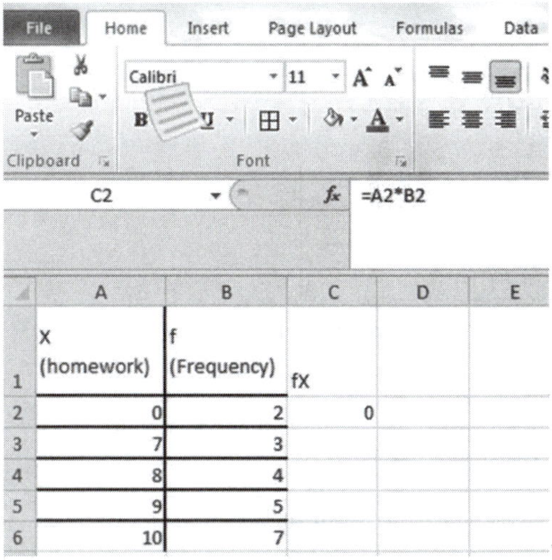

Then move the cursor to the lower right corner of **C2** until + shows up, left click and hold the mouse and drag it to **C6**, and then release the mouse. You will see the rest of the multiplication operations of *fX* automatically show up in **C3** to **C6**. Now you need to sum up all the multiplication operations from **C2** to **C6**. Move the cursor to **C7** and type "**=SUM(C2:C6)**" then hit **Enter**. The answer for the total homework grades is $\sum fX = 168$. In **C8**, type in "**=C7/B7**" to calculate mean $\bar{X} = \frac{\sum fX}{n} = \frac{168}{21} = 8$. Excel does not provide label for the answer so move the cursor to **B8** and label it as "**mean**". The mean of the homework grades is 8 as shown in Figure 3.6.

FIGURE 3.6 ● Programming $\sum fX$ and $\bar{X}$ in Excel

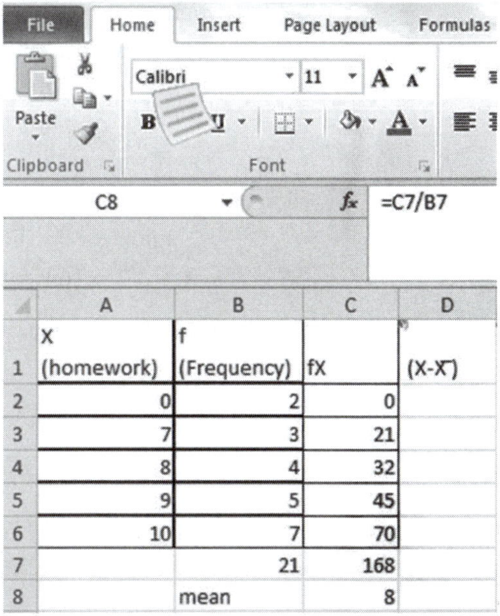

FIGURE 3.7 ● Programming $SS = \Sigma f(X - \overline{X})^2$ in Excel

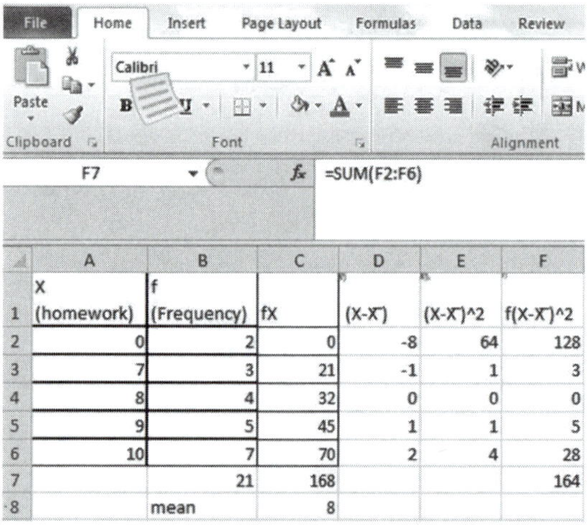

	A	B	C	D	E	F
1	X (homework)	f (Frequency)	fX	(X-X̄)	(X-X̄)^2	f(X-X̄)^2
2	0	2	0	-8	64	128
3	7	3	21	-1	1	3
4	8	4	32	0	0	0
5	9	5	45	1	1	5
6	10	7	70	2	4	28
7		21	168			164
8		mean	8			

The sample $SS = \Sigma(X - \overline{X})^2$ needs to be modified to $SS = \Sigma f(X - \overline{X})^2$ in a frequency table. Based on the order of operations, $(X - \overline{X})$, $(X - \overline{X})^2$, and $f(X - \overline{X})^2$ columns need to be created. Each column clearly represents one step in the order of operations. In **D1**, label the column as $(X - \overline{X})$. It is very tricky to get $\overline{X}$ in Excel, so you might want to use alternative label as (X-mean) for convenience. The label is for your information only, it does not contain any command. You can choose whatever you want to label each column. However, you need to learn the exact commands to conduct math operations properly. There is no creativity allowed in conducting math operations. Here are the exact commands. In **D2**, type "**=A2-C$8**", which shows the operation of first $(X - \overline{X})$. The **$** in the Excel provides the absolute cell reference to the value $\overline{X}$ in **C8**. Once the mean value is fixed, it is possible to copy the operation to produce the rest of $(X - \overline{X})$. Move the cursor to the lower right corner of **D2** until + shows up; then left click and hold the mouse and drag it to **D6** and release the mouse. You will see the rest of $(X - \overline{X})$ automatically show up in **D3** to **D6**. In **E1**, label the column as $(X - \overline{X})^2$. Or you may use **(X-mean)^2** as alternative label. In **E2**, type "**=D2^2**", which shows the operation of first $(X - \overline{X})^2$. Then move the cursor to the lower right corner of **E2** until + shows up, left click and hold the mouse, and drag it to **E6**; then release the mouse. You will see the rest of $(X - \overline{X})^2$ automatically show up in **E3** to **E6**. In **F1**, label it as $f(X - \overline{X})^2$. Or you can use **f(X-mean)^2** as alternative label. In **F2**, type "**=B2*E2**", which shows the operation of the multiplication of $f(X - \overline{X})^2$. Then move the cursor to the lower right corner of **F2** until + shows up, left click and hold the mouse, and drag it to **F6**; then release the mouse. You will see the rest of $f(X - \overline{X})^2$ automatically show up in **F3** to **F6**. In **F7**, type "**=SUM(F2:F6)**" to obtain $SS = \Sigma f(X - \overline{X})^2 = 164$ as shown in Figure 3.7.

Once you get $SS = 164$, you can calculate sample variance and standard deviation.

$$\text{Sample variance} = \frac{SS}{n-1} = \frac{164}{20} = 8.2$$

$$\text{Sample standard deviation} = s = \sqrt{\text{Variance}} = 2.9$$

Based on the round-off rule, the values of sample variance and sample standard deviation are reported at one more decimal place than in the original data.

2. The range for the rollover ratings on SUVs is the Maximum – the Minimum = 5 – 1 = 4.

You can follow the exact process as stated in Exercise Problem 1 to solve this problem. Now, solve the problem on your own before you look at the solutions.

To calculate *SS*, we need to know $\bar{X}$ first. In a frequency table, the mean formula is modified to $\bar{X} = \frac{\sum fX}{n} = \frac{173}{50} = 3.46.$

The *SS* formula in a frequency table is modified to $SS = \sum f(X - \bar{X})^2$. New columns of $(X - \bar{X})$, $(X - \bar{X})^2$, and $f(X - \bar{X})^2$ are created in Table 3.9a.

TABLE 3.9a ● Calculations for Rollover Ratings on SUVs

X (Number of Stars)	*f* (Frequency)	*fX*	$(X-\bar{X})$	$(X-\bar{X})^2$	$f(X-\bar{X})^2$
1	3	3	−2.46	6.0516	18.1548
2	7	14	−1.46	2.1316	14.9212
3	14	42	−0.46	0.2116	2.9624
4	16	64	0.54	0.2916	4.6656
5	10	50	1.54	2.3716	23.716
Total	50	173			64.42
	$\bar{X}$ = 3.46				

$$SS = 64.42$$

$$\text{Sample variance, } s^2 = \frac{SS}{n-1} = \frac{64.42}{49} = 1.31$$

$$\text{Sample standard deviation, } s = \sqrt{\text{Variance}} = \sqrt{1.31} = 1.14$$

What You Learned

The topics in Chapter 3 are variability measures. Variability measures demonstrate how dispersed or spread out individual values are. Variability measures quantify the differences among individual values in the distribution. Three variability measures are (1) range, (2) variance, and (3) standard deviation.

1. Range is the maximum minus the minimum in the distribution, which is highly sensitive to extreme values.
2. Variance and standard deviation are more reliable and more frequently used than range. Be careful in choosing the correct formula when calculating variance and standard deviation in a population versus in a sample.

$$\text{Population variance} = \sigma^2 = \frac{SS}{N}$$

$$\text{Population standard deviation} = \sigma = \sqrt{\frac{SS}{N}}$$

$$\text{Sample variance} = s^2 = \frac{SS}{df} = \frac{SS}{n-1}$$

$$\text{Sample standard deviation} = s = \sqrt{\frac{SS}{df}} = \sqrt{\frac{SS}{n-1}}$$

The concept of degrees of freedom is important when calculating sample variance and standard deviation. It will be repeatedly mentioned in other statistical analyses.

When using frequency tables to calculate variance and standard deviation, the formula $SS = \Sigma(X - \bar{X})^2$ needs to be modified to $SS = \Sigma f(X - \bar{X})^2$ to consider the impact of frequencies. The same modification also applies to population $SS = \Sigma f(X - \mu)^2$, when using frequency tables.

Key Words

Degrees of freedom: The degrees of freedom in a statistical procedure is the number of values involved in the calculation that can vary freely in the sample. 76

Mean squared deviation: The mean squared deviation is another term for the variance. 74

Range: The range is defined as the maximum minus the minimum in a variable. 72

Standard deviation: Standard deviation is a way to quantify individual differences, and it is mathematically defined as $\sigma = \sqrt{\text{Variance}}$ for a population or $s = \sqrt{\text{Variance}}$ for a sample. 73

Sum of squared deviations (*SS*): *SS* is the acronym for sum of squared deviations, population $SS = \Sigma(X - \mu)^2$, and sample $SS = \Sigma(X - \bar{X})^2$. *SS* is an important step in calculating variance and standard deviation. It is also called sum of squares. 74

Variability: Variability describes the extent to which observed values of a variable are dispersed, spread out, or scattered. 72

Variability measures: There are three commonly used measures for variability: (1) range, (2) variance, and (3) standard deviation. 72

Variance: Variance is a way to quantify individual differences, and it is mathematically defined as the mean squared deviation, population variance, $\sigma^2 = SS/N$, and sample variance $s^2 = SS/(n-1)$. 74

Learning Assessment

Multiple Choice Questions: Choose the Best Answer to Every Question

1. In a sample where the variance of variable *X* is 24, what is the standard deviation of *X*?
   a. 23
   b. 18.57
   c. 5.0
   d. 4.90

2. In a sample where the sum of squares (*SS*) of variable *X* is 112 and the sample size is 25, what is the variance of variable *X*?
   a. 4.96
   b. 4.67
   c. 4.48
   d. 4.12

3. In a population of 20 members, the sum of squares (*SS*) of variable *X* is 314, what is the standard deviation of variable *X*?
   a. 4.06
   b. 3.96
   c. 3.76
   d. 3.54

4. When there are 12 observations in a small sample, the degrees of freedom for the sample standard deviation is __________.
   a. 10
   b. 11
   c. 12
   d. 13

5. Variance is one of the measures for variability that requires the variable to be __________ scales of measurement.
   a. nominal, ordinal, interval, and ratio
   b. ordinal, interval, and ratio
   c. both interval or ratio
   d. only ratio

6. In a sample of 45 homework grades, the highest score is 10 and the lowest score is 5. The range for homework grades is __________.
   a. 5
   b. 6
   c. 22.5
   d. 23

7. For the scores 3, 3, 5, 7, 14, and 16, what is the correct answer for $\Sigma(X - \bar{X})$?
   a. −6
   b. 0
   c. +6
   d. 160

8. The relationship between variance and standard deviation is that
   a. standard deviation is smaller than variance
   b. standard deviation is larger than variance
   c. standard deviation is the square root of the variance
   d. standard deviation is equal to variance

9. What is the *SS* for the following sample data: 10, 7, 6, 10, 6, and 15?
   a. 60
   b. 36
   c. 100
   d. −25

10. Which of the following statements is correct?
    a. The degrees of freedom for sample variance are $n - 1$
    b. The degrees of freedom for population variance are $n - 1$
    c. The degrees of freedom for sample standard deviation are n
    d. *SS* means sum of standard deviations

11. The round-off rule states that the final values for measures of variation should be reported
    a. with the same decimal places as in the original data
    b. with one more decimal place than in the original data
    c. with two more decimal places than in the original data
    d. with three more decimal places than in the original data

12. Which one of the following measures of variability only involved extreme values of the sample?
    a. Range
    b. Variance
    c. Standard deviation
    d. Sum of squares

Free Response Questions

13. In a randomly selected sample of 50 women, shoe sizes are measured and reported in Table 3.10. What are the measures of variability for these 50 women's shoe sizes?

TABLE 3.10 ● Women's Shoe Sizes

X (Shoe Size)	*f* (Frequency)
5	2
6	5
7	14
8	16
9	7
10	5
11	1

14. If you think the drive-through service at fast-food restaurants is slow, you are not alone. A market research firm studied and reported fast-food drive-through time. Here is a sample of 55 randomly selected customers and their drive-through time as shown in Table 3.11. What are the measures of variability for these drive-through times?

TABLE 3.11 ● Fast-Food Drive-Through Time

X (Minutes)	*f* (Frequency)
1	2
2	6
3	20
4	17
5	7
6	3

15. In a randomly selected sample of 21 students who took statistics courses this semester, their midterm exam grades are reported in Table 3.12. Use Excel to calculate the measures of variability for these 21 students' exam grades.

TABLE 3.12 ● Students Exam Grades

Exam Grade
82
45
82
84
73
74
93
88
83
54
80
80
87
44
21
50
94
94
63
77
87

Answer to Pop Quiz Question

1. b

4

STANDARD Z SCORES

Learning Objectives

After reading and studying this chapter, you should be able to do the following:

- Convert raw scores to Z scores to get measures on a universal yard stick
- Identify unusual values or outliers using Z scores
- Calculate Z scores for population or sample data
- Explain how converting raw scores to Z scores does not change the shape of the distribution
- Apply the empirical rule to connect Z scores with probabilities in special cases

WHAT YOU KNOW

You learned measures of central tendency in Chapter 2 and measures of variability in Chapter 3. Central tendency measures reflect the center of the distribution, and variability measures demonstrate the individual differences. When data are normally distributed, the mean is the most commonly used measure for central tendency, and the standard deviation is the most commonly used measure of variability. In the presence of outliers, however, the mean is heavily influenced by them, and thus, it is not a suitable measure of central tendency. Moreover, you learned different formulas for population parameters versus sample statistics. When calculating the sample variance and standard deviation, the degrees of freedom is $df = n - 1$.

WHAT IS NEW

In Chapter 4, you will learn to use the mean and standard deviation together to create *Z scores*. *Z* scores are standard scores that describe the differences of individual values from the mean in units of standard deviation. Two main attributes of *Z* scores are direction and magnitude. Positive *Z* scores indicate that the values are above the mean, and negative *Z* scores indicate that the values are below the mean. The magnitude of a *Z* score is reflected by the absolute value of the *Z* score, which signifies the difference between the value and the mean. The higher the absolute value, the longer is the difference between the value and the mean. When $Z = 1$, it means that the value is 1 standard deviation above the mean. When $Z = -0.5$, it means that the value is 0.5 standard deviations below the mean. *Z* scores provide markers to identify the location of every value in the distribution. There are different formulas for population *Z* scores and sample *Z* scores, which are covered in this chapter.

STANDARD Z SCORES

When you were in high school preparing to apply to colleges, you might have taken some standardized tests such as the SAT or the ACT. Did you find the score reports confusing? If you did, you were not alone. Let's take a moment to discuss the makeup of SAT scores. Once you understand one standardized test, the same principle applies to other standardized tests. SAT section scores are reported on a scale from 200 to 800. The scaled SAT scores are theoretically designed to have a mean of 500 and a standard deviation of 100. However, the mean and standard deviation of the scaled scores in each section do not always turn out to be 500 and 100, respectively. Every year the College Board would calculate the mean and standard deviation of each section as shown in Table 4.1. Such numbers also vary from year to year. The College Board recorded and analyzed every test taker's score. There were 2,220,087 recorded SAT scores in 2019.

TABLE 4.1 ● 2019 SAT Mean and Standard Deviation in Evidence-Based Reading and Writing and Math Sections

Statistics	Evidence-Based Reading and Writing	Mathematics
N	2,220,087	2,220,087
Mean	531	528
Standard deviation	104	117

Note: Data retrieved from the College Board (2019). https://reports.collegeboard.org/pdf/2019-total-group-sat-suite-assessments-annual-report.pdf

Assume that you received the score report in 2019 with Evidence-Based Reading and Writing (ERW) = 670 and Math = 675 on the SAT. Which section was stronger compared with other students who took the same SAT exam? The answer is not as simple as directly comparing 670 and 675 to pick the higher value. Don't worry! I am here to help. The answer to this question will be provided in the first example of this chapter after discussing the concept of *Z* scores.

When you tried to compare values from different sections, the original values were not helpful in determining which one was your stronger section, even when both scores were reported on the same scale (ranging from 200 to 800). Both subsection scores had their own distributions with different means and standard deviations as shown in Table 4.1.

Here is another scenario. Assume that you also took the ACT and received 30 in English, 29 in Math, 31 in Reading, 29 in Science, and an ACT composite score of 29. According to the ACT National Norms, the means and standard deviations of the ACT subsection and composite score varied, as shown in Table 4.2.

You know that most colleges only require either ACT or SAT scores. You were probably wondering which set of the test scores has a better chance of getting you accepted in a selective university. The SAT scores were reported between 200 and 800, and the ACT scores were reported between 1 and 36. Clearly, the worst reported SAT scores were higher than the perfect ACT scores. The answer to which test score had a better chance of getting you accepted in a selective university would not be simply picking the higher reported scores between ACT and SAT. In applying to colleges, you especially need to figure out how your performance compared with other students who took the exams at the same time. *Z* score is the perfect tool to provide answers in such a situation. I bet that you wish you had learned *Z* scores in high school! No worry! Once you understand how to calculate *Z* scores, the same principle can be applied to all other standardized tests, including GRE scores, which may be required for many graduate schools.

***Z* scores** are standard scores that describe the differences of individual values from the mean in units of standard deviation. As long as the mean and standard deviation are known in a distribution, all original values (also known as raw scores) of a variable can be

TABLE 4.2 ● 2019–2020 National Norm of ACT Section Score and Composite Score

Item	English	Math	Reading	Science	Composite
Mean	20.2	20.5	21.3	20.8	20.9
Standard deviation	7.0	5.5	6.8	5.8	5.3

Note: Data retrieved from the ACT (2019). https://www.act.org/content/dam/act/unsecured/documents/MultipleChoiceStemComposite.pdf

converted into Z scores. It is useful to think of Z scores as a universal yardstick because all Z scores are reported only as the number of standard deviations above or below the mean without any reference to measurement unit (e.g., inches, pounds, dollar, minutes, or points). Z scores provide a rule of thumb to identify the location of a particular value in the distribution, including unusual and extreme values (i.e., outliers). The difference between unusual values and extreme values is that there are about 5% of values considered to be unusual in a distribution but they are known to exist. But extreme values (i.e., outliers) do not always exist in a distribution. When they do exist, less than 1% of the values in a distribution are considered outliers. Outliers have undue influence on the outcomes of statistical analysis. They need to be identified and treated with caution. The rule of thumb is that if a value X converts to a Z score and is less than 2 standard deviations from the mean, $|Z| < 2$, then X is an ordinary value. The rule of thumb to identify outliers is that if a value X converts to a Z score and is more than 3 standard deviations from the mean, $|Z| > 3$, X is an outlier. The Z values between 2 and 3 standard deviations away from the mean are unusual values, as shown in Figure 4.1.

Standard Z scores have three important attributes:

1. The sign + or – describes the score in relation to the mean. All positive Z scores are above the mean, and all negative Z scores are below the mean.
2. The magnitude of the Z score describes the distance between the value and the mean in number of standard deviation units.
3. When transforming all raw scores to Z scores, you are conducting a mathematical operation on the raw values. This mathematical operation does not change the shape of the distribution. The shape of the distribution remains the same as the original distribution. Z scores have mean = 0, standard deviation = 1, and variance = 1.

FIGURE 4.1 ● Ordinary Values, Unusual Values, and Outliers

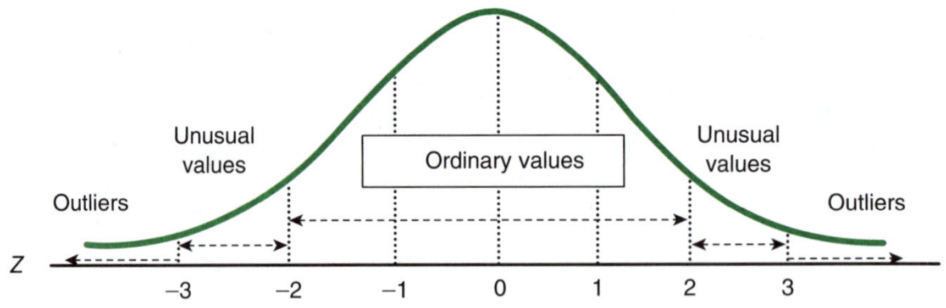

Z Scores for a Population

The Z-score formula for a population is $Z = \frac{(X - \mu)}{\sigma}$, where X is the value of the variable, μ the population mean, and σ, the population standard deviation. We get the numerator of the Z formula by calculating the distance between X and μ by doing a simple subtraction $(X - \mu)$. To complete the formula, you use the distance divided by σ. The result of the Z formula generates a standardized measure expressed in units of σ. You need to know the population mean and standard deviation before you can calculate Z scores. The SAT and the ACT are standardized aptitude tests used to predict how high school students will perform in colleges or universities. The testing industry keeps records of scores from everyone who took the exam and routinely reports means and standard deviations in different sections and composite scores and conducts comparisons across different groups. It is feasible to obtain the population mean and standard deviation for standardized tests such as the SAT and the ACT.

Z scores can be calculated to show the relative standing of the reported scores. Let me demonstrate the process of calculating Z scores with several examples.

EXAMPLE 4.1

Sarah received a SAT score report. Her score on ERW was 670 and on Math was 675. Which was Sarah's stronger section compared with other students who took the SAT exams? According to Table 4.1, SAT scores are normally distributed, and the population mean and standard deviation for SAT ERW are $\mu = 531$, $\sigma = 104$; and for Math, $\mu = 528$, $\sigma = 117$.

Using Z scores puts scores on a universal yardstick to figure out the relative standing of Sarah's scores in ERW and Math compared with other students who took the same exam.

a. Sarah's Z score in SAT ERW:

$$Z = \frac{X - \mu}{\sigma} = \frac{670 - 531}{104} = \frac{139}{104} = 1.34$$

b. Sarah's Z score in SAT Math:

$$Z = \frac{X - \mu}{\sigma} = \frac{675 - 528}{117} = \frac{147}{117} = 1.26$$

Z scores show the exact location of each individual score in each distribution. Both section scores are normally distributed. Z scores allow individual scores from different distributions to be placed on the same yardstick, as shown in Figure 4.2, where Z scores are overlaid with the population mean

(Continued)

(Continued)

and standard deviation. All scores are lined up by the mean and equally spaced by the standard deviation. Z scores have a mean = 0, and a standard deviation = 1. Therefore, in a Z scale, 1 standard deviation above the mean is expressed as 1, and for other scales, $\mu + \sigma$; 2 standard deviations above the mean is expressed as 2, and for other scales, $\mu + 2\sigma$; and 3 standard deviations above the mean is expressed as 3, and for other scales $\mu + 3\sigma$. Due to the symmetry of the normal distribution curve (i.e., the left side is the mirror image of the right side), the same principle applies to the left side of the curve. That is, in a Z scale, 1 standard deviation below the mean is expressed as −1, and for other scales, $\mu - \sigma$; 2 standard deviations below the mean is expressed as −2, and for other scales, $\mu - 2\sigma$; and 3 standard deviations below the mean is expressed as −3, and for other scales, $\mu - 3\sigma$, as shown in Figure 4.2.

FIGURE 4.2 ● *Z* Scale Overlaid With Population Parameters

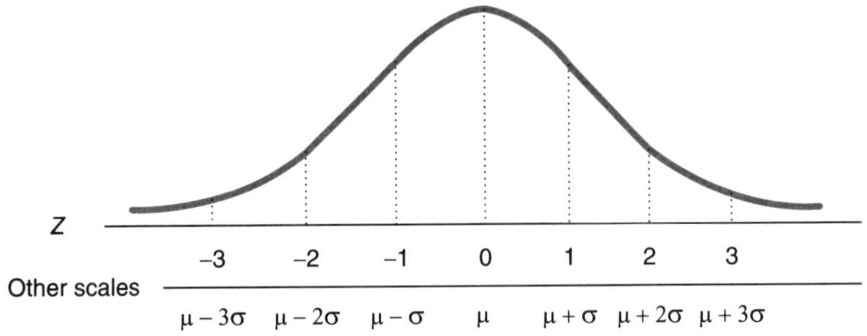

In standardized tests, higher Z scores mean better scores. Based on the location of the Z scores and the relative standing of the three Z scores, 1.34 > 1.26, Sarah's stronger section on the SAT was ERW when compared with other students who took the exams at the same time.

Z scores between 2 and −2 are considered to be ordinary. $Z = 2$ translates into the original scale as $\mu + 2\sigma$, and $Z = -2$ translates into the original scale as $\mu - 2\sigma$. In other words, most students' SAT ERW scores are between $531 - 2(104) = 323$ and $531 + 2(104) = 739$, SAT Math scores between $528 - 2(117) = 294$ and $528 + 2(117) = 762$. Scores outside these intervals are considered to be unusual.

Next, let's solve the problem of picking a better set of standardized test scores to boost Sarah's chances of getting admitted to a selective university.

EXAMPLE 4.2

Sarah also took the ACT last year. She scored 30 in English, 29 in Math, 31 in Reading, and 29 in Science. According to Table 4.2, the English scores have $\mu = 20.2$, $\sigma = 7.0$; Math, $\mu = 20.5$, $\sigma = 5.5$; Reading, $\mu = 21.3$, $\sigma = 6.8$; and Science, $\mu = 20.8$, $\sigma = 5.8$. Most colleges and universities only require either ACT scores or SAT scores for admission. Which set of scores would be a better choice to boost Sarah's chance of getting admitted into a selective university?

You already learned in Example 4.1 that picking a better set of standardized test scores differs from simply picking the higher reported scores. You learned that you need to convert reported scores to standard *Z* scores. Again, the *Z* scale provides a universal yardstick, allowing comparisons of the values of different variables with different distributions.

We already converted Sarah's SAT ERW score to a *Z* score = 1.34, and SAT Math score to *Z* = 1.26 in Example 4.1. Apply the same formula to convert Sarah's ACT scores to *Z* scores:

ACT English:

$$Z = \frac{X - \mu}{\sigma} = \frac{30 - 20.2}{7.0} = \frac{9.8}{7.0} = 1.40$$

ACT Math:

$$Z = \frac{X - \mu}{\sigma} = \frac{29 - 20.5}{5.5} = \frac{8.5}{5.5} = 1.55$$

ACT Reading:

$$Z = \frac{X - \mu}{\sigma} = \frac{31 - 21.3}{6.8} = \frac{9.7}{6.8} = 1.43$$

ACT Science:

$$Z = \frac{X - \mu}{\sigma} = \frac{29 - 20.8}{5.8} = \frac{8.2}{5.8} = 1.41$$

Based on the different *Z* scores from the ACT and SAT, Sarah's SAT *Z* scores ranged from 1.26 to 1.34 and ACT *Z* scores ranged from 1.40 to 1.55. Overall, Sarah scored better on the ACT than other students who took the same exams. Submitting ACT score reports instead of SAT would boost Sarah's chance of being admitted by a selective university. The basic principle of comparing the values of different variables with different distributions is to convert the reported scores to *Z* scores to allow meaningful comparisons on the universal yardstick, the *Z* scale.

Let's consider another example using *Z* scores but has nothing to do with standardized test scores.

EXAMPLE 4.3

According to current data from the National Center for Health Statistics, the height of adult women in the United States is normally distributed, with $\mu = 63.8$ inches and $\sigma = 4.2$ inches. Use this information to answer the following three questions.

a. What is the Z score for a woman with a height of 5 feet 3 inches?

b. How tall would a woman have to be to reach $Z = 2$?

c. How short would a woman have to be to have $Z = -1.25$?

Although 5 feet 3 inches is a common way to communicate height measures, it is necessary to convert all numbers into the same unit before calculating the Z scores. As you know, there are 12 inches in a foot.

a. $X = 5$ feet 3 inches $= 5 \times 12 + 3 = 63$ inches

$$Z = \frac{X - \mu}{\sigma} = \frac{63 - 63.8}{4.2} = \frac{-0.8}{4.2} = -0.19$$

An adult woman's height of 5 feet 3 inches puts her at 0.19 standard deviations below the mean.

b.

$$Z = \frac{X - \mu}{\sigma}$$

$$2 = \frac{X - 63.8}{4.2}$$

$$8.4 = X - 63.8$$

$$X = 72.2$$

A woman has to be 72.2 inches to reach $Z = 2$.

c.

$$Z = \frac{X - \mu}{\sigma}$$

$$-1.25 = \frac{X - 63.8}{4.2}$$

$$-5.25 = X - 63.8$$

$$X = 58.55$$

A woman has to be 58.55 inches to have a score of $Z = -1.25$.

EXAMPLE 4.4

Measuring a new born baby's height and weight has become a routine in every hospital. A new born baby boy weighs 7 pounds 2 ounces and is 21 inches tall. What are the new baby boy's Z scores in weight and height? Based on the CDC's growth chart for newborn boys, the mean weight $\mu = 3.53$ kilograms and $\sigma = 0.46$ kilograms, and the mean height $\mu = 49.99$ centimeters and $\sigma = 2.47$ centimeters. There are 16 ounces in a pound, 1 ounce equals 28.35 grams, and 1 inch equals 2.54 centimeters.

Height and weight are not only different measures with different distributions, they are also measured and reported by different measurement units. Make sure the correct μ and σ are used in the calculation process.

a. *Z* score for weight with all measures converted to grams:

$$7 \text{ lb } 2 \text{ oz} = 7 \times 16 + 2 \text{ oz} = 114 \text{ oz}$$

$$114 \text{ oz} = 114 \times 28.35 \text{ g} = 3{,}231.9 \text{ g}$$

$$3.53 \text{ kg} = 3.53 \times 1{,}000 \text{ g} = 3{,}530 \text{ g}$$

$$0.46 \text{ kg} = 0.46 \times 1{,}000 \text{ g} = 460 \text{ g}$$

$$Z = \frac{X - \mu}{\sigma} = \frac{3231.9 - 3530}{460} = \frac{-298.1}{460} = -0.65$$

b. *Z* score for height with all measures converted to centimeters:

$$21 \text{ in.} = 21 \times 2.54 \text{ cm} = 53.34 \text{ cm}$$

$$Z = \frac{X - \mu}{\sigma} = \frac{53.34 - 49.99}{2.47} = \frac{3.35}{2.47} = 1.36$$

According to the *Z* scores, this newborn boy's weight is 0.65 standard deviations below the mean, and the newborn boy's height is 1.36 standard deviations above the mean. Although height and weight come from different distributions, it is possible to compare the *Z* scores for height and weight. Accordingly, this boy weighs less but is taller than other newborn boys. He is a tall and slim newborn boy.

Why Statistics Matter

Children's height and weight measures vary naturally, but at times they can vary quite dramatically. The Centers for Disease Control and Prevention (CDC) has an extensive

(Continued)

(Continued)

data bank on health statistics. Children's growth charts on height and weight are part of the health statistics. The CDC publishes growth charts for infants between 0 and 36 months on its website, indicating the normal range of monthly growth. Babies grow at a fast pace within the first few months of birth. That is the reason to have a different growth chart for every month. Boys grow at a different pace from that of girls, so their growth charts are separate. By choosing the correct age and correct sex of the child on the growth charts, you can figure out roughly where a baby stands when compared with other babies of the same sex at the same age. These statistics are especially important to identify early health and growth problems in the instances where the baby's height or weight deviates significantly from the normal range.

EXAMPLE 4.5

In a family of five, the family members' ages are 17, 22, 26, 48, and 57. Convert every age into a Z score. What are the mean and standard deviation of the Z scores?

Populations don't always contain large number of people. The size of the population depends on the researchers' interests. My interest here is only on this particular family. Since there are five members in this family and all five members' ages are reported, it is a population. This example uses the same numbers as in Example 3.2, for which you have already calculated the mean and standard deviation.

$$\mu = \frac{\Sigma X}{N} = \frac{170}{5} = 34$$

$$\sigma = \sqrt{\frac{SS}{N}} = \sqrt{\frac{1222}{5}} = 15.6$$

You have everything you need to convert your raw score (X) into Z by using $Z = \frac{X - \mu}{\sigma}$. Then you can calculate the mean and standard deviation of Z the same way you calculated the mean and standard deviation of X as shown in Table 4.3.

$$\text{Mean of } Z \text{ scores}: \mu_Z = \frac{\Sigma Z}{N} = \left((-1.09) + (-0.77) + (-0.55) + 0.90 + 1.47\right)/5 = 0$$

$$SS \text{ of } Z = (-1.09)^2 + (-0.77)^2 + (-0.51)^2 + 0.90^2 + 1.47^2 = 5.00$$

$$\text{Standard deviation of } Z = \sigma_Z = \sqrt{\frac{SS}{N}} = \sqrt{\frac{5}{5}} = 1$$

TABLE 4.3 ● Family Members' Ages and Z Scores

X (Age)	$X - \mu$	$Z = (X - \mu)/\sigma$	$(Z - \mu_Z)^2$
17	−17	−1.09	1.18
22	−12	−0.77	0.59
26	−8	−0.51	0.26
48	14	0.90	0.81
57	23	1.47	2.16
170	0	0	5.00
ΣX		ΣZ	$\Sigma(Z - \mu_Z)^2$

This example is to demonstrate that when you convert every value in the population into Z scores, the Z scores have mean = 0 and standard deviation = 1. Z scores are customarily calculated to two places after the decimal point to conform to the way Z scores are reported in the Standard Normal Distribution Table, or the Z Table, which will be introduced in the next chapter.

Z Scores for a Sample

You just learned how to calculate Z scores using population mean and standard deviation. However, in reality, population means and standard deviations are usually unavailable. In this section, you will learn to calculate Z scores in a sample. When Z scores are calculated for a sample, both μ and σ are unknown, so the Z formula needs to be adjusted to $Z = \frac{X - \overline{X}}{s}$, where $\overline{X}$ is the sample mean, and s is the sample standard deviation. Sample Z scores can be useful tools to help you make sense of your daily experiences. An example should help make this clear.

EXAMPLE 4.6

The waiting time of a drive-through in a fast-food restaurant depends on the number of cars in front of you. The waiting time goes up and down throughout the day. A restaurant reported to have a mean waiting time of $\bar{X} = 15.5$ minutes and the standard deviation of $s = 6.3$ minutes based on a sample of 100 customers. Alex went to this restaurant and waited for 25 minutes to get her food. How is Alex's waiting time compared with other customers at this drive-through?

$$Z = \frac{X - \bar{X}}{s} = \frac{25 - 15.5}{6.3} = 1.51$$

Alex's waiting time was 1.51 standard deviations above the mean compared with other customers. In other words, Alex waited longer than the majority of customers at this drive-through. Maybe she should consider to visit the restaurant at a different time when it is not as busy.

The same principle for calculating *Z* scores for populations applies to *Z* scores for samples. The attributes of *Z* scores remain the same. Conducting a *Z* transformation does not change the shape of the distribution. The shape of the *Z*-score distribution is the same as that of the original distribution of the raw scores, and the *Z*-score distribution has mean = 0 and standard deviation = 1.

Let's take a look at a simple example to go through the process of converting raw scores to *Z* scores then calculate the mean and standard deviation of the *Z* scores from a sample. It is helpful to go through the calculation process with simple numbers, then you can apply the same process to more complicated numbers later.

EXAMPLE 4.7

A random sample of five students answered a survey about the number of hours spent studying the night before an exam. The answers were 1, 1, 3, 5, and 5 hours of studying time. Convert the hours into *Z* scores, and calculate the mean and standard deviation for the *Z* scores.

You need to find out the sample mean and standard deviation to convert *X* into *Z*.

$$\bar{X} = \sum X/n = (1+1+3+5+5)/5 = 15/5 = 3$$

$$SS = \sum (X - \bar{X})^2 = (1-3)^2 (1-3)^2 + (3-3)^2 + (5-3)^2 + (5-3)^2 = 16$$

$$\text{Variance } s^2 = SS/(n - 1) = 16/(5 - 1) = 4$$

$$\text{Standard deviation } s = \sqrt{\text{Variance}} = \sqrt{4} = 2$$

$$Z = \frac{X - \bar{X}}{s}$$

The results of this Z calculation are shown in the fourth column of Table 4.4.

$$\bar{Z} = \sum Z / n = ((-1)+(-1)+0+1+1)/5 = 0$$

Table 4.4 shows the step-by-step process of calculating the mean and standard deviation for the Z scores in Example 4.7.

TABLE 4.4 ● Step-by-Step Process for Calculating the Mean and Standard Deviation for Z Scores

X (Hours of Studying)	$X - \bar{X}$	$(X - \bar{X})^2$	$Z = (X - \bar{X}) / s$	$(Z - \bar{Z})^2$
1	−2	4	−1	1
1	−2	4	−1	1
3	0	0	0	0
5	2	4	1	1
5	2	4	1	1
Total		16	0	4
		SS of X		*SS* of Z

$$SS \text{ of } Z = (-1-0)^2 + (-1-0)^2 + (0-0)^2 + (1-0)^2 + (1-0)^2 = 4$$

$$\text{Variance of } Z = s^2 = SS/(n-1) = 4/(5-1) = 1$$

$$\text{Standard deviation of } Z = \sqrt{\text{Variance}} = \sqrt{1} = 1$$

This example demonstrates that Z scores from a sample have mean = 0 and standard deviation = 1. These attributes are true in a sample, just as they do in a population.

POP QUIZ

1. For a population with $\mu = 100$ and $\sigma = 10$, what is the X value corresponding to $Z = 0.75$?
   a. 92.5
   b. 97.5
   c. 105
   d. 107.5
2. What is the measurement unit of Z scores?
   a. The same as the measurement unit of the original scores
   b. The square of the measurement unit of the original scores
   c. The square root of the measurement unit of the original scores
   d. There is no measurement unit for Z scores.

EMPIRICAL RULE FOR VARIABLES WITH A NORMAL DISTRIBUTION

Although Z scale serves as a universal yardstick, not many people are well versed in it. You probably have not heard your friends express their test scores, height and weight measures, or income in Z scores. That's because most people don't know much about Z scores. Even Z scale is a universal yardstick, it is not commonly used in everyday conversations. If you don't believe me, you can start a social experiment that you only use Z scores to communicate about numbers to other people to see how they respond to you. To make the Z scale more accessible, statisticians created a computation rule to link Z scores with probabilities. Even most people don't fully understand the concept of probability, they have a rough idea of what probability means. The linkage between Z scores and probabilities can be demonstrated in the empirical rule. The **empirical rule** states that about 68% of the values fall within 1 standard deviation from the mean, about 95% of the values fall within 2 standard deviations from the mean, and about 99.7% of the values fall within 3 standard deviations from the mean, as shown in Figure 4.3. The empirical rule applies to all normally distributed variables.

Roughly 68% of the values fall within $-1 \leq Z \leq 1$ or $\mu \pm \sigma$

Roughly 95% of the values fall within $-2 \leq Z \leq 2$ or $\mu \pm 2\sigma$

Roughly 99.7% of the values fall within $-3 \leq Z \leq 3$ or $\mu \pm 3\sigma$

The empirical rule is also called the 68–95–99.7 rule. It helps people interpreting Z scores. Due to the fact that only 5% of the values fall outside 2 standard deviations from the mean, customarily, these 5% are viewed as unusual values. In Example 4.8, we will demonstrate how the empirical rule can help in interpreting Z scores.

FIGURE 4.3 ● Empirical Rule for Normally Distributed Variables

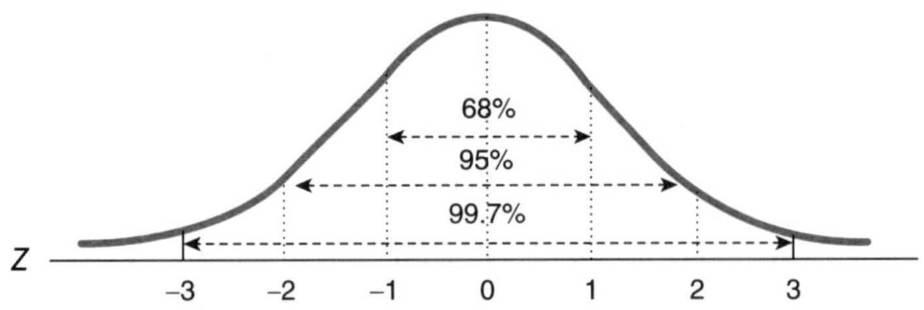

EXAMPLE 4.8

In the United States, adult men's height is normally distributed, with $\mu = 5$ feet 8 inches and $\sigma = 6$ inches. What percentage of men are taller than 6 feet 8 inches?

Let's convert the numbers to the same measure unit, inch, to figure out the Z score for 6 feet 8 inches.

$$\mu = 5 \text{ feet } 8 \text{ inches} = 5 \times 12 + 8 = 68 \text{ inches}$$

$$X = 6 \text{ feet } 8 \text{ inches} = 6 \times 12 + 8 = 80 \text{ inches}$$

$$Z = \frac{X - \mu}{\sigma} = \frac{80 - 68}{6} = \frac{12}{6} = 2$$

The question asks for the percentage of men taller than 6 feet 8 inches, so we need to figure out the probability that links with $Z > 2$. To answer this question, you have to combine your knowledge of the 60–95–99.7 rule and the attributes of Z scores. Let's list the key points of the knowledge that are helpful to solve this problem.

1. The entire area under the normal distribution contains all possible values so the total probability sums up to 1.
2. In a normal distribution, Z scores are symmetrical and the mean, median, and mode are located at the same spot. Since the distribution is symmetrical around this spot, 50% of the values are above the mean and 50% are below the mean.
3. According to the empirical rule, 95% of the values fall within $-2 \leq Z \leq 2$. The area between $Z = 0$ and $Z = 2$ is half of the 95%, due to the symmetry of the normal curve. Therefore, the percentage of men who are taller than 6 feet 8 inches is 100% – 50% – 47.5% = 2.5% as shown in Figure 4.4. Drawing the normal distribution with Z scale helps visualize the three key points listed about. Thus, only 2.5% of adult men are taller than 6 feet 8 inches.

FIGURE 4.4 ● Applying the Empirical Rule to Interpret $Z > 2$

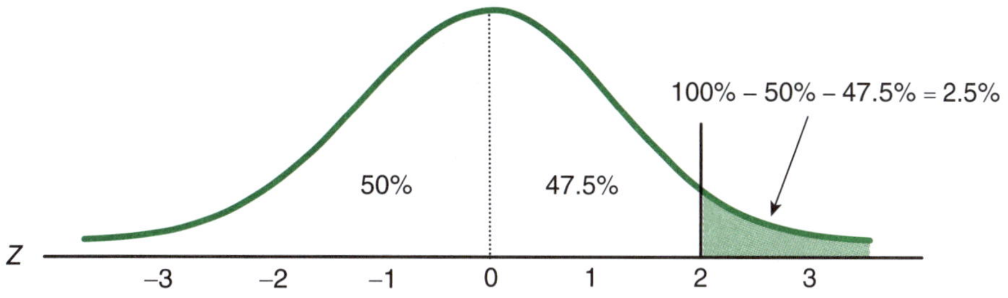

EXAMPLE 4.9

Tim's SAT Math score was 612. SAT Math scores are normally distributed, with $\mu = 497$ and $\sigma = 115$. What was the percentage of students who scored higher than Tim's score?

First, Tim's Math score needs to be converted to

$$Z = \frac{X - \mu}{\sigma} = \frac{612 - 497}{115} = 1.$$

Let's list the key points of the knowledge that are helpful to solve this problem.

1. The entire area under the normal distribution contains all possible values so the total probability sums up to 1.
2. In a normal distribution, Z scores are symmetrical, thus 50% of the values are above the mean and 50% are below the mean.
3. According to the empirical rule, 68% of the values fall within $-1 \leq Z \leq 1$. The area between $Z = 0$ and $Z = 1$ is half of the 68%, due to the symmetry of the normal curve.Therefore, the percentage of students who score above Tim's is 100% – 50% – 34% = 16% as shown in Figure 4.5. Drawing the normal distribution with Z scale helps visualize the three key points listed above. Thus, only 16% of students scored higher than Tim's. In other words, Tim's score was higher than 84% of the students who took the SAT Math at the same time.

FIGURE 4.5 ⬢ Applying the Empirical Rule to Interpret $Z > 1$

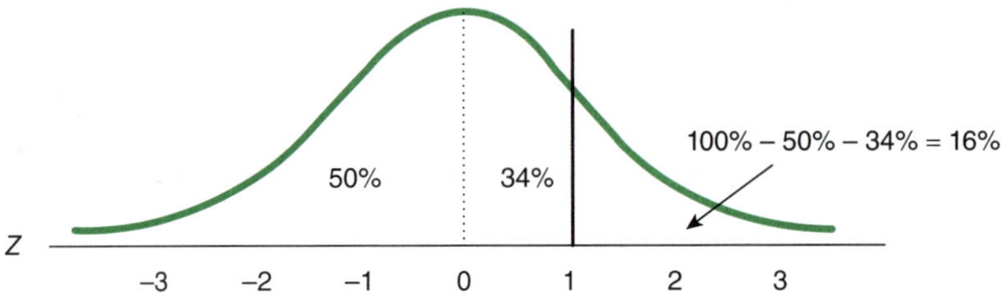

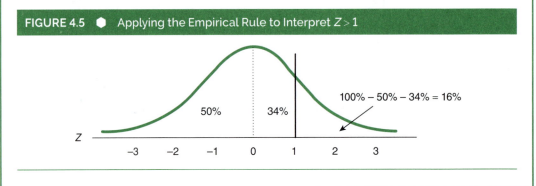

To reinforce this calculation process, here is one final example.

EXAMPLE 4.10

In the Ohio Division II high school boys' swimming competitions, the swimmers' times were recorded. The time (in seconds) to complete the 50-yards free style event (known simply as 50 Free) was normally distributed, with a mean of $\overline{X} = 25$ seconds and a standard deviation of $s = 2.5$ seconds. Josh swam a time of 20 seconds for the 50 Free event. What percentage of Division II high school boys swam faster than Josh in 50 Free?

The faster the swimmer, the less time it takes to complete the event. First, Josh's 50 Free time of 20 seconds needs to be converted to a Z score.

$$Z = \frac{X - \bar{X}}{s} = \frac{20 - 25}{2.5} = -2$$

Josh's time was 2 standard deviations below the mean. Let's list the key points of the knowledge that are helpful to solve this problem.

1. The entire area under the normal distribution contains all possible values so the total probability sums up to 1.
2. In a normal distribution, Z scores are symmetrical, thus 50% of the values are above the mean and 50% are below the mean.
3. According to the empirical rule, 95% of the values fall within $-2 \le Z \le 2$. The area between $Z = -2$ and $Z = 0$ is half of the 95%, due to the symmetry of the normal curve. Therefore, the percentage of Division II high school boys who swam faster than Josh is 100% – 50% – 47.5% = 2.5% as shown in Figure 4.6. Thus, 2.5% of Division II high schoolers boys could swim faster than Josh in 50 Free. In other words, Josh's 50 Free time was faster than 97.5% of the Division II high school boy swimmers.

FIGURE 4.6 ● Applying the Empirical Rule to Interpret $Z < -2$

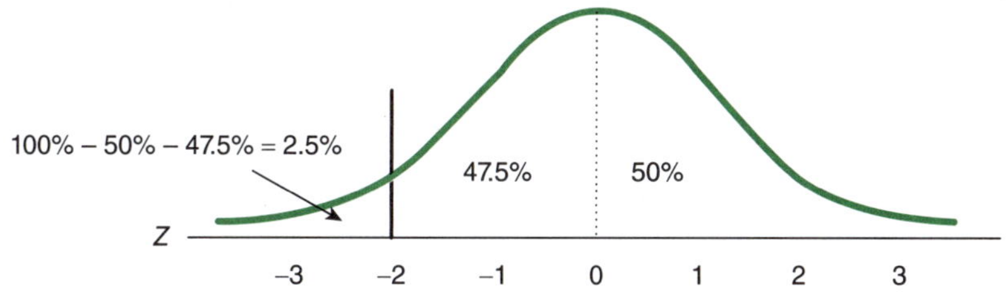

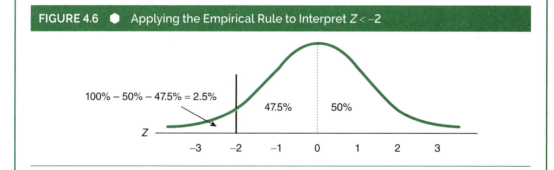

The empirical rule (68–95–99.7 rule) of normal distributions in combination with the symmetry of the normal distribution curve allows us to use probability to interpret Z scores clearly in a few special cases. To extend the connection between Z scores and probabilities beyond these special cases to all possible Z values, that is the topic of next chapter.

POP QUIZ

3. Based on the empirical rule, 99.7% of values are within 3 standard deviations from the mean, $-3 \leq Z \leq 3$. What percentage of values fall within $0 \leq Z \leq 3$?
   a. 49.9%
   b. 68%
   c. 95%
   d. 99.7%

4. Assume that the scores on a statistics exam were normally distributed. Bianca earned 74 points on the exam. The exam scores had $\overline{X} = 82$ and $s = 8$. What percentage of students scored higher than Bianca?
   a. 16%
   b. 32%
   c. 68%
   d. 84%

EXERCISE PROBLEMS

1. Body mass index (BMI) is a common way to measure the percentage of body fat and screen for obesity. The formula for BMI = Weight (in kilograms)/Height^2 (in square meters). If you are interested in finding out your BMI, you need to convert your weight to kilograms and your height to meters. Adult women and men have different BMI distributions. Men's BMI is normally distributed, with a mean $\mu = 27.8$ and a standard deviation $\sigma = 5.22$. Women's BMI is also normally distributed, with a mean $\mu = 27.3$ and a standard deviation $\sigma = 7.01$. If a man and a woman both have BMI = 34, compared with people of the same sex, whose BMI is farther from the mean?

2. The distribution of reaction time of young adults aged 20 to 25 was positively skewed, with a long tail to the right. In a random sample of 80 young adults, a mean of 400 milliseconds was reported. A Z score equal to 2 is reported for a participant with a reaction time of 500 milliseconds.
   a. What was the standard deviation of reaction time in this sample?
   b. If all the reaction times were converted to Z scores, what was the shape of the Z-score distribution?

3. Jimmy scored a 90 on an IQ test. The IQ test scores are normally distributed, with $\mu = 100$ and $\sigma = 10$. What percentage of people scored lower than Jimmy's score on the same IQ test?

Solutions

1. First, convert the original BMI = 34 to Z scores for men and women:

$$\text{Man's } Z = \frac{X-\mu}{\sigma} = \frac{34-27.8}{5.22} = 1.19$$

$$\text{Woman's } Z = \frac{X-\mu}{\sigma} = \frac{34-27.3}{7.01} = 0.96$$

The man's BMI = 34 converts to $Z = 1.19$; it is 1.19 standard deviations above the mean. The woman's BMI = 34 converts to $Z = 0.96$; it is 0.96 standard deviations above the mean as shown in Figure 4.7. Therefore, the man with BMI = 34 is farther from the mean compared with people of the same gender.

FIGURE 4.7 ● Locations of Two Calculated Z Scores Overlaid With the Original Scales

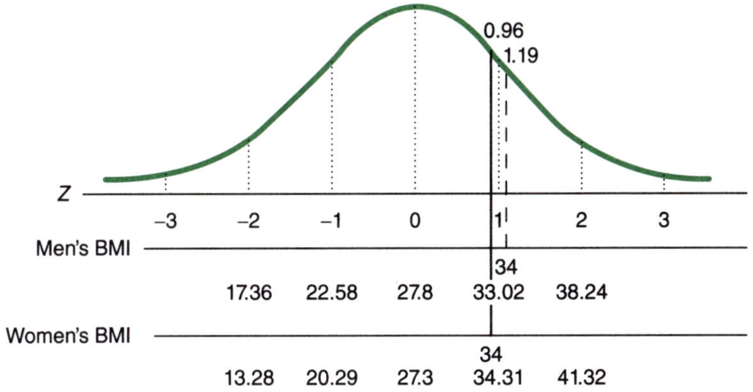

2. a. First, identify the formula for the Z-score conversion for a sample, $Z = (X - \bar{X})/s$. Then list all the numbers given in the problem statement into the formula:

$$Z = 2$$

$$X = 500$$

$$\bar{X} = 400$$

$$s = ?$$

You can solve one unknown with three values in this formula. Plug three values into the formula, and solve the fourth one.

$$2 = \frac{500 - 400}{s}$$
$$2s = 100$$
$$s = 50$$

The standard deviation of reaction time in this sample is 50 milliseconds.

2.b. The shape of the Z-score distribution is the same as that of the original reaction time. It is positively skewed, with a long tail to the right.

3. First, identify the formula for the Z-score conversion for a population $Z = (X - \mu)/\sigma$.

Then list all the numbers given in the problem statement into the formula:

$$X = 90$$
$$\mu = 100$$
$$\sigma = 10$$
$$Z = ?$$
$$Z = \frac{X - \mu}{\sigma} = \frac{90 - 100}{10} = -1$$

Jimmy's 90 converted to a $Z = -1$.

According to the empirical rule, the area between $-1 \le Z \le 1$ is 68%; therefore, the area between $Z = 0$ and $Z = -1$ is half of it: ½ × 68% = 34%.

Of the people who took the IQ test, 84% scored higher than Jimmy and 16% scored lower than Jimmy, as shown in Figure 4.8.

FIGURE 4.8 ● Applying the Empirical Rule to Interpret $Z < -1$

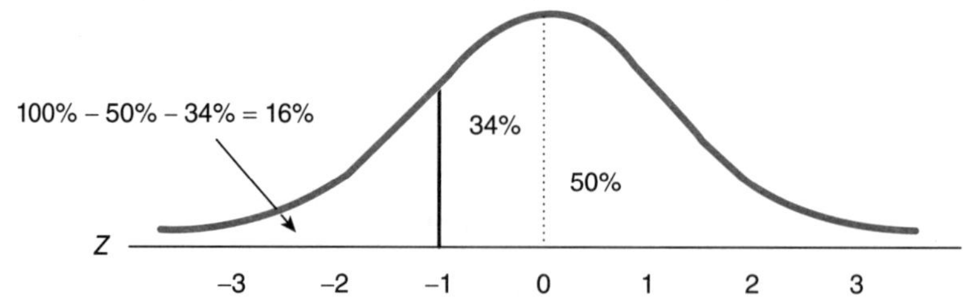

What You Learned

Chapter 4 focused on *Z* scores. You learned how to convert raw scores to *Z* scores using the mean and standard deviation of the distribution. The same principle can be applied to populations as well as samples. The distribution shape of the *Z* scores remains the same as that of raw scores. *Z* scores have a mean of 0 and a standard deviation of 1.

For populations, the *Z*-score formula is $Z = \frac{X - \mu}{\sigma}$.

For samples, the *Z*-score formula is $Z = \frac{X - \bar{X}}{s}$.

In both formulas, there are four values involved: (1) *X*, (2) *Z*, (3) the mean, and (4) the standard deviation of the distribution. You may solve one unknown with one equation. Therefore, as long as three values are provided, you can calculate the fourth one.

The empirical rule (68–95–99.7 rule) provides a rule of thumb for connecting *Z* scores and probabilities for normally distributed variables.

Key Words

Empirical rule: The empirical rule describes the following attributes for variables with a normal distribution: About 68% of the values fall within 1 standard deviation of the mean, about 95% of the values fall within 2 standard deviations of the mean, and about 99.7% of the values fall within 3 standard deviations of the mean. 108

***Z* scores:** *Z* scores are standard scores that describe the differences of individual raw scores from the mean in units of standard deviation. 97

Learning Assessment

Multiple Choice Questions: Choose the Best Answer in Every Question

1. For a population with $\mu = 100$ and $\sigma = 20$, what is the *X* value corresponding to $Z = 0.50$?
   a. 90
   b. 95
   c. 105
   d. 110

2. A sample of $n = 63$ scores has a mean of $\overline{X} = 45$ and a standard deviation of $s = 8$. In this sample, what is the Z score corresponding to $X = 39$?
   a. $Z = 0.75$
   b. $Z = 1.00$
   c. $Z = -0.75$
   d. $Z = -1.00$

3. A sample of $n = 45$ scores are transformed into Z scores. The mean for the 45 Z scores __________.
   a. is 0
   b. is 1
   c. is −1
   d. cannot be determined without more information

4. Under a normal distribution curve, the area covered by $\mu \pm 2\sigma$ is roughly __________% of the data.
   a. 50
   b. 68
   c. 95
   d. 99

5. For a sample with $\overline{X} = 36$, a score of $X = 40$ corresponds to $Z = 0.50$. What is the standard deviation for the sample?
   a. 2
   b. 4
   c. 8
   d. 16

6. A sample of $n = 78$ scores are transformed into Z scores. The standard deviation for the 78 Z scores __________
   a. is 0
   b. is 1
   c. is −1
   d. cannot be determined without more information.

7. In an IQ test with mean = 100 and standard deviation = 15, what score is 1.5 standard deviations above the mean ($Z = 1.50$)?
   a. 100
   b. 115
   c. 122.5
   d. 130

8. In a sample $n = 5$, all the raw scores are converted into Z scores. The first four Z scores are −1.19, −0.69, 0.75, and 1.12. What is the last Z score?
   a. −0.01
   b. −0.10
   c. 0.01
   d. 0.10

9. Which one of the following distributions where $X = 70$ would have the highest Z value?
   a. $\mu = 100$ and $\sigma = 60$
   b. $\mu = 100$ and $\sigma = 30$
   c. $\mu = 100$ and $\sigma = 15$
   d. $\mu = 100$ and $\sigma = 10$

10. Which one of the following distributions with $X = 60$ has the smallest Z score?
   a. $\mu = 50$ and $\sigma = 60$
   b. $\mu = 50$ and $\sigma = 30$
   c. $\mu = 50$ and $\sigma = 15$
   d. $\mu = 50$ and $\sigma = 10$

Free-Response Questions

11. Robin scored 40 points on an English test with $\bar{X} = 30$ and $s = 8$. He also scored a 50 on a math test with $\bar{X} = 55$ and $s = 5$. In which subject did Robin perform better than his classmates according to these test scores?

12. According to the College Board, the SAT ERW $\mu = 497$ and $\sigma = 115$, and the Math $\mu = 513$ and $\sigma = 120$. Ted's ERW score was 730, and his Math score was 760. Were Ted's scores considered unusual scores?

13. The national average of car insurance is \$1,427 per year and standard deviation is \$354. Alex paid \$1,781 a year for his car insurance. What percentage of people paid more than Alex did on their car insurance?

Answers to Pop Quiz Questions

1. d
2. d
3. a
4. d

5

BASIC PRINCIPLES OF PROBABILITY

Learning Objectives

After reading and studying this chapter, you should be able to do the following:

- Define probability terms such as simple event, event, and sample space
- Explain the addition rule, multiplication rule, and complementary rule in probability
- Identify and provide an example of a binomial probability distribution
- Explain how to construct a probability distribution table
- Define Z score and explain the relationship between probability and Z score in a normal distribution
- Describe the purpose of using the Z Table

WHAT YOU KNOW AND WHAT IS NEW

You learned frequency distribution tables in Chapter 2 as a way to organize and simplify numeric data by listing values in an ascending order and tallying the frequency for each value. You learned to calculate means and standard deviations using frequency tables in Chapter 3. You understand the importance of considering the impact of frequency when calculating the mean and standard deviation. You learned to transform raw scores to

Z scores by using the Z formula and to use the empirical rule to connect Z scores and probabilities in a few special cases in Chapter 4. The concepts you learned and the skills you mastered in those early chapters are the building blocks of statistics, and they continue to be relevant throughout the rest of this book.

The link between Z scores and probabilities makes Z scores more accessible and easier to interpret. In Chapter 5, you will learn the computation rule to connect Z scores with probabilities under normal distribution. First, you will learn the basic principles of probability in Chapter 5. *Probability* is a measure to quantify uncertainty. Probability is a new topic, and all its basic terms, such as *simple event*, *event*, and *sample space*, will be clearly defined. The mathematical operations (i.e., *addition rule*, *multiplication rule*, and *complementary rule*) of probability will be explained. The same principles for calculating mean and standard deviation from a frequency table can be applied in a probability distribution.

BASIC TERMS AND MATHEMATICAL OPERATIONS IN PROBABILITY

To understand the basic principles of probability, we need to start with the definitions of terms. It is somewhat ironic that probability, which is a measure to quantify uncertainty, usually makes students *feel* uncertain after learning about this topic. Hopefully, the examples that follow the definitions below will reinforce the meaning of each term and make the learning of this abstract concept more tangible.

Basic Terms in Probability

Learning a new mathematical topic, such as probability, is similar to learning a new language. There is a basic vocabulary that you have to know before you become fluent in the language. Understanding and memorizing the vocabulary is the first step to master the language. Here are the basic terms in probability.

Event: An **event** is defined as a set of outcomes from an experiment or a procedure (i.e., tossing a coin, throwing a die, going through a treatment, having a child, or answering a multiple-choice question).

Simple event: A **simple event** is defined as an elementary event that cannot be broken into simpler parts.

Sample space: A **sample space** is a complete list of all possible outcomes. The probabilities of all possible outcomes add to 1..

Notations in Probability

P stands for probability.

A or B denotes particular events.

P(A) stands for the probability of Event A happening.

$$P(\text{A}) = \frac{\text{Number of ways A happens}}{\text{Total number of all possible outcomes}}$$

Probability is expressed with numbers between 0 and 1, $0 \leq P(\text{A}) \leq 1$. When $P(\text{A}) = 0$, it means that Event A does not happen, and when $P(\text{A}) = 1$, it means that Event A definitely happens. Probability can never be greater than 1 or less than 0. When you answer a probability question with a negative number or a number greater than 1, your calculation is *absolutely* wrong. You have to redo your calculation!

Once you have the basic vocabulary in probability, we can proceed to discuss probability in everyday situations. When expectant parents are anticipating the birth of a baby, they wonder whether it is a boy or a girl. Most human births are single births, with one baby born per pregnancy. To keep things simple, we are limiting the discussion of probability to single births. The probability of having a boy is roughly .5, and the probability of having a girl is also .5. Let's use b for boy and g for girl. *P*(b) is the probability of having a boy, and *P*(g) is the probability of having a girl.

Here are probabilities from one single birth:

$$P(\text{b}) = .5$$

$$P(\text{g}) = .5$$

$$P(\text{b}) + P(\text{g}) = 1$$

The probability of having a boy or a girl in a single birth is pretty easy to calculate. There are two possible outcomes, and the sum of these two possible outcomes adds to 1. Having a boy is one possible event, and having a girl is the other possible event. Both events are examples of a simple event that cannot be broken down further. The sample space includes two possible outcomes of having a boy or having a girl. That is easy enough. Let's move on to discuss the probabilities of having two children.

When having two children, the parents might have two boys, two girls, or one boy and one girl. These are events. The event "one boy and one girl" can be broken further into having a boy first and a girl second or having a girl first and a boy second. The other two events (i.e., two boys or two girls) cannot be broken down any further. Table 5.1 lists the basic probability terms for having two children.

TABLE 5.1 ● Basic Probability Terms for Having Two Children

Procedure	Event	Simple Event
Having two children	Two boys	bb
	One boy and one girl	bg, gb
	Two girls	gg
Sample space		bb, bg, gb, gg

As you can see, the probabilities of having a boy or a girl when having two children are more complicated than the probability of doing so with only one child. To calculate the probability of having two children, we need to introduce mathematical operations of the probabilities.

Mathematical Operations of Probabilities

Table 5.1 shows that there are four simple events in the sample space of having two children. We need to calculate the probability of having two boys, $P(\text{bb})$; the probability of having one boy and one girl, $P(\text{bg})$ or $P(\text{gb})$; and the probability of having two girls, $P(\text{gg})$. Calculating these probabilities involves two new probability concepts: (1) *independent events* and (2) *multiplication rule for independent events*. A and B are two **independent events** when the occurrence of one event does not affect the probability of the occurrence of the other event. In the example of having two children, it means that the sex of the first child does not affect the sex of the second child.

The **multiplication rule for independent events** states that when Event A and Event B are independent, the probability of both events happening is $P(\text{AB}) = P(\text{A}) \times P(\text{B})$.

Let's think about the probability of having two boys. This can only happen when the first baby is a boy and the second baby is also a boy. The probability of the second baby's sex is independent of the probability of the first baby's sex. Thus, the probability of the first baby being a boy is $P(\text{b}) = .5$, and the probability of the second baby being a boy is $P(\text{b}) = .5$.

Therefore, the probability of having two boys is

$$P(\text{bb}) = P(\text{b}) \times P(\text{b}) = (.5) \times (.5) = .25$$

The probability of having two girls is, similarly,

$$P(\text{gg}) = P(\text{g}) \times P(\text{g}) = (.5) \times (.5) = .25$$

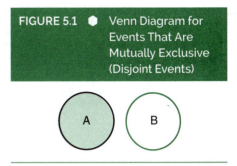

Now, let's figure out the probability of having one boy and one girl. This event can be broken down into two simple events: (1) having a boy first and then a girl (bg) and (2) having a girl first and then a boy (gb). Only one of these two events would occur in any family. We need to calculate *P*(bg or gb). In general, calculating *P*(A or B) involves understanding two new probability concepts: (1) *mutually exclusive events* and (1) *Addition Rule 1 of probability*. When A and B are **mutually exclusive events**, only one of the events can happen: If one happens, the other does not happen. They cannot both happen at the same time. There is no overlapping between Event A and Event B. When having one boy and one girl in a family, it means either having a boy first and then a girl (bg) or having a girl first and then a boy (gb), but two such events cannot happen at the same time. Mutually exclusive events are also called **disjoint events**. *Disjoint events* is simply another label for mutually exclusive events. A Venn diagram for events that are mutually exclusive (i.e., disjoint events) is shown in Figure 5.1.

FIGURE 5.1 ● Venn Diagram for Events That Are Mutually Exclusive (Disjoint Events)

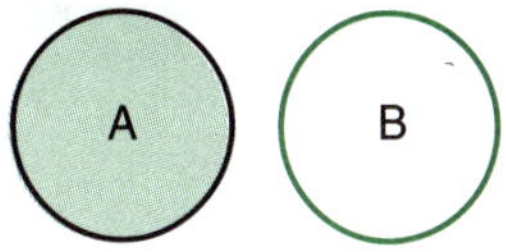

Addition Rule 1 of probability states that when Event A and Event B are mutually exclusive, the probability of either Event A or Event B happening is the sum of the probability of each event, $P(\text{A or B}) = P(\text{A} \cup \text{B}) = P(\text{A}) + P(\text{B})$

The event of having one boy and one girl includes two simple events, bg and gb. These two simple events are mutually exclusive because if one happens, the other does not, and they can't happen at the same time.

Having a boy first and a girl second: $P(\text{bg}) = P(\text{b}) \times P(\text{g}) = (.5) \times (.5) = .25$

Having a girl first and a boy second: $P(\text{gb}) = P(\text{g}) \times P(\text{b}) = (.5) \times (.5) = .25$

Therefore, the probability of having one boy and one girl is

$$P(\text{bg or gb}) = P(\text{bg}) + P(\text{gb}) = .25 + .25 = .50$$

Let's take this one step further and look into the probabilities of all possible outcomes of having three children. If you are thinking the more children you have, the more complicated the probability calculations will get. You are correct! Have no fear, I promise, we will stop the discussion at having three children in this chapter. Now, let's list all possible outcomes of having three children as shown in Table 5.2.

As mentioned earlier, the probability of having a boy or a girl in one birth is independent of having a boy or a girl in any other birth.

TABLE 5.2 ● Probabilities of Different Numbers of Boys or Girls in a Family With Three Children

Procedure	Event	Simple Event
Having three children	Three boys	bbb
	Two boys and one girl	gbb, bgb, bbg
	One boy and two girls	bgg, gbg, ggb
	Three girls	ggg
Sample space		bbb, gbb, bgb, bbg, bgg, gbg, ggb, ggg

The probability of having three boys is

$$P(\text{bbb}) = P(\text{b}) \times P(\text{b}) \times P(\text{b}) = (.5) \times (.5) \times (.5) = .125$$

The event of having two boys and one girl can be broken down into three simple events because the girl could be the first born, second, or the third. These events are mutually exclusive: gbb, bgb, and bbg.

$$P(\text{gbb}) = P(\text{g}) \times P(\text{b}) \times P(\text{b}) = (.5) \times (.5) \times (.5) = .125$$

$$P(\text{bgb}) = P(\text{b}) \times P(\text{g}) \times P(\text{b}) = (.5) \times (.5) \times (.5) = .125$$

$$P(\text{bbg}) = P(\text{b}) \times P(\text{b}) \times P(\text{g}) = (.5) \times (.5) \times (.5) = .125$$

Therefore, the probability of having two boys and one girl is

$$P(\text{gbb or bgb or bbg}) = P(\text{gbb}) + P(\text{bgb}) + P(\text{bbg}) = .125 + .125 + .125 = .375$$

The event of having one boy and two girls can also be broken down into three simple events, and these events are mutually exclusive: bgg, gbg, and ggb.

$$P(\text{bgg}) = P(\text{b}) \times P(\text{g}) \times P(\text{g}) = (.5) \times (.5) \times (.5) = .125$$

$$P(\text{gbg}) = P(\text{g}) \times P(\text{b}) \times P(\text{g}) = (.5) \times (.5) \times (.5) = .125$$

$$P(\text{ggb}) = P(\text{g}) \times P(\text{g}) \times P(\text{b}) = (.5) \times (.5) \times (.5) = .125$$

Therefore, the probability of having one boy and two girls is

$$P(\text{bgg or gbg or ggb}) = P(\text{bgg}) + P(\text{gbg}) + P(\text{ggb}) = (.125) \times (.125) \times (.125) = .375$$

The probability of having three girls is

$$P(\text{ggg}) = P(\text{g}) \times P(\text{g}) \times P(\text{g}) = (.5) \times (.5) \times (.5) = .125$$

Now, we have successfully calculated probability of different events in terms of having one child, two children, and three children. In all these events, for each birth, there are only two possible outcomes: a boy or a girl. The probability of having a boy stays at $P(\text{b}) = .5$ and the probability of having a girl also stays at $P(\text{g}) = .5$. Such conditions fit in with a particularly important probability distribution called a *binomial probability distribution*, which deserves further discussion.

Binomial Probability Distribution

A **binomial probability distribution** is a probability distribution applied to variables with only two possible outcomes (i.e., success vs. failure) in each trial. The binomial probability distribution has a fixed number of trials. The trials are independent of one another. The probability of a success and the probability of a failure remain the same in all trials. Linking the definition back to our previous example, having a child results in only two possible outcomes: a boy or a girl. To figure out the probability of every outcome, we need to know the number of children. Each birth is an independent event, in other words, the outcome of a previous birth will not affect the outcome of the next birth. The probability of having a boy and the probability of having a girl remain the same in all births.

In summary, a binomial probability distribution has the following four requirements:

1. The procedure has a fixed number of trials.
2. The trials are independent events. The probabilities of outcomes in one trial do not affect the probabilities of outcomes in the other trials.
3. There are only two possible outcomes for each trial: (1) success and (2) failure. These are commonly used labels. There is no value judgment involved. The outcomes can be head/tail or boy/girl.
4. The probability of a success and the probability of a failure remain the same in all trials.

Here is the basic notation vocabulary for binomial probability distributions:

S stands for success, and F stands for failure

$P(\text{S}) = p$, the probability of success in one trial

$P(\text{F}) = 1 - p = q$, the probability of failure in one trial

n = the number of trials

r = the number of successes in n trials

$P(r)$ = the probability of getting exactly r successes in n trials

Binomial probability formula:

$$P(r) = C_r^n p^r q^{(n-r)} = \frac{n!}{(n-r)!r!} p^r q^{(n-r)} \quad \text{for } r = 0,1,2,3,\ldots n$$

Let me break the binomial formula down and explain each component.

C_r^n is a mathematical operation to calculate the number of possible combinations to produce r number of successes out of n trials,

$$C_r^n = \frac{n!}{(n-r)!r!}$$

C stands for combination

$$n! = n \times (n-1) \times (n-2) \times (n-3) \cdots 3 \times 2 \times 1$$

$n!$ is pronounced as **n factorial**. It is a mathematical operation calculated as the multiplicative product of all positive integers less than or equal to n. 0! is defined as 1. You can calculate the factorial for positive integers. The value can get large pretty quickly. For example,

$$5! = 5 \times 4 \times 3 \times 2 \times 1 = 120$$

$$10! = 10 \times 9 \times 8 \times 7 \times 6 \times 5 \times 4 \times 3 \times 2 \times 1 = 3628800$$

Let's use simple numbers to practice the combination of r successes in n trials, C_r^n. If you produce three successes out of five trials, in how many different ways can it happen?

$$C_3^5 = \frac{n!}{(n-r)!r!} = \frac{5!}{(5-3)!3!} = \frac{5!}{2!3!} = 10$$

Let's apply the binomial probability formula to the example of having three children. The number of girls in a family with three children could be 0, 1, 2, or 3. Let's label having a girl as a success and having a boy as a failure. Again, there is no value judgment. Success and failure are commonly used labels to refer to two possible outcomes of one trial.

The probability of having no girl in three single births:

$n = 3$

$r = 0$

$p = .5$

$q = .5$

$$P(0) = \frac{3!}{(3-0)!0!}(.5)^0(.5)^3 = .125$$

The probability of having one girl in three single births:

$n = 3$

$r = 1$

$p = .5$

$q = .5$

$$P(1) = \frac{3!}{(3-1)!1!}(.5)^1(.5)^2 = .375$$

The probability of having two girls in three single births:

$n = 3$

$r = 2$

$p = .5$

$q = .5$

$$P(2) = \frac{3!}{(3-2)!2!}(.5)^2(.5)^1 = .375$$

The probability of having three girls in three single births:

$n = 3$

$r = 3$

$p = .5$

$q = .5$

$$P(3) = \frac{3!}{(3-3)!3!}(.5)^3(.5)^0 = .125$$

TABLE 5.3 ● Probability Distribution Table of the Number of Girls in a Family With Three Children

X (Number of Girls)	*P*(*X*)
0	.125
1	.375
2	.375
3	.125

Applying the binomial probability formula, we easily get the probabilities of having no, one, two, and three girls without analyzing and listing the entire sample space. Table 5.3 is the binomial probability distribution table of the number of girls in a family with three children.

If you survey 10 families with three single births, you might not get exactly the same probabilities. The attributes obtained from a small sample size (i.e., $n < 30$) tend to fluctuate depending on which families are included in the sample. If you obtain information from every family with three children, you get the probability distribution of the entire population. You are more likely to get the probability distribution as described in Table 5.3. Probability distributions are used to describe populations. Let's proceed with calculating the population mean, variance, and standard deviation with this example. Here are formulas for the population parameters of a probability distribution.

Population mean for a probability distribution: $\mu = \Sigma[XP(X)]$.

Population mean is calculated as the sum of the values of X multiplied by its associated probability.

Population variance for a probability distribution: $\sigma^2 = [(X-\mu)^2 P(X)]$.

Population variance is calculated as the sum of squared deviations multiplied by its associated probability.

Population standard deviation for a probability distribution:

$$\sigma = \sqrt{\text{Variance}} = \sqrt{\Sigma[(X-\mu)^2 P(X)]}.$$

As always, population standard deviation is equal to the square root of the population variance.

The reason to discuss the binomial probability distribution is that it has specially simple formulas to calculate μ, σ^2, and σ as shown below.

$$\mu = np$$

$$\sigma^2 = npq$$

$$\sigma = \sqrt{npq}$$

Let's use both sets of formulas to calculate μ, σ^2, and σ by using the numbers in Table 5.3. The first set of formulas apply to all probability distribution, and the specially simple formulas only apply to binomial probability distribution. According to the order of operations in the general probability formula, $\mu = \Sigma[XP(X)]$, the multiplication between X and $P(X)$ inside the parentheses brackets needs to be done first. Add up every $XP(X)$ to obtain $\Sigma[XP(X)]$ at the end. According to the order of operations in the formula for $\sigma^2 = \Sigma[(X-\mu)^2 P(X)]$, three more columns need to be calculated: $(X-\mu)$, $(X-\mu)^2$, and $(X-\mu)^2 P(X)$. Then, every $(X-\mu)^2 P(X)$ is added up to obtain $\Sigma[(X-\mu)^2 P(X)]$ at the end.

Each column in Table 5.3a demonstrates one step in the order of operations to get the μ, and σ^2.

$$\text{Mean} = \mu = \Sigma[XP(X)] = 1.5$$

$$\text{Variance} = \sigma^2 = \Sigma[(X-\mu)^2 P(X)] = 0.75$$

$$\text{Standard deviation} = \sigma = \sqrt{\text{Variance}} = \sqrt{\Sigma[(X-\mu)^2 P(X)]} = \sqrt{0.75} = 0.87$$

TABLE 5.3a ● Mean and Variance of the Probability Distribution of the Number of Girls in a Family With Three Children

X (Number of Girls)	$P(X)$	$XP(X)$	$X-\mu$	$(X-\mu)^2$	$(X-\mu)^2 P(X)$
0	.125	0	−1.5	2.25	0.28125
1	.375	0.375	−0.5	0.25	0.09375
2	.375	0.75	0.5	0.25	0.09375
3	.125	0.375	1.5	2.25	0.28125
Total		$\mu = 1.5$			$\sigma^2 = 0.75$

These general probability distribution formulas are exactly the same as the formula you learned in Chapter 3 when using a frequency distribution table for a population.

Population mean for a frequency table: $\mu = \Sigma fX/N$

For every X, the associated probability of X is $P(X) = f/N$; therefore,

$$\mu = \frac{\Sigma fX}{N} = \Sigma[XP(X)]$$

For the same reason, $P(X) = f/N$, the population variance for a frequency table:

$$\sigma^2 = \frac{SS}{N} = \Sigma\frac{f(X-\mu)^2}{N} = \Sigma[(X-\mu)^2 P(X)]$$

Population standard deviation: $\sigma = \sqrt{\text{Variance}}$

Let's verify the answers calculated from the general probability distribution formulas with the specially simplified formulas for the binomial probability distribution.

$$\mu = np = 3(.5) = 1.5$$

$$\sigma^2 = npq = 3(.5)(.5) = 0.75$$

$$\sigma = \sqrt{npq} = \sqrt{0.75} = 0.87$$

The binomial probability distribution is a special type of probability distribution. Its simplified formulas generate the same answers for μ, σ^2, and σ as a regular probability distribution. Once you replace $P(X) = f/N$, the formulas are mathematically identical to the ones you used in Chapter 3 for the frequency distribution of a population.

Binomial probability distributions have wide applications because many things in life can be classified into success or failure. For example, a set of product quality test outcomes can be classified as successes or failures. The probability of success and the probability of failure don't always split evenly at 50:50. Let's go over an example where the probability of success and the probability of failure are not evenly split.

EXAMPLE 5.1

A statistics professor uses quizzes to evaluate students' learning. There are five multiple-choice questions in a quiz. Each question has four answer options: a, b, c, and d. What is the probability of a student simply guessing blindly and getting the correct answers on four out of the five questions?

There are five questions, $n = 5$; the student gets four questions correct, $r = 4$.

The probability of success for each question is $p = 1/4 = .25$.

The probability of failure for each question is $q = 1 - p = .75$.

$$P(4) = C_r^n p^r q^{(n-r)} = C_4^5 (.25)^4 (.75)^{(5-4)} = \frac{5!}{(5-4)!4!}(.25)^4(.75)^1 = 5(.25)^4(.75)^1 = .0146$$

As you can tell, blindly guessing four out of the five questions correctly is a low-probability event. This student is very lucky to get such an outcome by guessing blindly. Probability is often linked with luck. Our subjective perceptions do not always match with calculated probabilities. People have the tendency to perceive that positive things are more likely to happen and negative things are less likely to happen than the calculated probabilities. For example, people who buy lottery tickets tend to think that they will be the big winners someday. No one ever wakes up in the morning and thinks that they will get into a car accident because of driving while texting. However, the probability of getting into a car accident when driving distracted is much higher than the probability of being the jackpot winner.

POP QUIZ

1. Which one of the following is not a required attribute of a binomial probability distribution?
   a. The procedure has a fixed number of trials.
   b. The trials are independent of one another.
   c. There are only two possible outcomes for each trial: success and failure.
   d. The probability of a success is $p = .5$, and the probability of a failure is $q = .5$.

2. When Events A and B are mutually exclusive, what is $P(A \cup B)$?
   a. $P(A) + P(B)$
   b. $P(A) - P(B)$
   c. $P(A) \times P(B)$
   d. $\frac{P(A)}{P(B)}$

The Dice Game

Probabilities are also commonly applied in dice games. A die is a three-dimensional cube with six sides; *dice* refers to two or more such objects. Each side has a different number of dots ranging from 1 to 6.

EXAMPLE 5.2

When you roll a die, what are the probabilities of getting 0, 1, 2, 3, 4, 5, 6, and 7 dots?

In Chapter 2, we discussed the probabilities of obtaining various numbers of dots by throwing one die, as shown in Table 5.4. It is clear that it is not possible to get 0 dots or 7 dots by throwing a single die. When some events are definitely not happening, their probabilities are 0. Therefore, $P(0) = 0$ and $P(7) = 0$. The rest of the probabilities are $P(1) = P(2) = P(3) = P(4) = P(5) = P(6) = 1/6 = .167$. The probability of getting 1, 2, 3, 4, 5, or 6 dots by throwing one die is identical across possible outcome. Such a distribution is labeled as a uniform distribution,

TABLE 5.4 ● Probabilities of Obtaining Various Numbers of Dots by Throwing One Die

X (Number of Dots)	Probability $P(X)$
1	1/6 = .167
2	1/6 = .167
3	1/6 = .167
4	1/6 = .167
5	1/6 = .167
6	1/6 = .167
Total	1

EXAMPLE 5.3

What is the probability of throwing a single die to get a number smaller than 4 or an odd number?

This is a probability question of either Event A or Event B, expressed as $P(A \cup B)$, or "the probability of A union B." First, we need to verify if Events A and B are mutually exclusive. Event A contains numbers smaller than 4, that is, {1, 2, 3}; Event B contains odd numbers, {1, 3, 5}. Event A and Event B are not mutually exclusive. Events A and B overlap with each other. Therefore, we need to introduce the **Addition Rule 2 of Probability**.

Addition Rule 2 of Probability: When Event A and Event B are not mutually exclusive, the probability of either Event A or Event B happening, or both, is the sum of the probabilities of each event minus the probability of the overlap between the two events, $P(A \cup B) = P(A) + P(B) - P(A \cap B)$.

The overlap of the two events is expressed as $P(A \cap B)$, or "Event A intersects Event B." In the example, Event A is a number smaller than 4, A = {1, 2, 3}, and Event B is an odd number, B = {1, 3, 5}. The overlap between Event A and Event B is $(A \cap B)$ = {1, 3}. $(A \cap B)$ means that both events are true at the same time. The numbers are both smaller than 4 *and* odd numbers.

A Venn diagram for Events A and B when they are not disjoint is shown in Figure 5.2. The light blue circle with a solid black outline represents the probability of Event A, *P*(A); the white circle with a black dashed outline represents the probability of Event B, *P*(B); and the darker blue portion represents the probability of both Events A and B, $P(A \cap B)$.

FIGURE 5.2 ● Venn Diagram for Events That Are Not Disjoint

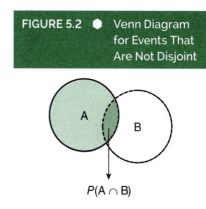

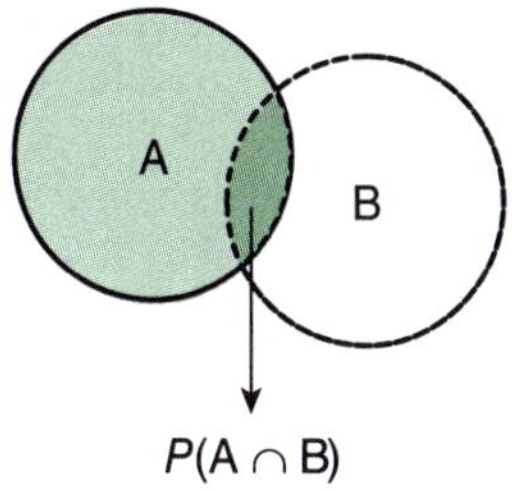

$$P(A \cup B) = P(A) + P(B) - P(A \cap B).$$

The reason we subtract the overlap is to eliminate the double counting in order to keep an accurate count of all simple events.

Event A = {1, 2, 3}, and its probability is $P(\text{A}) = 3/6 = .5$

Event B = {1, 3, 5}, and its probability is $P(\text{B}) = 3/6 = .5$

The overlapping part of Events A and B is {1, 3}, and its probability is $P(A \cap B) = 2/6 = .33$. This probability gets counted twice, once with Event A and again with Event B.

$$P(\text{A} \cup \text{B}) = P(\text{A}) + P(\text{B}) - P(\text{A} \cap \text{B}) = .5 + .5 - .33 = .67$$

The probability of throwing a die to get a number smaller than 4 or an odd number is .67.

We can also verify the calculated probability by listing all the simple events.

The sample space for throwing a die is {1, 2, 3, 4, 5, 6}.

Event A covers numbers smaller than 4; A = {1, 2, 3}Event B covers odd numbers; B = {1, 3, 5}

$(A \cup B)$ means that either Event A or Event B happens. The numbers can be smaller than 4 or odd numbers; $(A \cup B)$ = {1, 2, 3, 5}, and $P(A \cup B) = \frac{4}{6} = .67$.

EXAMPLE 5.4

What are the mean, variance, and standard deviation of the total numbers of dots from throwing two dice?

First, it is useful to figure whether the number of dots from two dice fits a general probability distribution or a binomial probability distribution. Every time you throw a die, you are getting one out of six possible outcomes so it does not fit with the binomial probability distribution that only allows two possible outcomes for each trial. We discussed the probabilities of obtaining specific numbers of dots by throwing two dice in Chapter 2. The total numbers of dots from throwing two dice and their probabilities are shown in Table 5.5. These probabilities allow us to put together a **probability distribution** table for the total numbers of dots from two dice. There are three requirements for constructing a probability distribution table:

1. The variable X is a discrete, random, and numerical variable. The number of possible values in X is finite. Each value is associated with a probability.
2. The probability of each value needs to be within 0 and 1, inclusive. This is expressed as $0 \leq P(X) \leq 1$.
3. The sum of the probabilities of all the values is 1. It is expressed as $\Sigma P(X) = 1$.

TABLE 5.5 ● Total Numbers of Dots From Throwing Two Dice and Their Probabilities

X (Number of Dots)	$P(X)$ (Probability)
2 (1,1)	1/36 = .028
3 (1,2), (2,1)	2/36 = .056
4 (1,3), (2,2), (3,1)	3/36 = .083
5 (1,4), (2,3), (3,2), (4,1)	4/36 = .111
6 (1,5), (2,4), (3,3), (4,2), (5,1)	5/36 = .139
7 (1,6), (2,5), (3,4), (4,3), (5,2), (6,1)	6/36 = .167
8 (2,6), (3,5), (4,4), (5,3), (6,2)	5/36 = .139
9 (3,6), (4,5), (5,4), (6,3)	4/36 = .111
10 (4,6), (5,5), (6,4)	3/36 = .083
11 (5,6), (6,5)	2/36 = .056
12 (6,6)	1/36 = .028
Total	1.001

The total numbers of dots from two dice could be 2, 3, 4, 5, 6, 7, 8, 9, 10, 11, and 12. When two dice are thrown, the values of the total number of dots are random, discrete, and numerical. There are 11 possible outcomes and each outcome is associated with a probability, and the probability of each outcome is between 0 and 1. The total probability is $\Sigma P(X) = 1$. This example fulfills all three requirements of a probability distribution.

The $\Sigma P(X)$ (or the sum of all probabilities) equals 1.001 instead of 1, which is attributable to rounding.

Population mean for a probability distribution: $\mu = \Sigma[XP(X)]$.

Population variance for a probability distribution: $\sigma^2 = \Sigma[(X - \mu)^2 P(X)]$

Population standard deviation for a probability distribution: $\sigma = \sqrt{\text{Variance}} = \sqrt{\Sigma[(X - \mu)^2 P(X)]}$

Let's document the step-by-step process by adding columns to Table 5.5. The new columns are shown in Table 5.5a. According to the order of operations in the formula $\mu = \Sigma[XP(X)]$, the multiplication between X and $P(X)$ inside the brackets needs to be done first. Add up every $XP(X)$ to obtain $\Sigma[XP(X)]$ at the end. According to the order of operations in the formula $\sigma^2 = \Sigma[(X - \mu)^2 P(X)]$, three more columns need to be created: $X - \mu$, $(X - \mu)^2$, and $(X - \mu)^2 P(X)$. Then add up every $(X - \mu)^2 P(X)$ to obtain $\Sigma[(X - \mu)^2 P(X)]$ at the end.

TABLE 5.5a ⬢ Mean, Variance, and Standard Deviation of the Total Numbers of Dots From Two Dice and Their Probabilities

X (Number of Dots)	$P(X)$ (Probability)	$XP(X)$	$X - \mu$	$(X - \mu)^2$	$(X - \mu)^2 P(X)$
2	.028	0.056	-5.0	25	0.702
3	.056	0.168	-4.0	16	0.899
4	.083	0.332	-3.0	9	0.750
5	.111	0.555	-2.0	4	0.447
6	.139	0.834	-1.0	1	0.141
7	.167	1.169	0.0	0	0.000
8	.139	1.112	1.0	1	0.137
9	.111	0.999	2.0	4	0.441
10	.083	0.83	3.0	9	0.744
11	.056	0.616	4.0	16	0.893
12	.028	0.336	5.0	25	0.698
Total	1.001	$\mu = 7.0$			$\sigma^2 = 5.9$

(Continued)

(Continued)

According to Table 5.5a,

$$\text{Mean} = \mu = \Sigma[XP(X)] = 7.0$$

$$\text{Variance} = \sigma^2 = \Sigma[(X-\mu)^2 P(X)] = 5.9$$

$$\text{Standard deviation} = \sigma = \sqrt{\text{Variance}} = \sqrt{\Sigma[(X-\mu)^2 P(X)]} = \sqrt{5.9} = 2.4$$

The mean number of dots from two dice is 7, and the standard deviation is 2.4.

Gambling payouts are tied to the probabilities of events. The rule is that the smaller the probabilities, the larger the payouts. That is the reason why throwing two dice landing seven dots will not win any payout.

POP QUIZ

3. Which one of the following attributes is not required for constructing a probability distribution table?
   a. The variable X is a discrete, random, numerical variable. The number of possible values of X is finite. Each value is associated with a probability.
   b. The probability of each value needs to be between 0 and 1, inclusive. This is expressed as $0 \leq P(X) \leq 1$.
   c. The sum of the probabilities of all the values equals 1. It is expressed as $\Sigma P(X) = 1$.
   d. The probability remains the same for every value.
4. What is the probability of throwing two dice to get the total number of dots greater than 4 and odd numbers?
   a. .250
   b. .333
   c. .444
   d. .667

LINKAGE BETWEEN PROBABILITY AND *Z* SCORE IN A NORMAL DISTRIBUTION

In a normal distribution, the area under the curve of a probability distribution equals 1 because it covers all possible values of Z. When you obtain a specific Z score, you can identify the location of that Z score in the distribution. Using the Z score as a boundary, the area under the curve of a probability distribution can be divided. For example, for

$Z > 2$, the probability $P(Z > 2)$ can be identified. I want to make this perfectly clear $P(Z > 2)$ is the same as $P(Z \geq 2)$ because probability is calculated as an area under the normal distribution curve, whether the particular Z value is included does not change the calculation of the area. The empirical rule (68–95–99.7 rule) provides a rough linkage between probabilities and Z scores under the normal distribution: 68% of the values are within $-1 \leq Z \leq 1$, 95% are within $-2 \leq Z \leq 2$, and 99.7% are within $-3 \leq Z \leq 3$, as discussed in Chapter 4. We can explore the linkage between probabilities and Z scores in more detail with a **probability density function.** A probability density function describes the probability for a continuous, random variable to be in a given range of values. When the normal distribution is expressed by Z scores, the probability density function formula is as given below:

$$y = \frac{1}{\sigma\sqrt{2\pi}} e^{-\frac{1}{2}\left(\frac{x-\mu}{\sigma}\right)^2}$$

Under the standard normal distribution, $\mu = 0$ and $\sigma = 1$, the function can be simplified to

$$y = \frac{1}{\sqrt{2\pi}} e^{-\frac{z^2}{2}}$$

$$e \cong 2.71828\ldots$$

$$\pi \cong 3.14159\ldots$$

The formula looks scary, but the good news is that you don't actually have to do any calculation using this formula. Some very considerate and smart people have created the Standard Normal Distribution Table (Z Table) to provide precise link between probabilities and Z values, as shown in Appendix A. All you need to do is learn how to use the Standard Normal Distribution Table. Many different ways to present the Z Table have shown up in various statistics textbooks. The Z Table ranges from one to eight pages, with different instructions on how to use it. For this book, I chose to use a two-page Z Table with one page of instructions. The first page of the Z Table consists of positive Z values, and the second page consists of negative Z values. It is important that you follow the instructions on how to use the Z Table. Keep in mind that Z (capital Z) stands for a collection of Z values and z (lowercase z) stands for a particular z value specified in a problem statement.

Here is a list of what you need to know about the Z Table:

1. All Z values are reported to two places after the decimal point, and all probabilities are reported to four places after the decimal point.
2. The column represents Z values to one place after the decimal point, and the row represents Z values' second place after the decimal point.

3. The intersection between the column and the row shows the area to the left of the specified z value or the probability of all Z values smaller than the specified z value, which is expressed as $P(Z < z)$.

4. If the problem statement asks for the probability of Z values larger than the specified z value, which is expressed as $P(Z > z)$, it can be calculated as 1 minus the probability listed in the Z Table.

Once you learn how to use the Z Table, your understanding of the link between probabilities and Z scores will go beyond the limited, special Z values as stated in the empirical rule in Chapter 4. You will be able to figure out the probability associated with any range of Z values. Let's use examples to illustrate the use of the Z Table. You will learn to figure out the probability to the left of a particular z value, $P(Z < z)$; the probability to the right of a particular z value, $P(Z > z)$; or the probability in between two specified z values, $P(z_1 < Z < z_2)$. Let's use four simple questions in Example 5.5 to illustrate how to solve the probability when given particular z value(s).

EXAMPLE 5.5

Find the probabilities associated with the specified Z values.

a. What is the probability that your workout time is less than 1.53 standard deviations above the mean?

b. What is the probability of being shorter than 0.25 standard deviations below the mean in height?

c. What is the probability that your screen time is higher than 1.78 standard deviations above the mean for people your age?

d. What is the probability of being within 0.5 standard deviations of the mean in the running speed?

Answers:

a. It is important to learn to mathematically express the probability of less than 1.53 standard deviations above the mean as $P(Z < 1.53)$.

Draw and identify this area in the standard normal distribution curve. The drawing does not need to be precise, but it helps identify any additional steps required to figure out the probability as shown in Figure 5.3.

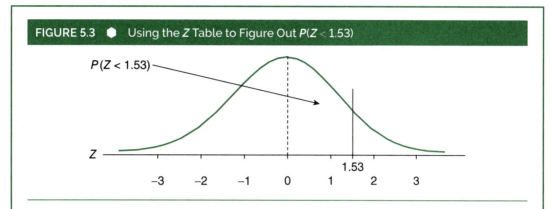

FIGURE 5.3 ● Using the Z Table to Figure Out $P(Z < 1.53)$

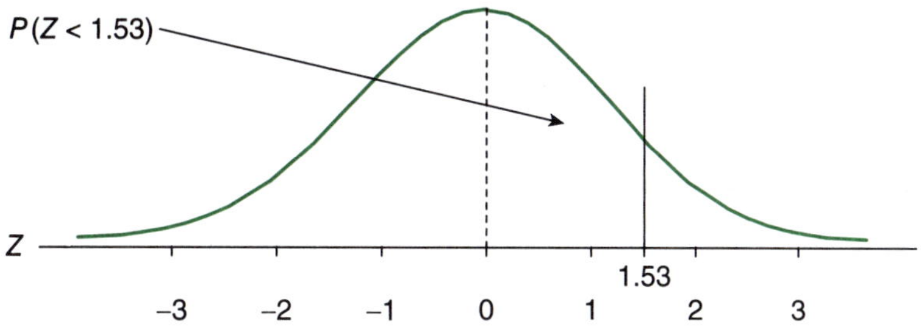

The Z Table provides the probability to the left of the specified Z values. According to the Z Table, $P(Z < 1.53) = .9370$; no additional calculation is needed. The probability of having a workout time less than 1.53 standard deviations above the mean is .9370.

Alternatively, you can use the Excel function to find $P(Z < 1.53)$. To get the probability of the area to the left of the specified z value is the same as finding the percentile rank for the z value. In a blank cell in Excel where you want the answer to appear, type "=**NORM.S.DIST(1.53, TRUE)**", and hit **Enter**. This function finds the percentile rank for the specified z value. You will see the answer .9370.

b. It is important to learn to mathematically express the probability of being shorter than 0.25 standard deviations below the mean as $P(Z < -0.25)$. Then draw and identify this area in the standard normal distribution curve as shown in Figure 5.4.

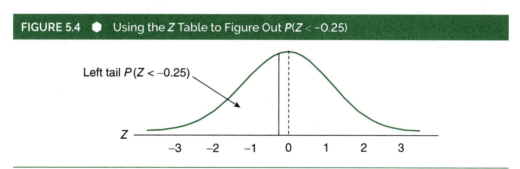

FIGURE 5.4 ● Using the Z Table to Figure Out $P(Z < -0.25)$

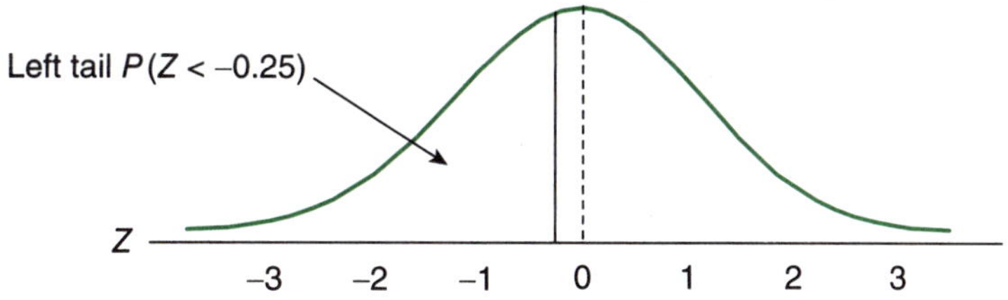

The Z Table provides $P(Z < z)$. Therefore, the answer for $P(Z < -0.25)$ can be found directly in the Z Table: $P(Z < -0.25) = .4013$.

Alternatively, you can use the Excel function to find $P(Z < -0.25)$. To get the probability of the area to the left of the specified z value is the same as finding the percentile rank for the z value. In a blank cell in Excel where you want the answer to appear, type "=**NORM.S.DIST(-0.25, TRUE)**", and hit **Enter**. This function finds the percentile rank for the specified z value. You will see the answer .4013.

(Continued)

(Continued)

c. The mathematical way to express the probability of a score higher than 1.78 standard deviations above the mean is $P(Z > 1.78)$. Once you have stated this, draw and identify this area in the standard normal distribution curve. The drawing helps you identify any additional steps required to figure out the probability as shown in Figure 5.5.

FIGURE 5.5 ● Using the *Z* Table to Figure Out *P*(*Z* > 1.78)

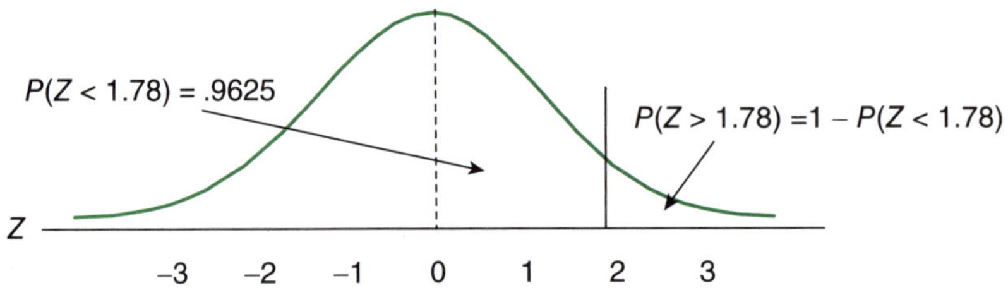

The *Z* Table provides $P(Z < z)$. It specifies the area to the left of the specified *Z* value. However, Example 5.5c asks for $P(Z > 1.78)$, the area to the right of the *z* value. The entire area under the curve is 1. Therefore, the area to the right of the *z* value equals 1 minus the area to the left of the *z* value; that is, $P(Z > 1.78) = 1 - P(Z < 1.78) = 1 - .9625 = .0375$. This answer tells you that the time you spent on the screen of your smart devices is higher than 96.25% of people your age and only 3.75% of them spent more time on the screen than you. It is time to consider spending more time doing other things.

Alternatively, you can use the Excel function to find $P(Z < 1.78)$. The same way as the *Z* Table only provide the probability to the left of the specified *z* value, the Excel function only provides the probability of the area to the left of the specified *z* value. In a blank cell in Excel where you want the answer to appear, type "=**NORM.S.DIST(1.78, TRUE)**", and hit **Enter**. You will see the answer .9625. Because everything under the probability curve is 1, you need to use 1 – .9625 = .0375 to obtain the probability of the area to the right of the specified *z* value.

d. It is important to learn to express the probability of being within 0.5 standard deviations from the mean as $P(-0.5 < Z < 0.5)$. Once you have expressed this, draw and identify this area in the standard normal distribution curve. The drawing does not need to be precise, but it helps identify any additional steps required to figure out the probability as shown in Figure 5.6.

FIGURE 5.6 ● Using the *Z* Table to Figure Out *P*(-0.5 < *Z* < 0.5)

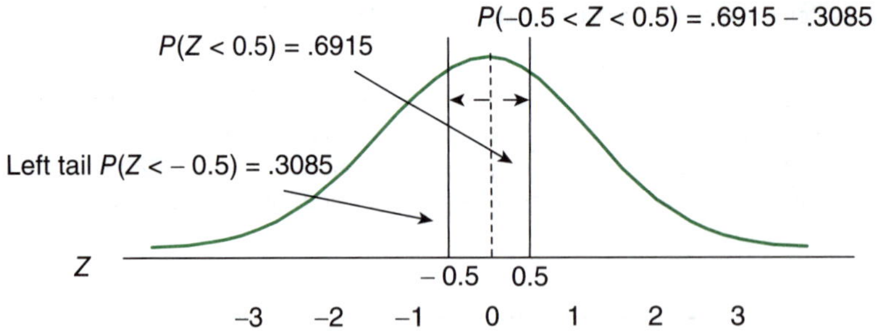

The Z Table provides $P(Z < z)$.

$$P(Z < -0.5) = .3085$$
$$P(Z < 0.5) = .6915$$

Therefore, the area in the middle, $P(-0.5 < Z < 0.5)$ is the difference between $P(Z < 0.5)$ and $P(Z < -0.5)$. It is calculated as $.6915 - .3085 = .3830$.

You have learned to use Z values to identify probabilities in the previous four questions. Since there is one-on-one unique connection between any z value and the probability, we can reverse the process, which means when given a probability, you can figure out the z value. In the next three questions, you will learn to use the known probabilities to figure out the Z values. When the probability provided in the question does not match with the detailed four places after the decimal point as listed in the Z Table, I suggest that you use the closest probability you can find in the Z Table to solve the problem.

EXAMPLE 5.6

Find the Z values associated with the specified probabilities.

a. What is the Z value that sets the top 1% apart from the rest of the distribution?
b. What Z values set the boundaries of the middle 50% from the rest of the distribution?
c. Which Z value sets the fastest 5% of swimmers' times in a 100-meter freestyle race apart from the rest of the distribution?

Answers:

a. To solve this problem, you should start drawing the figure of the top 1% at the right side of the distribution as shown in Figure 5.7. Due to the fact that everything under the curve is 1, the area to the left side of the z value is (1 – the area at the right side). The Z Table only provides the probabilities for $P(Z < z)$, therefore, the easiest way to express the z value that sets the top 1% apart from the rest of the distribution mathematically is $P(Z < z) = .99$. Next step is to search the Z Table to identify a Z value that gives a $P(Z < z)$ as close to .99 as possible. The probability closest to .99 in the Z Table is .9901, and the Z value associated with this probability is 2.33, as shown in Figure 5.7. $Z = 2.33$ sets the top 1% apart from the rest of the distribution.

Alternatively, you can use the Excel **NORM.S.INV(probability)** function to find the specific z value given the probability. The Excel function only provides the probability of the area to the left of the specified z value. To figure out the z value that separates the top 1% from the rest of the distribution, you need to

(Continued)

(Continued)

use .99 as the probability to the left of the z value. In a blank cell in Excel where you want the answer to appear, type "**=NORM.S.INV(0.99)**", and hit **Enter**. You will see the answer 2.33.

FIGURE 5.7 ● Using the Z Table to Figure Out the Z Value That Separates the Top 1% From the Rest of the Distribution

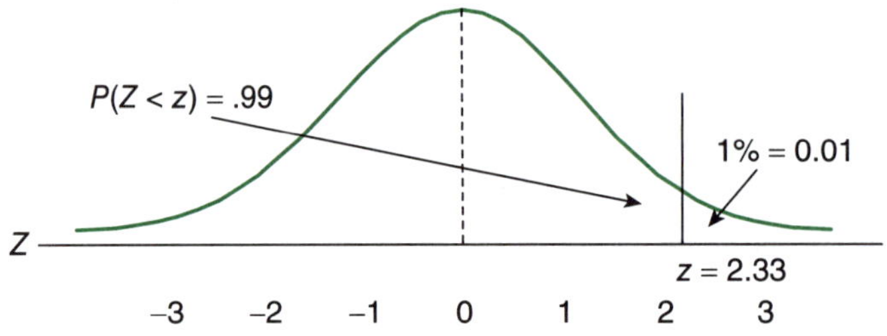

b. The mathematical way to express the Z values that set the boundaries of the middle 50% from the rest of the distribution is $P(-z < Z < z) = .50$. Due to the symmetrical attributes of the Z Table, the left side is the mirror image of the right side. The middle 50% sits in the center of the distribution, with 25% above the mean and 25% below the mean. The Z values that set the boundaries have the same magnitude but with different signs—one is positive and the other is negative. Once you have expressed this, draw and identify this area in the standard normal distribution curve.

$$\text{Left tail } P(Z < -z) = .25$$

Search the Z Table to identify a Z value that gives a $P(Z < -z)$ as close to .25 as possible. The probability in the Z Table closest to .25 is .2514, and the Z value associated with this probability is −0.67, as shown in Figure 5.8.

Alternatively, you can use the Excel **NORM.S.INV(probability)** function to find the specific z value given the probability. In a blank cell in Excel where you want the answer to appear, type "**=NORM.S.INV(0.25)**", and hit **Enter**. You will see the answer −0.67.

FIGURE 5.8 ● Using the Z Table to Figure Out $P(-z < Z < z) = .50$

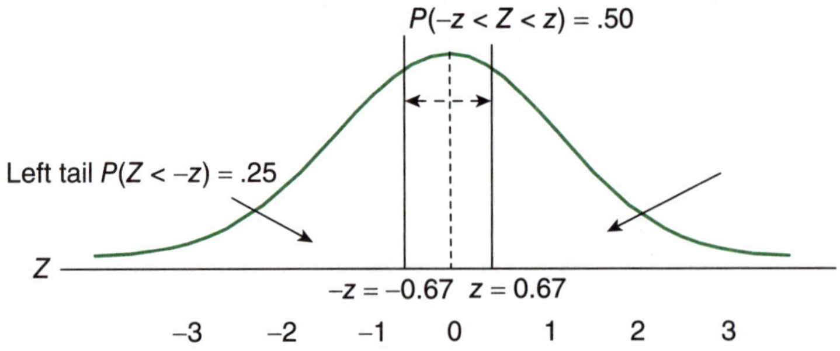

Using the symmetry of the Z curve, you know that the other Z value above the mean is 0.67.

Therefore, the Z values that set the boundaries of the middle 50% from the rest of the distribution are −0.67 and 0.67.

c. It is important to mathematically express the probability correctly. The faster the swimmer, the less time she or he takes to complete an event. Therefore, the fastest 5% of swimmers' times is at the left tail, $P(Z < -z) = .05$. Once this has been expressed, draw and identify this area in the standard normal distribution curve as shown in Figure 5.9.

FIGURE 5.9 ⬢ Using the Z Table to Figure Out $P(Z < -z) = .05$

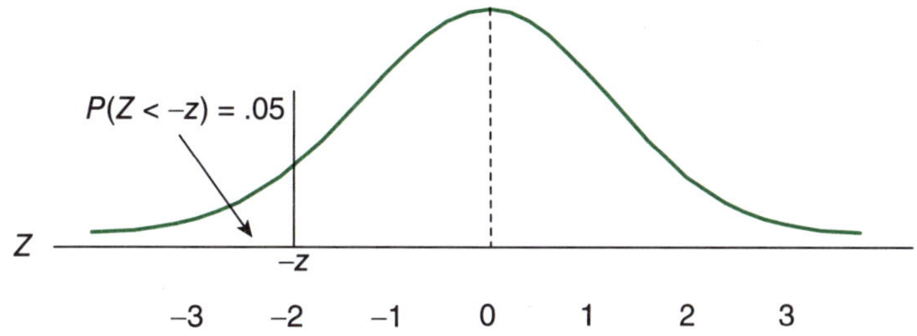

There are two probabilities with equal distance to .05 in the Z Table: .0505 and .0495, and the corresponding Z values are −1.64 and −1.65, respectively. Either answer is fine. However, I recommend using the z value with the larger absolute value to set a higher standard for future statistical testing purposes. Some statistics textbooks even go for interpolation to get a Z value of −1.645 and then round it to −1.65. The top 5% of swimmers' times are faster than 1.65 standard deviations below the mean.

Alternatively, you can use the Excel **NORM.S.INV(probability)** function to find the specific z value given the probability. In a blank cell in Excel where you want the answer to appear, type "=**NORM.S.INV(0.05)**", and hit **Enter**. You will see the answer −1.64485. You might see minor differences in the results between using the Z Table versus the Excel functions due to rounding.

POP QUIZ

5. The connection between the probabilities and Z scores is made possible by
   a. the multiplication rule of probability.
   b. the Addition Rule 1 of probability.
   c. the Addition Rule 2 of probability.
   d. the probability density function.

(Continued)

(Continued)

6. Which one of the following attributes applies to the Z Table?
   a. All possible Z values under the normal distribution curve have a probability equal to 1.
   b. If you cut the Z Table in the middle, the left side is a mirror image of the right side.
   c. $P(Z > z) = 1 - P(Z < z)$.
   d. All of the above attributes apply to the Z Table.

PROBABILITIES, Z SCORES, AND RAW SCORES

You have learned the connection between Z values and probabilities in a normal distribution from the previous section. However, it is unusual to hear someone say that they scored 1.57 standard deviations above the mean in a Sociology or a Psychology final exam. The link between probabilities and Z scores in a normal distribution needs to be extended further to raw scores to make this connection fully practical and useful. Such a connection between raw scores and probabilities also helps with interpreting raw scores. This is the reason why standardized test scores come with a percentile rank, which is the probability of other test scores lower than that particular score. The connection between raw scores and Z scores is through the Z formulas you learned in Chapter 4. Let's review the Z formulas. Repetition is the best way to learn statistics.

The Z formula for a population is $Z = (X - \mu)/\sigma$. When Z scores are calculated for a sample, both μ and σ are unknown, so the Z formula needs to be adjusted to $Z = (X - \bar{X})/s$. Therefore, the link between probabilities and raw scores can be established by a two-step process:

1. Transform raw scores (X) to Z scores by using the Z formula:

$$Z = \frac{X - \mu}{\sigma} \text{ or } Z = \frac{X - \bar{X}}{s}$$

2. Use the Z Table to connect the Z scores with probabilities.

Let's use examples to demonstrate this two-step process.

EXAMPLE 5.7

Incubation period is the time lapse between the exposure to a virus and symptoms onset. The incubation period of COVID-19 was normally distributed with a $\bar{X} = 7.8$ days and the standard deviation, s = 2.2. Brendan is a healthy 35-year-old who attended a large birthday party with many people, 9 days later he felt sick and tested positive for COVID-19. What was the percentage of COVID-19 patients who had shorter incubation period than Brendan's?

First, convert the raw score to a *z* score based on the *Z* formula. Then use the *Z* Table to figure out the probability to interpret this situation.

$$Z = \frac{(X - \bar{X})}{s} = \frac{(9 - 7.8)}{2.2} = 0.55$$

$$P(Z < 0.55) = .7088$$

Draw and identify this area under the standard normal distribution curve. According to the *Z* Table, the probability of $Z < 0.55$ is .7088 as shown in Figure 5.10.

FIGURE 5.10 ● Using the *Z* Table to Figure Out $P(Z < 0.55)$

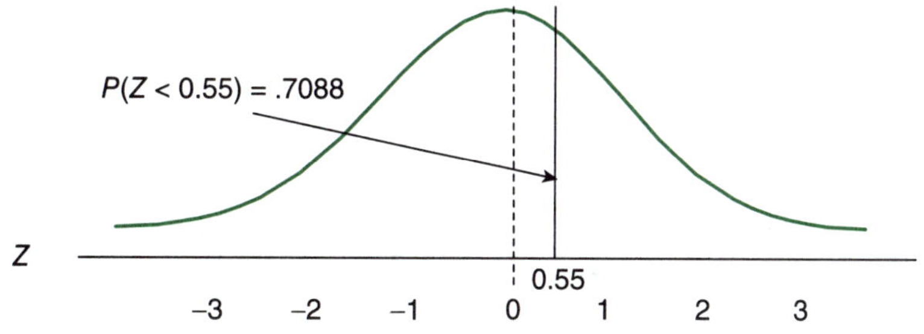

Different people responded to virus differently. The incubation period for COVID-19 ranged from 1 day to more than 14 days. Brendan's 9-day incubation period was longer than 70.88% of COVID-19 patients. There were still 29.12% COVID-19 patients who had longer incubation period than 9 days. It is wise to forego large gatherings and exercise good hygiene and keep social distancing whenever possible to avoid the exposure and spread of COVID-19.

The two-step procedures that first link the probabilities with *Z* scores then link *Z* scores with raw scores provide the exact location where the value is in relation to the rest of the distribution. This is a good way to interpret raw scores. Psychologists and other social scientists constantly develop new measures to quantify social or psychological attributes according to certain rules. In the next example, we will look at a newly developed measure to quantify "life satisfaction."

EXAMPLE 5.8

Satisfaction with life is a subjective feeling that psychologists have tried to measure over the past several decades. Researchers usually ask participants to indicate their levels of agreement with statements such as "I feel I can take on anything in life," "I am happy about everything in my life," or "I have a cheerful influence on other people." Assume that life satisfaction scores are normally distributed, with $\bar{X} = 35.5$ and $s = 8.7$. What is the probability of randomly selecting an individual whose life satisfaction score is higher than 50?

The question asks for $P(X > 50)$. You know that there is no direct link between raw scores and probability. The raw score, 50, needs to be transformed into a Z score. The Z Table provides the probability of Z scores lower than a specified z score, $P(Z < z)$. Then $P(Z > z)$ can be calculated as $1 - P(Z < z)$. Here is the two-step process.

Step 1:

$$Z = \frac{X - \bar{X}}{s} = \frac{50 - 35.5}{8.7} = 1.67$$

Step 2: Draw and identify this area under the standard normal distribution curve, $P(Z > 1.67)$ as shown in Figure 5.11.

According to the Z Table, $P(Z < 1.67) = .9525$

$$P(X > 50) = P(Z > 1.67) = 1 - P(Z < 1.67) = 1 - .9525 = .0475$$

FIGURE 5.11 ● Using the *Z* Table to Figure Out *P* (*Z* > 1.67)

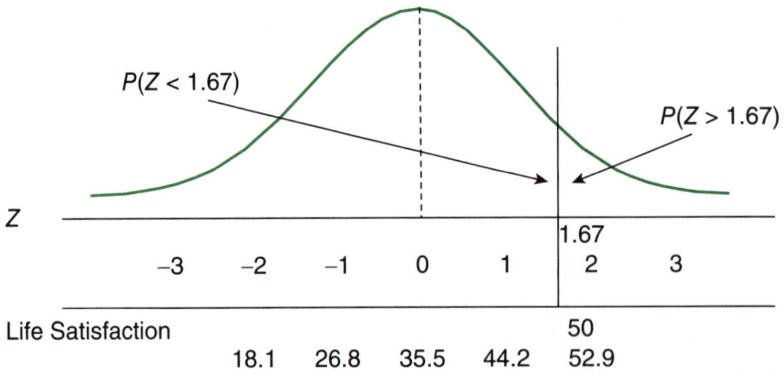

The probability of randomly selecting an individual whose life satisfaction score is greater than 50 is .0475, or 4.75%. This is a small probability event ($P = .0475$), which means that not many people have life satisfaction scores higher than 50.

Converting raw scores into Z scores provides for easy interpretation of a particular value on the universal yardstick (Z scale). Such a process applies not only to social or psychological measures but also to physical measures such as BMI. The universal yardstick provides a good reference of what is being considered as normal by the society. The ideal of "being normal" usually is expressed by the middle 95% of a distribution.

EXAMPLE 5.9

Adult women's BMIs are normally distributed, with a mean $\mu = 27.1$ and a standard deviation $\sigma = 5.43$. What are the low end and the high end for the middle 50% of adult women's BMIs?

The connection between raw scores and probabilities is indirect. It needs to go through a two-step process. In this example, where the probability is known, you need to figure out the Z scores. First, the connection between the Z score and the probability can be obtained from the Z Table. Second, raw scores can be calculated from the Z formula. Draw and identify this area under the standard normal distribution curve, $P(-z < Z < z) = .50$ as shown in Figure 5.12.

FIGURE 5.12 ● Using the Z Table to Figure Out $P(-z < Z < z) = .50$

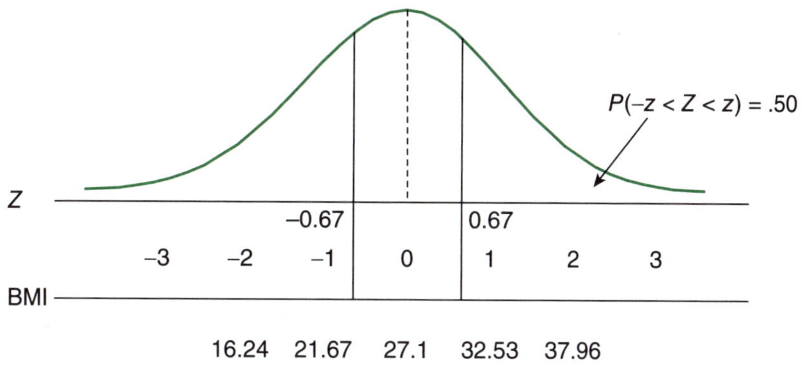

Use the following characters of Z Table to solve this problem: (1) Everything under the standard normal curve is 1 and (2) the Z curve is symmetrical whereby the Z values that separate the middle 50% also divide the two tails evenly.

Left tail $P(Z < -z)$ = Right tail $P(Z > z) = 1 - P(Z < z) = .25$

Therefore, $P(Z < -z) = .25$, and $P(Z < z) = .75$

(Continued)

(Continued)

The Z Table provides the probability to the left of a specific z value, $P(Z < z)$; thus, we need to identify the left tail, $P(Z < -z)$, as close to .25 as possible or $P(Z < z)$ as close to .75 as possible. The Z Table provides $P(Z < -z)$; when the probability is .25, the Z value is −0.67. Or the Z Table provides $P(Z < z)$; when the probability is .75, the Z value is 0.67. The Z values set the boundaries of the middle 50% as −0.67 and 0.67.

Then the raw scores can be calculated from the Z formula. The low end of the middle 50% of adult women's BMI:

$$Z = \frac{X - \mu}{\sigma}$$

$$-0.67 = \frac{X - 27.1}{5.43}$$

$$X = 23.46$$

The high end of the middle 50% of adult women's BMI:

$$Z = \frac{X - \mu}{\sigma}$$

$$0.67 = \frac{X - 27.1}{5.43}$$

$$X = 30.74$$

The low end of the middle 50% of adult women's BMI is 23.46, and the high end is 30.74. This range covers 50% of adult women's BMI.

Why Statistics Matter

With a preset standard, such as the middle 95% or middle 68%, on the Z scale, medical professionals can effectively apply and communicate the medical test results (e.g., blood pressure, blood cholesterol level, blood sugar, heart rate, or blood white cell count) to flag the patients whose test results were outside the normal ranges for proper treatments.

POP QUIZ

7. Assume that a health insurance company conducted a free health screening for its customers between 35 and 65 years old. During this health screening, the total blood cholesterol was measured for the 2,800 customers who participated. The total blood cholesterol level has $\bar{X} = 158.3 \text{ mg / dl}$ and $s = 40.7 \text{ mg/dL}$. High total cholesterol is a risk factor for heart diseases. The insurance company decided to use $Z > 1.25$ as a red flag to send its customers educational information on keeping their cholesterol under control. What is the cutoff of the total cholesterol level for customers to receive this educational information, and what proportion of the customers will receive this information?

EXERCISE PROBLEMS

1. The total screen time spent on computers, smart devices, video games, and TV/movies tend to increase with age among children. However, higher screen time is associated with lower psychological well-being such as less curiosity, lower self-control, and worse emotional stability. The negative effects of screen time is more serious for preschoolers (ages 2–5). In a large national survey on screen time, researchers found that the daily screen time for 2- to 5-year-old children was normally distributed with a $\bar{X} = 2.28 \text{ hours}$ and $s = 1.72$ hours. What proportion of preschoolers have daily screen time less than 2 hours?
2. Adult males' body mass index (BMI) is normally distributed with $\mu = 26.1$ and $\sigma = 5.45$. What proportion of adult males have a BMI between 21 and 39?
3. SAT Evidence-Based Reading and Writing scores are normally distributed, with $\mu = 497$ and $\sigma = 115$. SAT Math scores are normally distributed, with $\mu = 513$ and $\sigma = 117$. A prestigious university only accepts students who score above the 90th percentile on SAT ERW and SAT Math. What SAT scores are needed to be accepted in this university?

Solutions

1. The question asks for $P(X < 2)$. Convert $X = 2$ to a Z score, using the Z formula.

$$P(X < 2) = P\left(Z < \frac{(2 - 2.28)}{1.72}\right) = P(Z < -0.16) = .4364$$

The proportion of preschoolers whose daily screen time less than 2 hours is 43.64%. Parents with preschoolers need to limit their preschoolers' daily screen time to uphold their psychological well-beings.

2. The question asks for $P(21 < X < 39)$. There is no direct connection between raw scores and probability under the normal distribution. The two-step process is needed to solve the problem.
   a. Transform raw scores (X) to Z scores by using the Z formula:

$$P(21 < X < 39) = P\left(\frac{21-26.1}{5.45} < Z < \frac{39-26.1}{5.45}\right) = P(-0.94 < Z < 2.37)$$

   b. Use the Z Table to connect Z scores with probabilities.

Figure out the probability between these two specified Z values: $P(-0.94 < Z < 2.37)$.

Draw and identify this area under the standard normal distribution curve, $P(-0.94<Z<2.37)$ as shown in Figure 5.13.

$P(-0.94 < Z < 2.37)$ can be calculated as the difference between these two probabilities $P(Z < 2.37) - P(Z < -0.94) = .9911 - .1736 = .8175$.

FIGURE 5.13 ● Using the Z Table to Figure Out $P(-0.94 < Z < 2.37)$

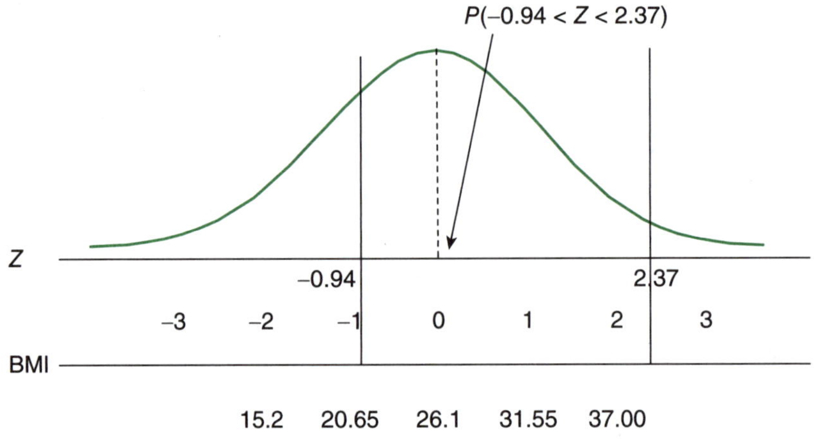

The answer is that 81.75% of adult males have a BMI between 21 and 39. This exercise demonstrates that you can figure out the probability within any two specified raw scores by using the two-step process to connect raw scores with Z scores and then Z scores with probabilities.

3. The question asks for the raw score for the 90th percentile on SAT Evidence-Based Reading and Writing as well as the 90th percentile on SAT Math.

$$P(Z < z) = .90$$

Draw and identify this area under the standard normal distribution curve, $P(Z < z) = .90$. According to the Z Table, the probability closest to .90 is .8997, which corresponds to $Z = 1.28$ as shown in Figure 5.14.

FIGURE 5.14 ● Using the Z Table to Figure Out $P(Z < z) = .90$

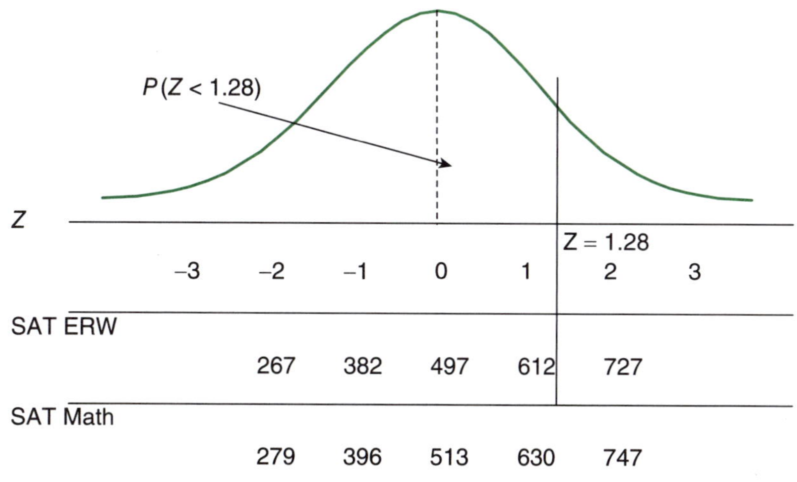

SAT Evidence-Based Reading and Writing:

$$Z = \frac{X - \mu}{\sigma}$$

$$1.28 = \frac{X - 497}{115}$$

$$X = 1.28 \times 115 + 497 = 644.2$$

SAT Math:

$$Z = \frac{X - \mu}{\sigma}$$

$$1.28 = \frac{X - 513}{117}$$

$$X = 1.28 \times 117 + 513 = 662.76$$

Students need to score higher than 644.2 on Evidence-Based Reading and Writing and higher than 662.76 on Math to be accepted by this university.

What You Learned

You have learned the basic vocabulary in probability: event, simple event, and sample space. It is important to review the definitions of these terms in the Key Words.

Binomial probability formula:

$$P(r) = C_r^n p^r q^{(n-r)} = \frac{n!}{(n-r)!r!} p^r q^{(n-r)} \quad \text{for } r = 0,1,2,3,\ldots,n$$

where

C = combination

p = probability of success in one trial

$q = 1 - p$ = probability of failure in one trial

n = number of trials

r = number of successes in n trials

$P(r)$ = probability of getting exactly r successes in n trials

Binomial probability distribution is a special probability distribution. The calculation of μ, σ^2, and σ in a binomial probability distribution is simplified.

$$\mu = np$$

$$\sigma^2 = npq$$

$$\sigma = \sqrt{npq}$$

The formulas for calculating the population mean, variance, and standard deviation from a probability distribution are the same as those using a frequency distribution table.

Population mean for a frequency table: $\mu = \sum \frac{fX}{N}$

For every X, the associated probability of X is $P(X) = f/N$; therefore,

$$\mu = \sum \frac{fX}{N} = \Sigma[XP(X)]$$

For the same reason, $P(X) = f/N$, the population variance for a frequency table:

$$\sigma^2 = \frac{SS}{N} = \frac{\Sigma f(X - \mu)^2}{N} = \Sigma[(X - \mu)^2 P(X)]$$

Population standard deviation: $\sigma = \sqrt{\text{Variance}} = \sqrt{\frac{\Sigma f(X - \mu)^2}{N}} = \sqrt{\Sigma[(X - \mu)^2 P(X)]}$

Three attributes of the Standard Normal Distribution Table (*Z* Table) are as follows:

1. All *Z* values are reported to two places after the decimal point, and all probabilities are reported to four places after the decimal point.
2. The column represents *Z* values to one place after the decimal point, and the row represents *Z* values' second place after the decimal point.
3. The intersection between the column and the row shows the area to the left of the specified *z* value, or the probability of all *Z* values smaller than the specified value, *z*, which is expressed as $P(Z < z)$.

The connection between probabilities and raw scores can be established by a two-step process:

1. Transform raw scores (*X*) to *Z* scores by using the *Z* formula:

$$Z = \frac{X - \mu}{\sigma} \text{ or } Z = \frac{X - \bar{X}}{s}$$

2. Use the *Z* Table to connect *Z* scores with probabilities under the normal distribution.

Key Words

Addition Rule 1 of probability: When Event A and Event B are mutually exclusive, the probability of either Event A or Event B happening is the sum of the probabilities of each event: $P(\text{A or B}) = P(\text{A} \cup \text{B}) = P(\text{A}) + P(\text{B})$. 123

Addition Rule 2 of probability: When Event A and Event B are not mutually exclusive, the probability of either Event A or Event B, or both, happening is the sum of the probabilities of each event minus the overlapping part of the two events: $P(\text{A or B}) = P(\text{A} \cup \text{B}) = P(\text{A}) + P(\text{B}) - P(\text{A} \cap \text{B})$. 133

Binomial probability distribution: A binomial probability distribution is a probability distribution that applies to variables with only two possible outcomes (i.e., success vs. failure) in each trial. The trials are independent of one another. The probability of a success and the probability of a failure remain the same in all trials. 125

Disjoint events: Disjoint events are mutually exclusive events. There are no overlapping parts between the events. 123

Event: An event is defined as a set of outcomes from an experiment or a procedure. 120

Independent events: Two events A and B are independent when the occurrence of one event does not affect the probability of the occurrence of the other event. 122

Multiplication rule for independent events: When Event A and Event B are independent, the probability of both events happening is $P(AB) = P(A) \times P(B)$. 122

Mutually exclusive events: When Event A and Event B are mutually exclusive, it means that only one of the events can happen: If one happens, the other does not happen. 123

***n* factorial:** *n* factorial is calculated as the multiplicative product of all positive integers less than or equal to n, $n! = n \times (n-1) \times (n-2) \times (n-3) \cdots 3 \times 2 \times 1$. 0! is defined as 1. 126

Probability density function: A probability density function describes the probability for a continuous, random variable to be in a given range of values. 137

Probability distribution: There are three requirements for a probability distribution: 134

1. The variable X is a discrete, random, and numerical variable. The number of possible values of X is finite. Each value is associated with a probability.
2. The probability of each value needs to be within 0 and 1, inclusive. It is expressed as $0 \le P(X) \le 1$.
3. The sum of the probabilities of all the values equals 1. It is expressed as $\Sigma P(X) = 1$.

Sample space: A sample space is a complete list of all possible outcomes. 120

Simple event: A simple event is defined as an elementary event that cannot be broken into simpler parts. 120

Learning Assessment

Multiple Choices: Circle the Best Answer to Every Question

1. For a normal distribution, $P(Z > 1.5) = .0668$. What is $P(Z < -1.5)$?
   a. .0668
   b. .9332
   c. .9772
   d. .0228
2. What is the answer for $P(Z < 0.25)$?
   a. .25
   b. .75

c. .5987

d. −.4013

3. What is the answer for $P(-2.01 < Z < 1.45)$?

a. .0963

b. .9043

c. .8389

d. .1611

4. What is the probability that the total number of dots from throwing two dice is eight?

a. 3/36

b. 4/36

c. 5/36

d. 6/36

5. Which of the following is true?

a. $P(Z > 2.0) > P(Z > 1.0)$

b. $P(Z > -2.0) > P(Z > -1.0)$

c. $P(Z > 0) > P(Z < 0)$

d. $P(Z < -2.0) > P(Z < -1.0)$

6. What is the answer for $P(-2.05 < Z < -1.14)$?

a. .0202

b. .0668

c. .1069

d. .1271

Free Response Questions

7. The distribution of adult females' BMI is normal, with $\mu = 27.1$ and $\sigma = 4.5$. What proportion of the adult females have a BMI between 21 and 31?

8. What is the probability of throwing two dice and getting the total number of dots as even numbers greater than six?

9. A cardiologist studies heart rate variability. Heart rates are normally distributed, with a mean of 68 beats per minute and a standard deviation of 19 among a sample of 100 patients. What is the probability of randomly selecting a patient to obtain a heart rate less than 60?

10. The ACT composite score is normally distributed, with a mean of 21.8 and a standard deviation of 5.4. What is the ACT composite score that separates the top 30% from the rest of the distribution?

Answers to Pop Quiz Questions

1. d
2. a
3. d
4. c
5. d
6. d
7. $Z = 1.25$ converts to a raw score with $\bar{X} = 158.3\text{mg/dl}$ and $s = 40.7$ mg/dl.

$$Z = \frac{X - \bar{X}}{s}$$

$$1.25 = \frac{X - 158.3}{40.7}$$

$$X = 209.2$$

Thus, customers will receive the educational information if their total cholesterol level is higher than 209.2.

$$P(Z > 1.25) = 1 - P(Z < 1.25) = 1 - .8944 = .1056$$

Thus, 10.56% of the customers will receive the educational information on keeping total cholesterol under control.

6

THE CENTRAL LIMIT THEOREM

Learning Objectives

After reading and studying this chapter, you should be able to do the following:

- Define sampling error, and explain why it is an unavoidable natural occurrence in empirical research
- Identify the principal characteristics of the central limit theorem
- Explain the law of large numbers in calculating probability
- Describe how a Z test is conducted, and interpret the results of a Z test

WHAT YOU KNOW AND WHAT IS NEW

You have learned the connection between raw scores and Z scores via the Z formula in Chapter 4, and the connection between Z scores and probabilities via Z Table in Chapter 5. You know how to transform a raw score into a Z score, figure out its exact location in the distribution, and find out the connection between the probability and the specified Z score. All of this information continues to be relevant in Chapter 6, where you will learn the most fundamental probability theory, the *central limit theorem*.

In a large population, we can randomly select a sample of n values and calculate the mean of these n values. We can do this again by taking another sample of n values and calculating the mean. The mean we get from the second sample may be different from the mean from the first sample simply because different values are selected in each sample. The central limit theorem is the theoretical description of the sampling distribution of the sample means. It also describes how sample means with different sample sizes behave in relation to the population mean, so that we can draw accurate conclusions when conducting hypothesis tests. Hypothesis testing forms the logic of decision-making rules in testing the strength of evidence provided by sample statistics, which is the topic of Chapter 7. This demonstrates the accumulative nature of learning statistics. Concepts that you learned from earlier chapters are necessary and relevant in later chapters. Once you learn the relationship between sample means and population mean, you will learn how to use sample statistics to make inferences about the population parameters. You can calculate the probability of a sample statistic being located within a certain distance from the population parameter. As you will see, Chapter 6 is mostly a logical extension of Chapter 5 with slight modifications.

SAMPLING ERROR

A population is defined as an entire collection of everything or everyone that researchers are interested in studying or measuring. A sample is defined as a subset of the population from which measures are actually obtained. The definitions of sample and population are detailed in Chapter 1. Such fundamental concepts deserve to be repeated. Frankly, repetition is the key to learn statistics. If you have any difficulty in a particular statistical concept the first time you read or hear it, studying it several times will help.

There are many possible samples within a population. No matter how hard you try, how perfectly you plan, and how well you achieve the random sampling, your sample statistics still do not match population parameters exactly. A **sampling error** is defined as a situation in which a sample is randomly selected and a natural difference, divergence, distance, or error occurs between the sample statistic and the corresponding population parameter. Such differences are really not "errors" because no mistakes are made in the calculation process. They are likely due to random sampling variations. Sample variations are attributed to individual differences from the participants who are included in different samples. There are many kinds of individual differences such as age, height, weight, speed, intelligence, physical strength, artistic skills, creativity, exam scores, and so on. For example, if you randomly select a sample of 25 people who work remotely from the population of all remote-working people, you can calculate their average time spent on daily video meetings. Let's label the mean as $\overline{X}_1$, repeat the procedure to get a second sample of 25 people whose mean daily video meeting time is $\overline{X}_2$, repeat the procedure to get a third sample of 25 people whose mean daily video meeting time is $\overline{X}_3$, and you may keep on doing this for k times. These sample means $\overline{X}_1, \overline{X}_2, \overline{X}_3, \ldots \overline{X}_k$ might not be exactly the same as the population mean. The difference between a sample mean and the population mean is called a sampling error. Sampling errors are unavoidable natural fluctuations due to the fact that different individuals are included in different samples.

If sampling errors always exist, one way or another, the sample statistics are not going to be exactly the same as the population parameters. Why bother to go through all the careful planning and meticulous research designs to ensure the quality of data collection? The key to that question is that sampling errors are random errors. As long as they are random errors, they can be explained by probability theory.

There are many different kinds of errors that can happen when conducting research. Sampling error is just one of them. Besides sampling errors, there are nonsampling errors. Nonsampling errors are mostly avoidable and preventable human errors. Such human errors include survey questions written with ambiguous wording provoking different interpretations from respondents, respondents offering socially desirable but not necessarily honest answers, data coding errors, data entry errors, nonresponses, inappropriate statistical analyses, and/or wrong conclusions. Such human errors are not random errors. Nonrandom errors could not be explained by probability theory. Careful planning, meticulous research design, and proper training in statistical analysis are specifically developed to minimize nonsampling errors because they are not fixable or interpretable by probability theory. When a sample contains nonsampling errors, it is inappropriate to use sample statistics to make inferences about the population parameters.

POP QUIZ

1. Which one of the following sampling procedures makes an effort to minimize nonsampling errors?
   a. TV shows asked the audience to call in to vote for their favorite contestants.
   b. Pollsters asked voters how they cast their ballots at the exit of a voting location.
   c. The ratings collected by ratemyprofessors.com.
   d. University students were assigned a number; then, 10% of the students were selected by a random number generator to provide opinions on the campus parking services.

SAMPLING DISTRIBUTION OF THE SAMPLE MEANS AND THE CENTRAL LIMIT THEOREM

To understand the central limit theorem, you have to assume that a sample of a particular size, n, is randomly selected from a population. Then, the same procedure is repeated k times to create k samples of the same size from the same population. To illustrate the process, the outcomes are summarized in Table 6.1.

TABLE 6.1 ● Randomly Selecting a Sample With *n* Cases, Then Repeating the Same Procedure *k* Times

Sampling	Sample Size	Sample Mean
Sample 1	n	$\bar{X}_1$
Sample 2	n	$\bar{X}_2$
Sample 3	n	$\bar{X}_3$
Sample 4	n	$\bar{X}_4$
⋮	n	⋮
Sample k	n	$\bar{X}_k$

The sample mean ($\bar{X}$) is calculated for each of the samples. Therefore, we have $\bar{X}_1$ from the first sample, $\bar{X}_2$ from the second sample, and all the way to $\bar{X}_k$ from the kth sample. This hypothetical distribution of all sample means selected from a given population (i.e., $\bar{X}_1, \bar{X}_2, \ldots \bar{X}_k$) is defined as the **sampling distribution of the sample means**. These sample means might take on different values due to sampling error. When you calculate the mean of all sample means, $\mu_{\bar{X}}$, the answer is the population mean, μ. You can easily calculate the standard deviation of the sample means, which is given a special label, $\sigma_{\bar{X}}$, as the **standard error of the mean**. The sample means are normally distributed as long as the population is normally distributed or the sample size is greater than 30. All these characteristics of sample means can be expressed as $N(\mu, \sigma_{\bar{X}})$, which is defined as a normal distribution with a mean of μ and standard deviation of $\sigma_{\bar{X}}$. The standard error of the mean is calculated by the population standard deviation divided by the square root of the sample size, $\sigma_{\bar{X}} = \sigma / \sqrt{n}$. Judging by the formula, you can see the larger the sample size, the smaller the standard error of the mean. It demonstrates that using a larger sample size produces a more stable sample mean with less variability and closer to the population mean. We can understand how sample means behave in relation to the population mean. In reality, the population mean is unknown, and we don't have the time, money, or energy to do repeated sampling an unlimited number of times. Therefore, all we have is the information from one sample. Understanding the sampling distribution of the sample means allows us to use the sample statistic to gain useful insights into the unknown population parameter.

The **central limit theorem** is a mathematical proposition that describes important characteristics of the sampling distribution of the sample means: distribution shape, central tendency, and variability. The essential characteristics of the central limit theorem are listed below:

1. The sampling distribution of the sample means becomes more and more normally distributed as the sample size increases. When sample size $n > 30$, the

sampling distribution of means is approximately normally distributed regardless of the shape of the original population distribution.

2. If the population is normally distributed, the sampling distribution of the sample mean is also normally distributed regardless of the sample size.
3. The mean of all sample means (also called the expected value of the mean) is the population mean, $\mu_{\bar{X}} = \mu$.
4. The standard error of the mean is the standard deviation of all sample means, $\sigma_{\bar{X}} = \sigma / \sqrt{n}$.

In summary, the central limit theorem states that for a sample size = n, the distribution of sample means is normally distributed with a mean $\mu_{\bar{X}} = \mu$ and standard deviation $\sigma_{\bar{X}} = \sigma / \sqrt{n}$. This statement is true when the sample size is large, $n > 30$, or when the population is normally distributed. As the sample size increases, the standard error of the mean decreases. In other words, large samples tend to generate smaller standard errors of the mean; thus, the sample means tend to cluster closer to the population mean than small samples. The relationship between the sample size and the standard error of the mean is reversely related. However, the relationship between the population standard deviation and the standard error of the mean is positively related. The larger the standard deviation of the population, the larger is the standard error of the mean.

EXAMPLE 6.1

SAT Math scores are normally distributed with $\mu = 513$ and $\sigma = 120$. Researchers randomly select a sample with a particular sample size and then repeat the procedure many times. The sample mean is calculated for each sample. The standard error of the mean is the standard deviation of all sample means.

a. What is the standard error of the mean for a sample size of 4?
b. What is the standard error of the mean for a sample size of 25?
c. What is the standard error of the mean for a sample size of 100?

Answers:

$$\text{Standard error} = \sigma_{\bar{X}} = \frac{\sigma}{\sqrt{n}}.$$

a. When $n = 4$, $\sigma_{\bar{X}} = \frac{\sigma}{\sqrt{n}} = \frac{120}{\sqrt{4}} = 60$

b. When $n = 25$, $\sigma_{\bar{X}} = \frac{\sigma}{\sqrt{n}} = \frac{120}{\sqrt{25}} = 24$

c. When $n = 100$, $\sigma_{\bar{X}} = \frac{\sigma}{\sqrt{n}} = \frac{120}{\sqrt{100}} = 12$

There is an inverse relationship between the sample size and the standard error of the mean. As the sample size increases, the standard error of the mean decreases. This is the reason why researchers prefer using sample statistics collected from large sample sizes to small sample sizes in making inferences about the population parameters.

POP QUIZ

2. According to the central limit theorem, which one of the following sampling distributions of the mean might not be normally distributed?
   a. When samples of a particular sample size, $n > 30$, are randomly selected from a normally distributed population.
   b. When samples of a particular sample size, $n > 30$, are randomly selected from a population that is not normally distributed.
   c. When samples of a particular sample size, $n \leq 30$, are randomly selected from a normally distributed population.
   d. When samples of a particular sample size, $n \leq 30$, are randomly selected from a population that is not normally distributed.

RELATIONSHIPS BETWEEN SAMPLE MEANS AND THE POPULATION MEAN

You learned the Z formula in Chapter 4, allowing you to convert a raw score X to a Z score by using $Z = (X - \mu) / \sigma$. The basic principle of the Z formula is $Z = (\text{Raw score} - \text{Mean})/\text{Standard deviation}$. Now, apply the same principle to convert a sample mean, $\bar{X}$, to a Z value. You learned that the mean of sample means $\mu_{\bar{X}}$ is μ and the standard deviation of sample means, also called the standard error of the mean, is $\sigma_{\bar{X}} = \sigma / \sqrt{n}$, so converting a sample mean to a Z value becomes $Z = \dfrac{\bar{X} - \mu}{\sigma / \sqrt{n}}$. The Z value provides the exact location of a particular sample mean in the distribution of all sample means. This process allows us to conduct a **Z test** between a sample mean and the population mean. Z tests measure the difference between a sample statistic and its hypothesized population parameter in units of standard error. Let's use examples to illustrate how to conduct a Z test.

EXAMPLE 6.2

Adult male heights are normally distributed with $\mu = 69.0$ inches and $\sigma = 6.0$ inches. What is the probability of randomly selecting a sample of 16 adult males whose mean height is less than 67 inches?

The process of comparing a sample mean to the population mean is slightly different from comparing one individual score to the population mean. According to the question, $P(\bar{X} < 67)$ is the correct mathematical expression. There is no direct connection between $\bar{X}$ and probability, so $\bar{X}$ needs to be converted into a Z value by using $Z = \frac{\bar{X} - \mu}{\sigma / \sqrt{n}}$. The secret of successfully conducting a Z test is to place the numeric values stated in the problem in their rightful spots in the formula.

$$Z = \frac{\bar{X} - \mu}{\frac{\sigma}{\sqrt{n}}} = \frac{67 - 69}{\frac{6}{\sqrt{16}}} = \frac{-2}{1.5} = -1.33$$

$$P(\bar{X} < 67) = P(Z < -1.33)$$

Draw the normal distribution curve and identify $P(Z < -1.33)$ under the standard normal distribution curve.

The standard error of the mean $= \sigma_{\bar{X}} = \frac{\sigma}{\sqrt{n}} = \frac{6}{\sqrt{16}} = 1.5$ as shown in Figure 6.1 on the adult male height scale.

FIGURE 6.1 ● Using the *Z* Table to Figure Out *P*(*Z* < –1.33)

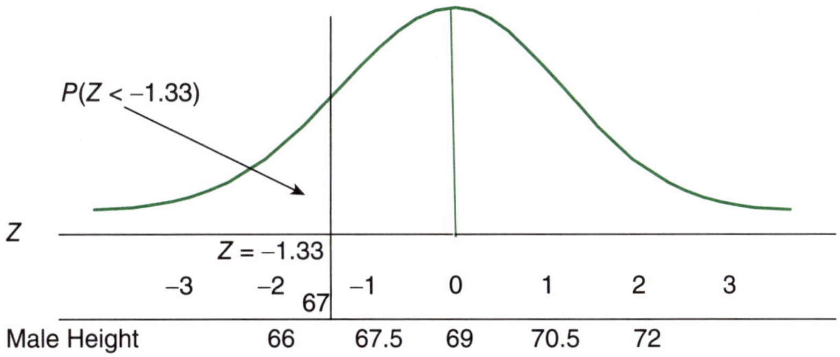

The left tail can be directly obtained from the Z Table.

$$P(\bar{X} < 67) = P(Z < -1.33) = .0918$$

Alternatively, you can use the Excel function to find $P(Z < -1.33)$. To get the probability of the area to the left of the specified z value is the same as finding the percentile rank for the z value. In a blank cell in

(Continued)

(Continued)

Excel where you want the answer to appear, type "**=NORM.S.DIST(-1.33, TRUE)**", and hit **Enter**. You will see the answer 0.091759.

The probability of randomly selecting 16 adult males whose mean height is less than 67 inches is .0918 or 9.18%. This example clearly illustrates the differences between what you learned in Chapter 4 converting a raw score X to a Z score by using $Z = (X - \mu)/\sigma$ versus what you learned in this chapter converting a sample mean, $\bar{X}$, to a Z value by using $Z = \frac{\bar{X} - \mu}{\sigma/\sqrt{n}}$. Simply remember that you apply the same principle of calculating the Z score by putting the mean of all the sample means and the standard deviation of the sample means (i.e., the standard error of the mean) in the formula when dealing with comparing a sample mean with a population mean. You should also notice that the standard deviation of all sample means (i.e., the standard error of the mean $\sigma_{\bar{X}} = \frac{\sigma}{\sqrt{n}}$) is much smaller than the population standard deviation, σ.

EXAMPLE 6.3

ACT composite scores are normally distributed with $\mu = 21.0$ and $\sigma = 5.4$. A sample of nine students is randomly selected. What is the probability that the sample mean is greater than 25?

The question asks for $P(\bar{X} > 25)$. There is no direct connection between $\bar{X}$ and probability, so $\bar{X}$ needs to be converted into a Z value by using $Z = \frac{\bar{X} - \mu}{\sigma/\sqrt{n}}$.

$$Z = \frac{\bar{X} - \mu}{\frac{\sigma}{\sqrt{n}}} = \frac{25 - 21}{\frac{5.4}{\sqrt{9}}} = \frac{4}{1.8} = 2.22$$

Draw and identify this area under the standard normal distribution curve, $P(Z > 2.22)$.

The standard error of the mean $= \sigma_{\bar{X}} = \frac{\sigma}{\sqrt{n}} = \frac{5.4}{\sqrt{9}} = 1.8$ as shown in Figure 6.2 on the ACT composite scale.

FIGURE 6.2 ● Using the Z Table to Figure Out $P(Z > 2.22)$

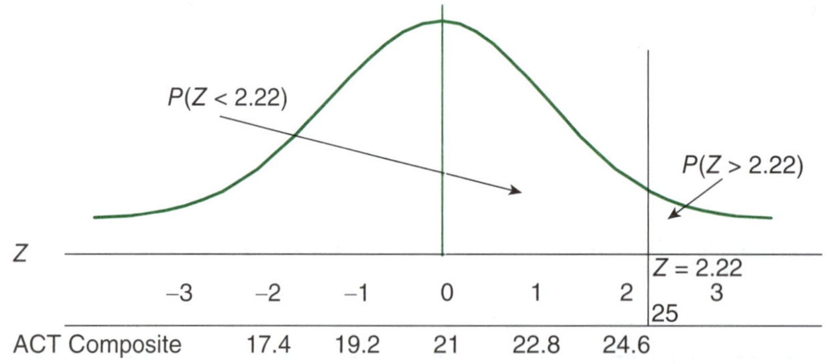

The Z Table only provides the area to the left of a particular z value, $P(Z < z)$. The area to the right of that z value can be calculated by $P(Z > z) = 1 - P(Z < z)$.

$$P(\bar{X} > 25) = P(Z > 2.22) = 1 - P(Z < 2.22) = 1 - .9868 = .0132$$

Alternatively, you can use the Excel function to find $P(Z < 2.22)$. The same way as the Z Table only provide the probability to the left of the specified z value, the Excel function only provides the probability of the area to the left of the specified z value. In a blank cell in Excel where you want the answer to appear, type "**=NORM.S.DIST(2.22, TRUE)**", and hit **Enter.** You will see the answer 0.9868. Because everything under the probability curve is 1, you need to use 1 – 0.9868 to obtain the probability of the area to the right of the specified z value. From now on, using the same principle, you can identify the probability either to the left, $P(Z < z)$ or to the right of a specified z value, $P(Z > z)$.

The probability of randomly selecting nine students whose mean ACT composite score is greater than 25 is .0132. This is a low probability event. You might have noticed that it is more common to get an individual ACT score greater than 25 than getting a randomly selected sample of nine students whose sample mean is greater than 25. That is because the standard error of the mean is only 1.8 for a sample of nine students. The sample means are more likely to be clustered around the population mean than individual scores. The difference between 25 and 21.0 is 2.22 times the standard error of the mean. The link between the Z value and the probability tells us that it is an unusual situation.

EXAMPLE 6.4

The total cholesterol levels for women aged 20 to 30 years are normally distributed with $\mu = 190$ mg/dl and $\sigma = 40$ mg/dl. A sample of 25 women in this age-group is randomly selected. What is the probability that the mean of these 25 women's total cholesterol is between 170 and 200?

According to the question, $P(170 < \bar{X} < 200)$ is the correct mathematical expression. The connection between sample means and probabilities is not direct. Apply the two-step process: first, use the Z formula to connect the raw values with the Z values. Second, use the Z Table to connect the Z values with the probabilities. The probability in between two Z values are the difference of their probability values listed in the Z Table.

$$P(170 < \bar{X} < 200) = P\left(\frac{170-190}{\frac{40}{\sqrt{25}}} < Z < \frac{200-190}{\frac{40}{\sqrt{25}}}\right)$$

$$= P(-2.50 < Z < 1.25)$$

$$= .8944 - .0062 = .8882$$

Draw and identify this area under the standard normal distribution curve, $P(-2.50 < Z < 1.25)$, as shown in Figure 6.3. The standard error of the mean $= \sigma_{\bar{X}} = \frac{\sigma}{\sqrt{n}} = \frac{40}{\sqrt{25}} = 8$ as shown in the cholesterol scale.

(Continued)

(Continued)

FIGURE 6.3 ● Using the Z Table to Figure Out $P(-2.50 < Z < 1.25)$

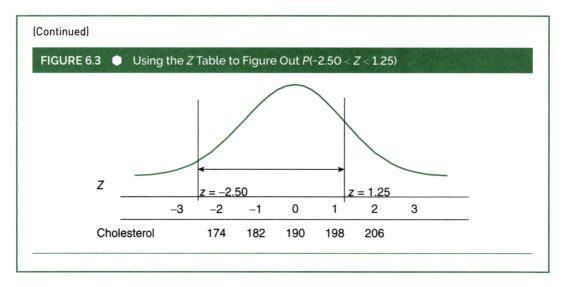

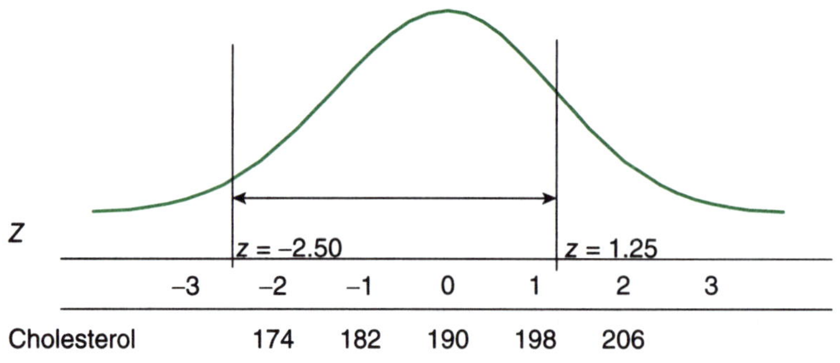

The Z tests compare an observed sample mean with its known or hypothesized population mean in the units of standard error of the mean, $\sigma_{\bar{X}} = \sigma / \sqrt{n}$. The larger the sample size, the smaller the standard error of the mean. Large sample sizes produce stable sample statistics, and they tend to cluster closer to the population parameters than small sample sizes. That is the reason why sample size is one of the main characteristics of a research study and requires careful advanced planning. If the sample size is too large, it wastes time and resources to collect data. If the sample size is too small, it produces unstable and unreliable statistics that might not be representative of the population parameters.

Social scientists create a standard procedure to test the strength of empirical evidence. For example, the Z tests compare an observed sample mean with its known or hypothesized population mean in the units of standard error. Then, they use the Z Table to figure out the probability associated with the calculated Z value. Such a probability forms the basis to evaluate the strength of empirical evidence.

POP QUIZ

3. In the Z formula for an individual raw score X and a Z formula for a sample mean, which part of the formula remains the same?
   a. Population mean, μ
   b. Sample mean, $\bar{X}$
   c. Raw score, X
   d. Standard error of the mean, $\sigma_{\bar{X}} = \frac{\sigma}{\sqrt{n}}$

THE LAW OF LARGE NUMBERS

The **law of large numbers** is another probability theorem that describes the relationship between sample means and population means. The law of large numbers describes the results of performing the same experiment multiple times: As the number of trials n approaches infinity, the observed sample mean approaches the population mean. Assume that X is the number of successes in n independent trials with a probability of success in each trial, p. The population mean is also called the expected value of X, which is expressed as $E(X) = \mu = np$.

Let's use an example to explain the law of large numbers. When you toss a fair coin, the probability of getting a head is .5, and the probability of getting a tail is .5. If you toss 10 coins at the same time, the number of heads you might get is 0, 1, 2, 3, 4, 5, 6, 7, 8, 9, or 10 in any particular trial. The results might not be exactly split 50–50 as the expected value of heads = np = 10(.5) = 5 in every trial because there are natural fluctuations due to chance. You may continue tossing these 10 coins 100 times, 1,000 times, or 10,000 times if you have time in your hand. As the number of trials approaches infinity, the observed sample mean approaches the population mean. The law of large numbers is intuitive because it demonstrates, in the long run, that natural fluctuations will get evened out. Thus, the observed sample mean will equal the population mean (or the expected value) eventually.

The law of large numbers is a simple theorem. Nothing can explain the process better than actually doing such an exercise. Of course, you will not reach the infinite number of trials. However, you will observe the fluctuations through a relatively small number of trials and observe the outcomes over a large number of trials. The last problem in the "Exercise Problems" is a class exercise on the law of large numbers.

POP QUIZ

4. The law of large numbers describes the relationship between the sample means and the population mean as which of the following numbers approaches infinity?
   a. The number of successes
   b. The population mean
   c. The number of trials
   d. The sample mean

Author's Aside

All gambling games involve probability. If you apply the knowledge of the law of large numbers to gambling, you will see that it is always better to be the owner of a casino than be a gambler. The law of large numbers and gambling rules both favor "the house" (the casino itself). Individual gamblers might beat the house sometimes, but in the long run, the house always wins. This is why casinos offer cheap bus tours, cheap cocktails, and even a small amount of free tokens to attract as many gamblers as possible. From the casino owners' point of view, the more the gamblers in the casinos, the more profits the house make.

EXERCISE PROBLEMS

1. Assume teenagers' screen time is normally distributed with a mean $\mu = 3.2$ hours and a standard deviation $\sigma = 1.1$.
   a. What is the probability of randomly selecting a group of five teenagers whose average screen time is greater than 3.5 hours?
   b. What is the probability of randomly selecting a group of 50 teenagers whose average screen time is greater than 3.5 hours?
   c. Explain the impact of sample size on the results.
2. The SAT ERW score is normally distributed with $\mu = 497$ and $\sigma = 115$. What is the probability of randomly selecting a group of 25 high school seniors to obtain their average SAT ERW score less than 450?
3. The Logistics Director of a company that manufactures personal protective equipment (PPE) tries to figure out the probability of fitting 110 boxes of PPEs into a small cargo plane to expedite the shipping so hospitals can receive PPEs as soon as possible. Each box contains various items with various quantities. Without knowing the weight of every box, the Logistics Director decides to calculate the probability of getting 110 boxes on this plane without overloading it. Based on the company's records, the weight of the boxes is normally distributed with $\mu = 38.69$ pounds and $\sigma = 14.78$ pounds. The small plane has a cargo capacity of 4,500 pounds. What is the probability that this small cargo plane can carry all 110 boxes of PPEs within its cargo capacity?
4. *The law of large numbers class exercise.* This exercise is designed to show natural fluctuations in a small number of trials and a convergent between the sample mean and the population mean when averaging the outcomes of a large number of trials. Every student in the class is given the assignment to record the number of heads by tossing 10 coins at the same time. Repeat the process 20 times and report the number of heads for each trial using Table 6.2. Compile the report for the entire class using Table 6.2 in Excel. I had 25 students in the class, thus created $25 \times 20 = 500$ trials of tossing 10 coins. Calculate the number of heads out of 10 coins with the number of trials obtained from the entire class. What is the average number of heads of tossing 10 coins over 20 trials for each student? What is the average number of heads for the entire class of 25 students' tossing 10 coins over 20 trials? How do these two means compare with the expected value $E(X) = \mu = np = 10(.5) = 5$?

TABLE 6.2 ● Number of Heads When Tossing 10 Coins at the Same Time

Trial	1	2	3	4	5	6	7	8	9	10	11	12	13	14	15	16	17	18	19	20
Student1																				

Solutions

1. The sample statistics calculated from a large sample produced more stable results than those calculated from a small sample.

   a. The standard error of the mean is $\sigma_{\bar{X}} = \frac{\sigma}{\sqrt{n}}$.

   $$\text{When } n = 5, \sigma_{\bar{X}} = \frac{\sigma}{\sqrt{n}} = \frac{1.1}{\sqrt{5}} = 0.492$$

According to the question, $P(\bar{X} > 3.5)$ is the correct mathematical expression. There is no direct connection between sample means and probabilities. The $\bar{X}$ needs to be converted into a Z score by using $Z = \frac{\bar{X} - \mu}{\frac{\sigma}{\sqrt{n}}}$.

$$P(\bar{X} > 3.5) = P\left(Z > \frac{3.5 - 3.2}{\frac{1.1}{\sqrt{5}}}\right) = P(Z > 0.61) = 1 - .7291 = .2709$$

Draw and identify this area under the standard normal distribution curve, $P(Z > 0.61)$ as shown in Figure 6.4. The $\sigma_{\bar{X}} = \frac{\sigma}{\sqrt{n}} = \frac{1.1}{\sqrt{5}} = 0.492$ is shown in the screen time scale. When randomly selecting five teenagers, the probability of their average screen time greater than 3.5 hours is .2709.

FIGURE 6.4 ● Using the Z Table to Figure Out $P(Z > 0.61)$

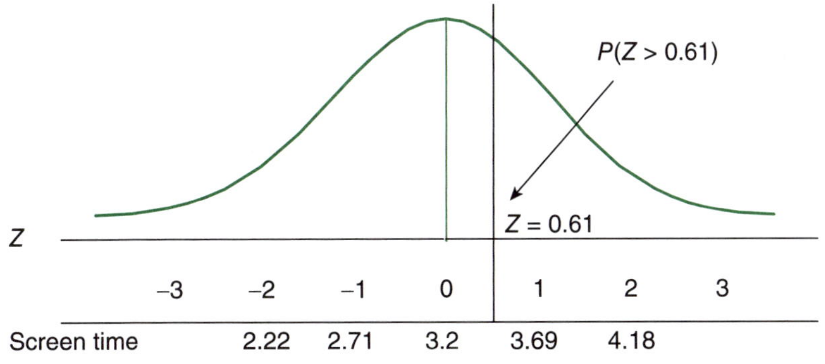

b. The standard error of the mean is $\sigma_{\bar{X}} = \frac{\sigma}{\sqrt{n}}$.

When $n = 50, \sigma_{\bar{X}} = \frac{\sigma}{\sqrt{n}} = \frac{1.1}{\sqrt{50}} = 0.156$

According to the question, $P(\bar{X} > 3.5)$ is the correct mathematical expression. There is no direct connection between sample means and probabilities. The $\bar{X}$ needs to be converted into a Z score by using. $Z = \frac{\bar{X} - \mu}{\sigma / \sqrt{n}}$.

$$P(\bar{X} > 3.5) = P\left(Z > \frac{3.5 - 3.2}{\frac{1.1}{\sqrt{50}}}\right) = P(Z > 1.92) = 1 - .9726 = .0274$$

Draw and identify this area under the standard normal distribution curve, $P(Z > 1.92)$ as shown in Figure 6.5. The $\sigma_{\bar{X}} = \frac{\sigma}{\sqrt{n}} = \frac{1.1}{\sqrt{50}} = 0.156$ is shown in the screen time scale.

When randomly selecting 50 teenagers, the probability of their average screen time greater than 3.5 hours is .0274.

FIGURE 6.5 ● Using the Z Table to Figure Out $P(Z > 1.92)$

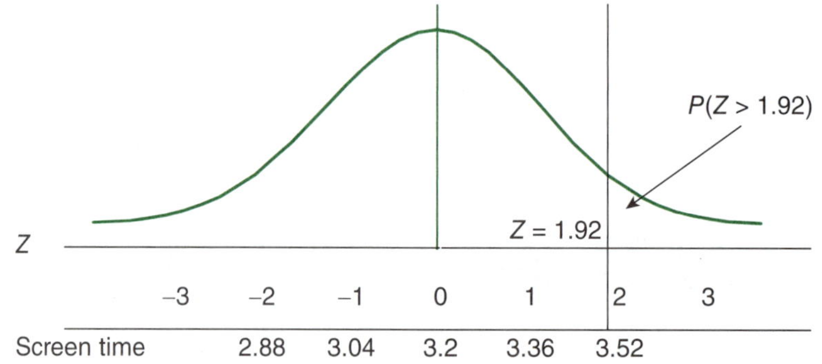

c. Larger sample size leads to smaller standard error of the mean; when $n = 5$, $\sigma_{\bar{X}} = 0.492$ and when $n = 50$, $\sigma_{\bar{X}} = 0.156$. In both hypothetical samples, their means were identical ($\bar{X} = 3.5$), and the population mean was the same ($\mu = 3.2$); however, the different standard error of the mean, $\sigma_{\bar{X}}$, generated different calculated Z values. The probability of randomly selecting five teenagers with an average screen time greater than 3.5 hours was higher than randomly selecting 50 teenagers with such a result. The results with higher probabilities mean that they are more likely to happen due to pure chance.

2. The correct mathematical expression according to the question is

$$P(\bar{X} < 450) = P\left(Z < \frac{(450-497)}{\frac{115}{\sqrt{25}}}\right) = P(Z < -2.04) = .0207$$

Draw and identify this area under the standard normal distribution curve, $P(Z < -2.04)$ as shown in Figure 6.6. The $\sigma_{\bar{X}} = \frac{\sigma}{\sqrt{n}} = \frac{115}{\sqrt{25}} = 23$ is shown in the SAT scale. The probability of randomly selecting 25 high school seniors to obtain a mean SAT ERW score less than 450 is .0207. This is a small probability event.

FIGURE 6.6 ● Using the Z Table to Figure Out P(Z < -2.04)

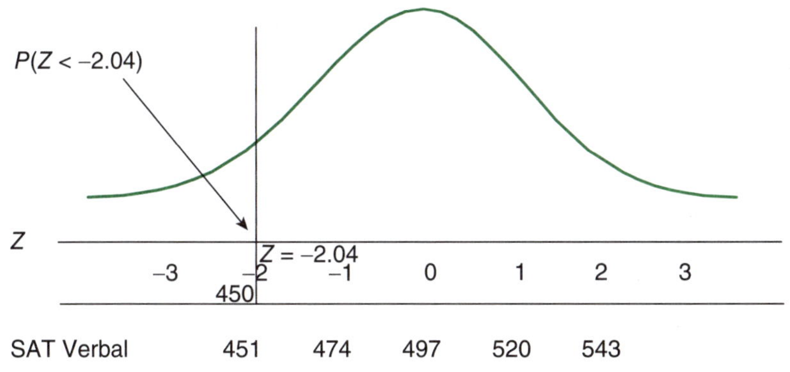

3. The Logistic Director is calculating the probability of getting all 110 boxes of PPE on a small cargo plane with a payload capacity of 4,500 pounds. Although the weight of every box is unavailable, the goal is to getting the boxes on the plane without overloading it. Therefore, the average weight of the 110 boxes should be less than $\frac{4500}{110} = 40.91$. The mathematical solution of this problem is $P(\bar{X} < 40.91)$. There is no direct connection between sample mean and probability, so $\bar{X}$ needs to be converted to a Z value.

$$P(\bar{X} < 40.91) = P\left(Z < \frac{(40.91-38.69)}{\frac{14.78}{\sqrt{110}}}\right) = P(Z < 1.57) = .9418$$

Draw and identify this area under the standard normal distribution curve, $P(Z < 1.57)$ as shown in Figure 6.7.

FIGURE 6.7 Using the Z Table to Figure Out $P(Z < 1.57)$

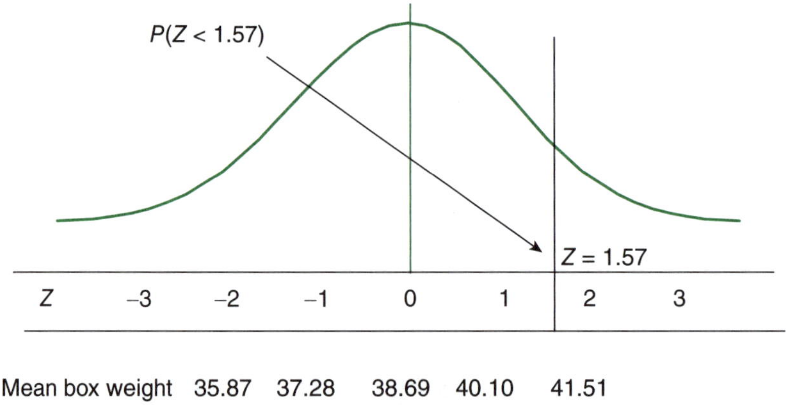

The probability of getting 110 boxes on the small cargo plane without overloading it is .9418. It is highly likely that all 110 boxes will be within the cargo capacity of the plane. Therefore, based on the knowledge of the distribution of box weight without knowing every individual box weight, the Logistics Director estimated the probability of all 110 boxes of PPEs will fit in this small cargo plane is .9418 and it is highly likely that 110 boxes will be delivered to hospitals quickly.

4. The solution for this question depends on the actual data collected from the entire class.

Here is an example of 25 students who used Table 6.2 to record number of heads when tossing 10 coins at the same time, and repeat the process 20 times. The results are shown in Table 6.3. All students' records were merged to create Table 6.3. First, the average number of heads from Student 1's 20 trials of tossing of 10 coins was calculated, and it was 5.3 heads. Then, the mean for every student's 20 trials is calculated and reported in the last column. You can see the variation of number of heads was large for individual trials. It could range from 0 to 10 heads. However, when the mean was calculated over 20 trials, it ranged from 3.9 to 6.4. Then, the mean of all 25 students' 20 trials (i.e., effectively, mean for 500 trials of tossing 10 coins) was 4.95, which was very close to $E(X) = \mu = np = 10(.5) = 5$. This process demonstrated that when the number of trials got larger and larger, the sample mean got closer and closer to the population mean. If you are not satisfied with this result from 500 trials, you may keep on increasing the number of trials until it approaches infinity. It will take you forever to get there.

TABLE 6.3 ⬢ 25 Students Toss 10 Coins for 20 Trials

Trial	1	2	3	4	5	6	7	8	9	10	11	12	13	14	15	16	17	18	19	20	$\bar{X}$
Student 1	3	6	4	5	8	8	4	6	3	2	9	5	6	4	7	8	2	7	2	7	5.3
Student 2	3	7	4	6	2	8	5	6	7	3	4	7	3	5	6	5	2	7	4	5	4.95
Student 3	3	6	5	4	3	2	8	4	6	5	3	4	7	6	5	2	7	4	6	5	4.75
Student 4	5	6	5	4	5	4	6	4	2	7	4	7	5	6	4	3	2	5	6	4	4.7
Student 5	6	4	7	5	8	8	7	4	7	6	8	6	5	5	7	3	6	9	7	4	6.1
Student 6	5	4	2	8	6	4	8	3	5	2	4	6	4	4	5	2	5	3	4	4	4.4
Student 7	4	3	5	4	3	5	8	8	7	3	4	4	5	6	6	3	7	5	3	7	5
Student 8	3	5	5	7	6	8	2	6	4	6	6	2	7	3	5	8	4	3	4	4	4.9
Student 9	6	7	5	6	5	4	6	4	2	5	4	7	3	2	8	5	2	7	4	2	4.7
Student 10	4	5	6	6	4	6	5	8	5	5	6	6	6	5	2	3	4	5	4	6	5.05
Student 11	5	5	5	7	7	5	6	5	4	5	3	8	6	5	5	5	4	5	6	5	5.3
Student 12	4	4	5	4	5	4	4	3	6	4	5	6	5	5	2	6	4	4	5	7	4.6
Student 13	4	4	6	3	1	5	6	3	4	4	4	5	7	6	2	3	2	4	5	6	4.2
Student 14	6	3	5	7	4	4	5	5	7	7	4	5	4	5	6	7	5	4	5	7	5.25
Student 15	3	4	7	3	6	4	6	5	4	6	6	4	8	4	5	2	7	3	8	4	4.95
Student 16	5	4	2	4	5	4	8	3	5	7	3	4	6	3	4	5	5	4	4	3	4.4
Student 17	5	6	8	7	4	5	5	4	3	5	5	3	4	6	2	4	3	6	6	3	4.7
Student 18	8	3	6	3	4	6	3	4	5	7	6	2	3	8	4	6	7	2	5	4	4.8
Student 19	6	5	5	6	5	5	5	5	7	5	4	5	4	6	4	5	5	5	6	5	5.15
Student 20	6	3	1	6	2	7	4	2	6	2	5	5	5	6	6	4	3	4	8	4	4.45
Student 21	7	5	7	4	6	5	5	5	7	4	4	2	5	4	5	3	6	2	6	5	4.85
Student 22	4	5	6	7	3	8	10	9	4	7	6	5	9	5	8	6	4	10	4	8	6.4
Student 23	7	8	4	5	6	9	10	5	6	3	2	4	6	5	8	9	5	6	7	8	6.15
Student 24	3	2	6	4	4	1	3	5	7	2	4	5	3	4	4	2	6	3	6	4	3.9
Student 25	5	4	4	6	2	5	5	8	4	7	4	5	3	7	5	7	7	4	3	2	4.85
Total																					4.95

Tossing a coin generates two possible outcomes: head or tail. Each trial is independent with other trials. The probability of success and failure stays the same throughout the trials. Therefore, the mean and standard deviation of the results of tossing 10 coins can be calculated with the binomial probability distribution formulas where $\mu = np$ and $\sigma = \sqrt{npq}$. For each individual trial of tossing 10 coins, $\mu = np = 10\,(.5) = 5$ and $\sigma = \sqrt{npq} = \sqrt{10(.5)(.5)} = 1.58$. For average of 20 trials, $\mu = 5$ and the standard error of the mean is $\sigma_{\bar{X}} = \frac{\sigma}{\sqrt{n}} = \frac{1.58}{\sqrt{20}} = 0.35$. The distribution of number of heads can be illustrated in Figure 6.8 using Z scale, X scale, and $\bar{X}$ scale. You can also understand why the number of heads varied more in individual trials than in the average of 20 trials.

FIGURE 6.8 ⬢ *Z* Scale, *X* Scale and $\bar{X}$ Scale on the Number of Heads in Tossing 10 Coins

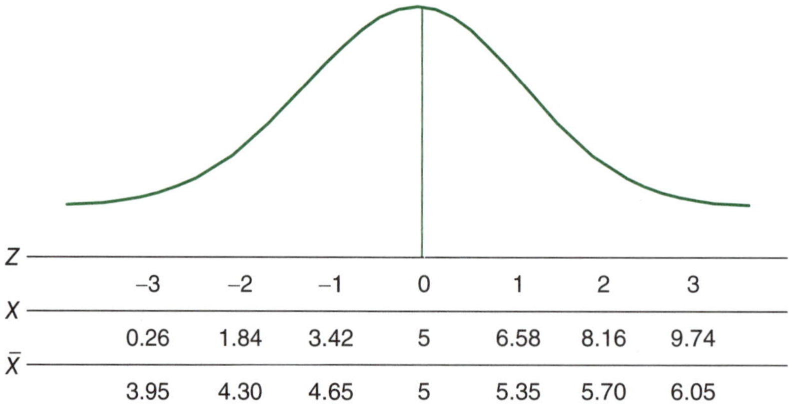

What You Learned

In this chapter, you have learned the central limit theorem, the law of large numbers, and how to apply the same principle you learned in the Z formula to convert a sample mean to a Z value by using $Z = \frac{\bar{X} - \mu}{\sigma / \sqrt{n}}$ to figure out the exact location of a particular sample mean in the distribution of all sample means. Knowing how to use the Z Table to connect Z values and probabilities remains important for both Chapters 5 and 6. Overall, you've learned that when you repeat the same random sampling procedure to produce many sample means, the sample means form a normal distribution with a mean of μ and standard deviation of $\sigma_{\bar{X}} = \sigma/\sqrt{n}$ as long as the samples are selected from a normally distributed population or the sample size is large (i.e., $n > 30$). Sampling errors produce random errors. As long as the errors are random, the relationship between the sample means and the population mean can be explained by the probability theory. The location of the sample mean can be identified exactly in the distribution. The probability associated with the Z value serves as a basis to evaluate the relationship between the sample mean and the population mean.

Key Words

Central limit theorem: The central limit theorem states that when the same random sampling procedure is repeated to produce many sample means, the sample means form a normal distribution with a mean of μ and a standard deviation of $\sigma_{\bar{X}} = \sigma/\sqrt{n}$ as long as the samples are selected from a normally distributed population or the sample size is large (i.e., $n > 30$). 160

Law of large numbers: The law of large numbers is another probability theorem describing the relationship between sample means and population means. When the number of trials approaches infinity, the observed sample mean approaches the population mean. 167

Sampling distribution of the sample means: The sampling distribution of the sample means is defined as the distribution of all means from samples of the same sample size, n, randomly selected from the population (i.e., $\bar{X}_1, \bar{X}_2, \bar{X}_3, \ldots$). 160

Sampling error: A sampling error is defined as when a sample is randomly selected, a natural divergence, difference, distance, or error occurs between sample statistics and population parameters. 158

Standard error of the mean: The standard error of the mean is defined as the standard deviation of all sample means. Its calculation is the standard deviation of the population divided by the square root of the sample size, $\sigma_{\bar{X}} = \sigma/\sqrt{n}$. 160

***Z* test:** A Z test measures the difference between a sample mean and its population mean in units of standard error of the mean, $Z = \frac{\bar{X} - \mu}{\sigma/\sqrt{n}}$. 162

Learning Assessment

Multiple Choice: Circle the Best Answer in Every Question

1. Assume that IQ test scores are normally distributed with a mean of 100 and standard deviation of 10. What is the probability of randomly selecting an individual with an IQ greater than 130?
   a. .9987
   b. .0013
   c. .9772
   d. .0228

2. Assume that adult heights forms a normal distribution with a mean, $\mu = 68.34$ inches and a standard deviation, $\sigma = 6.57$ inches. What is the probability of randomly selecting a group of 25 people and their average height is taller than 5 feet 11 inches?
   a. .9783
   b. .9772
   c. .0228
   d. .0217

3. In any normal distribution, the middle 50% is bound by __________.
   a. $-0.5 < Z < 0.5$
   b. $-0.67 < Z < 0.67$
   c. $-1.5 < Z < 1.5$
   d. $-1.67 < Z < 1.67$

4. The standard deviation of all sample means is also called
   a. the standard error of the mean
   b. the sampling error
   c. the central limit theorem
   d. the sum of squared deviations

5. What is the probability of randomly selecting a sample of $n = 20$ people with an average happiness score greater than 60? Assume that happiness scores are normally distributed with a $\mu = 65$ and $\sigma = 15$?
   a. .0681
   b. .0823
   c. .9177
   d. .9319

6. The number of days of hospital stay after a particular procedure is a skewed distribution. Researchers randomly select 40 patients after this procedure. Assuming that the sampling process can be repeated many times, the shape of the sampling distribution of sample means is
   a. the same shape as that of the population
   b. normally distributed
   c. negatively skewed
   d. positively skewed

7. A population is normally distributed with $\mu = 80$ and $\sigma = 8$. The sampling distribution of sample means for sample size $n = 4$ selected from this population would have a standard error of the mean of __________.
   a. 80
   b. 8
   c. 4
   d. 2

8. Which of the following samples would have the smallest standard error of the mean?
   a. $n = 25$ from a population with $\sigma = 10$
   b. $n = 25$ from a population with $\sigma = 20$
   c. $n = 100$ from a population with $\sigma = 10$
   d. $n = 100$ from a population with $\sigma = 20$

9. A random sample of $n = 14$ scores is obtained from a normal population with $\mu = 19.87$ and $\sigma = 4.12$. What is the probability of randomly selecting a sample with a mean less than 21?
   a. .1587
   b. .3085
   c. .6915
   d. .8485

10. A sample is obtained from a population with $\mu = 100$ and $\sigma = 10$. Which of the following sample means would produce a *Z* score closest to zero?
   a. A sample of $n = 25$ with $\overline{X} = 102$
   b. A sample of $n = 100$ with $\overline{X} = 102$
   c. A sample of $n = 25$ with $\overline{X} = 104$
   d. A sample of $n = 100$ with $\overline{X} = 104$

11. What are the *Z* values that set the middle 99% apart from the rest of the distribution?
   a. $Z < 1.96$
   b. $Z < 2.58$
   c. $|Z| < 1.96$
   d. $|Z| < 2.58$

Free Response Questions

12. A new water taxi has a maximal load capacity of 3,200 pounds. Adult males' weight is normally distributed with $\mu = 182.5$ pounds and $\sigma = 40.8$ pounds and adult females' weight is normally distributed with $\mu = 165.2$ pounds and $\sigma = 45.6$ pounds. The safety certificate sets the maximum number of people at 16 (including the driver). It is not possible to know how many male and female passengers will get on the water taxi for every trip. On average, males weigh more than females. To calculate the worst case scenario (e.g., the highest probability to overload the water taxi), it is prudent to assume that all passengers and the driver are male in this calculation. What is the probability that this water taxi overloads when carrying 15 male passengers and a male driver?

13. An internet car insurance company advertises that it takes $\mu = 15$ minutes to get an insurance quote to save you money. The company's records show that the standard deviation for the time spent on

the site to get insurance is $\sigma = 5$ minutes. The company randomly selects a sample of 16 customers and records their time spent on the site to get their insurance quotes. What is the probability that the sample mean of this randomly selected 16 customers' online time to get an insurance quote is longer than 17 minutes?

14. A food processing company labels the net weight of a bag of frozen chicken vegetable dumplings as $\mu = 16$ ounces. According to the company's records, the standard deviation of the net weights of the frozen chicken vegetable dumplings is $\sigma = 1.5$ ounces. What is the probability of randomly selecting a sample of nine bags of the frozen chicken vegetable dumplings with the average net weights less than 15 ounces?

Answers to Pop Quiz Questions

1. d
2. d
3. a
4. c

7

HYPOTHESIS TESTING

Learning Objectives

After reading and studying this chapter, you should be able to do the following:

- Define the terms *Type I error* and *Type II error*, and explain their meanings in hypothesis tests
- Identify and describe the four steps in conducting a hypothesis test
- Explain the importance of the null hypothesis and the alternative hypothesis in conducting a hypothesis test
- Compare and contrast a one-tailed test and a two-tailed test
- Describe the relationship between the α level and the rejection zone when conducting a hypothesis test
- Explain why rejecting a null hypothesis or failing to reject a null hypothesis are mutually exclusive and collectively exhaustive

WHAT YOU KNOW AND WHAT IS NEW

You have learned the connection between probability and the sampling distribution of the sample means in Chapter 6. Understanding that connection allows you to compare a particular sample mean with a known population mean when the population standard deviation, σ, is known. In conducting a comparison between a sample mean and a population mean, researchers have to make a critical judgment on "are the means the same or different?" Scientists have developed a standardized procedure called *hypothesis testing* to make such a critical judgment.

Hypothesis testing is a process that researchers use to test a claim about a population. Hypothesis testing usually involves four steps.

Step 1: Start with explicitly stating the pair of hypotheses: the null hypothesis and the alternative hypothesis.

Step 2: Identify the rejection zone for the hypothesis test.

Step 3: Calculate the appropriate test statistic.

Step 4: Make the correct conclusion.

Obviously many basic terms involved in these four steps need to be clearly defined first. These terms may sound very technical, scientific, and unfamiliar; therefore, an analogy to the litigation process will be used to illustrate the hypothesis testing procedure because almost every student seems somewhat familiar with the litigation process due to the popularity of many legal dramas either on TV series or popular movies.

TYPE I ERROR AND TYPE II ERROR

In a criminal case, both the defense attorneys and the prosecutors can gather evidence to support their arguments on the defendant's alleged crime. Similarly, scientists have to collect data to test whether their experiments generate the effects stipulated by their hypotheses. Hypotheses are basically "educated guesses" in layman's terms. Let's take some time to learn the precise definition of hypotheses and their functions in scientific studies.

Researchers declare the purpose of their study by using the pair of hypotheses: null hypothesis and alternative hypothesis. The **null hypothesis, H_0**, is the "no effect" statement that indicates whatever effect researchers want to establish with empirical data does not exist. Researchers hope to use data to nullify or reject the null hypothesis. When the null hypothesis is rejected, the researchers turn to the alternative hypothesis. The **alternative hypothesis, H_1**, states the effect that researchers are interested in establishing; thus, H_1 is also referred to as the research hypothesis. The null hypothesis and alternative hypothesis are *mutually exclusive* statements describing population parameters. "Mutually exclusive" was a key term first introduced in Chapter 5; it means H_0 does not overlap with H_1. Only one of these two statements can be true. When one of them is true, the other is not true. In reality, the effect either exists or does not exist. At the end of a hypothesis test, we find out whether the null hypothesis is rejected by the data or not.

You might be confused at this point. Why bother to state a no effect hypothesis and then try to reject it? The reason to use the null hypothesis as the starting point of any empirical study is to assume that the study has no effect unless there is strong evidence to reject or dispute the no effect hypothesis. Such a process is established by a standard scientific procedure to guard against anyone falsely claiming a treatment effect when such an effect does not exist.

For example, in the wake of the Ebola outbreak in West Africa, many online venders claimed that their products can cure Ebola. These products ranged from vitamin C, silver, dark chocolate, cinnamon bark, and oregano to snake venom. The Food and Drug Administration (FDA) and Federal Trade Commission (FTC) had to warn these online companies to stop fraudulent claims immediately or face potential legal actions. It is illegal to market dietary supplements to claim that they can cure human diseases. New drugs may not be legally introduced or delivered into interstate commerce without prior approval from the FDA. FDA approval of new drugs requires vigorous clinical studies to validate their effects. Such a process is designed to protect the general public from falling victim to the modern snake oil scams. Con artists are very skilled at creating fear to make a profit. Falsely claimed treatment effects of dietary supplements plague the internet every day. Some of the false claims include curing cancers, AIDS, and Ebola; reversing the aging process; and losing weight. That is the reason why scientifically proven clinical studies are required to start with the no effect hypothesis; then strong evidence is provided to dispute the no effect hypothesis. That means no effect exists until scientists can gather strong evidence to prove otherwise.

We can further illustrate this point by drawing an analogy to the legal system in the United States. In the legal system, the defendant is presumed innocent until proven guilty. In other words, the starting point of criminal trials is the presumption of innocence. It is up to the prosecutors to present evidence beyond a reasonable doubt to convince the jury otherwise. If there is not enough evidence to convince the jury beyond a reasonable doubt that the defendant had committed the alleged crime, the defendant must be found "not guilty." The presumption of innocence is best described as an assumption of innocence that is indulged in the absence of contrary evidence. Hypothetically, in an extreme case, neither the prosecutor nor the defendant's lawyer offers a shred of evidence to prove or disprove the defendant's guilt. No one in the courtroom has any idea about what happened. In this case, using U.S. legal conventions, the verdict must be "not guilty."

The same logic applies to hypothesis testing. The starting point of a hypothesis test is to assume that the new treatment has no effect (H_0 is true). When no evidence or weak evidence is provided, the default conclusion should be "fail to reject H_0." It is up to the researchers to provide strong evidence to dispute or reject H_0.

The decisions from hypothesis tests can be either correct or incorrect, exactly the same way as the decisions from a jury trial can be just or unjust. Let's use the legal system to examine the possible outcomes of a jury trial. Table 7.1 shows all possible outcomes from different combinations of two possible scenarios of the unknown reality: (1) the

TABLE 7.1 ● Possible Outcomes of a Jury Decision

		Unknown Reality	
		Defendant Did Not Do It	Defendant Did It
Jury decision	Guilty	Mistake I	Correct decision
	Not guilty	Correct decision	Mistake II

defendant did not commit the crime or (2) the defendant did commit the crime; and two possible jury decisions: (1) guilty or (2) not guilty.

The cells inside Table 7.1 represent four different combinations of the unknown reality and the jury decisions. The table clearly shows that there are two correct decisions: (1) when the defendant did not commit the crime and the jury found the defendant not guilty and (2) when the defendant actually committed the crime and the jury found the defendant guilty. There are two different types of mistakes. Mistake I means when the defendant did not commit the crime but the jury found the defendant guilty. Mistake II means when the defendant committed the crime but the jury found the defendant not guilty.

Mistakes incur costs to the society and individuals alike. Which type of mistake is worse? Is Mistake I (i.e., putting innocent defendants in prison for crimes they did not commit) worse than Mistake II (i.e., letting guilty criminals go free)? Students usually had heated debates over different answers to this question. One of the most eloquent answer was, 'Mistake I is worse than Mistake II because Mistake I has the same problem as in Mistake II that the guilty person was not held responsible for the crime, plus an innocent person was put in jail for it. Clearly, Mistake I is worse than Mistake II.' Although the table shows two types of mistakes and two types of correct decisions, fortunately, they are not at a 50–50 split. There are quality control standards in the legal system to make sure that the number of correct decisions outweighs the number of mistakes.

Now apply the same logic to hypothesis testing. There are two possible scenarios of the unknown reality: (1) H_0 is true and (2) H_0 is false. There are also two possible research decisions: (1) reject H_0 or (2) fail to reject H_0. Table 7.2 shows all possible combinations of the unknown reality and research decisions.

In Table 7.2, I want to focus on the two types of mistakes: (1) Type I error and (2) Type II error. **Type I error** occurs when researchers decide to reject H_0 but the H_0 is true. **Type II error** occurs when researchers fail to reject H_0 but the H_0 is false. In plain English, a Type I error occurs when the researchers falsely claim that there is an effect but the effect actually does not exist. Type II error occurs when the researchers fail to

TABLE 7.2 ● Possible Outcomes of a Hypothesis Test

Research Decision	Unknown Reality	
	H_0 Is True	H_0 Is False
Reject H_0	Type I error	Correct decision
Fail to reject H_0	Correct decision	Type II error

claim an effect but the effect actually exists. Under the do-no-harm principle, scientists collectively agree that Type I error is worse than Type II error.

To ensure the quality of scientific research, scientists collectively agree to keep Type I error under control. In the example of the FDA, the cost of Type I error is to potentially expose the public to useless treatments that might potentially harm their health. Traditionally, the acceptable risk of committing Type I error, α is set to be at a .05 level for psychological and other social scientific research because psychologists and social scientists collectively agree that if 1 in 20 studies generates a bogus effect, it is an acceptable risk. Type II error, on the other hand, does not incur such risk to the public. The probability of committing Type II error is called β. Researchers simply fail to claim an effect when an effect actually exists. In such situations, researchers can always start over and conduct more studies later. Sooner or later, someone is going to discover the overlooked effect.

Now, let's discuss the correct decisions. First, when H_0 is true, and the researchers fail to reject H_0, this is a correct decision. The probability of such a correct decision can be calculated as $(1 - \alpha)$. This probability $(1 - \alpha)$ is also called the confidence level, which will be introduced in Chapter 8. Second, when H_0 is false and the researchers reject H_0, it is again a correct decision. The probability of such a correct decision is calculated as $(1 - \beta)$, which is called power. **Power** is defined as the sensitivity to detect an effect when an effect actually exists. The power of the hypothesis test is the probability of correctly rejecting H_0 when H_0 is false.

Power is an important statistical concept and is influenced by the following three factors:

1. Sample size, n
2. The probability of committing Type I error, α
3. Effect size

Sample size and power are positively correlated. All else being equal, when the sample size is larger, power is bigger. When the sample size is smaller, power is smaller. Large sample size makes a standard error small. When a standard error is small, it is more likely to detect an effect when it actually exists. Sample size is the most influential factor in calculating statistical power because sample size is under the control of researchers. Researchers can plan ahead to figure out how big the sample size needs to be to reach a desirable power level.

The probability of committing Type I error, α, and power are also positively correlated. When α is higher, power is higher. When α is lower, power is lower. The rationale is that when the Type I error is under strict control, the risk of falsely claiming an effect is low; however, the sensitivity to detect an effect is also low. A Type I error is to reject H_0, when H_0 is true. A Type II error is to fail to reject H_0 when H_0 is false. It is obvious that there is a trade-off relationship between the probability of committing a Type I error, α, and the probability of committing a Type II error, β. As one goes up, the other has to come down. Therefore, when α goes up, researchers decide to increase the risk of Type I error; then β will go down, and $(1 - \beta)$ will go up. Just remember that α and β move in opposite directions, but α and $(1 - \beta)$ move in the same direction.

The **effect size** is positively related with power. The effect size is a standardized measure of the difference between the sample statistic and the hypothesized population parameter in units of standard deviation. We will have detailed discussion on effect size in Chapter 8. Effect size makes it possible to compare results across different studies. A large effect size means that there is a substantial difference between the sample statistic and the hypothesized population parameter. When standardized effect size is larger, power is bigger. When standardized effect size is smaller, power is smaller. Therefore, a large effect size makes it easy to reject H_0. A small effect size makes it difficult to reject H_0.

A summary table of the probabilities associated with all possible outcomes from conducting a hypothesis test is shown in Table 7.3. The probability of committing a Type I error is α, and the probability of correctly failing to reject H_0 when H_0 is true is $(1 - \alpha)$. As we discussed in Chapter 5, these two events are mutually exclusive and they complement each other because the probabilities of these two events add to 1. Similarly, the probability of committing a Type II error is β and the probability of correctly rejecting H_0 when H_0 is false is $(1 - \beta)$, which is also called the statistical power. The statistical power and the probability of committing the Type II error are mutually exclusive and complementary events. The statistical power is positively correlated with the sample size, the probability of committing a Type I error, and its effect size.

Author's Aside

The calculation of power is more complicated than you may like to pursue in an introductory statistics course. If you are interested in calculating power in different situations, you may refer to Cohen (1988) and Park (2010).

TABLE 7.3 ● Probability of Type I Error, Type II Error, and Power

Research Decision	Unknown Reality	
	H_0 Is True	H_0 Is False
Reject H_0	α	$1 - \beta$
Fail to reject H_0	$1 - \alpha$	β

POP QUIZ

1. What is the correct definition of Type I error?
   a. Reject H_0 when H_0 is true.
   b. Fail to reject H_0 when H_0 is true.
   c. Reject H_0 when H_0 is false.
   d. Fail to reject H_0 when H_0 is false.

2. What is the correct definition of Type II error?
   a. Reject H_0 when H_0 is true.
   b. Fail to reject H_0 when H_0 is true.
   c. Reject H_0 when H_0 is false.
   d. Fail to reject H_0 when H_0 is false.

3. Which of the following variables is positively correlated with the statistical power?
   a. The probability of committing a Type I error, α
   b. The sample size, n
   c. The effect size
   d. All of the above three variables are positively correlated with the statistical power.

THE FOUR-STEP PROCEDURE TO CONDUCT A HYPOTHESIS TEST

Hypothesis testing is a standardized process to test the strength of scientific evidence for a claim about a population. Conducting a hypothesis test involves the following four-step procedure. All four steps are explained and elaborated in the following sections.

Step 1: Explicitly state the pair of hypotheses.

Step 2: Identify the rejection zone for the hypothesis test.

Step 3: Calculate the test statistic.

Step 4: Make the correct conclusion.

Step 1: Explicitly State the Pair of Hypotheses

Hypothesis tests are conducted to test the strength of scientific evidence for a claim about a population. Research purposes are explicitly expressed by the pair of hypotheses. H_0 is the "no effect" (null) hypothesis, and H_1 is the research (alternative) hypothesis, which researchers turn to when H_0 is rejected by the data. H_1 states the effect that researchers are interested in. Both hypotheses describe characteristics of the population. Hypothesis tests can be nondirectional or directional. Let's use an example to explain nondirectional versus directional hypotheses.

The average IQ score for preschoolers aged 3 to 5 years is $\mu = 100$. Researchers investigated the maternal drinking on IQ of preschoolers by comparing a sample of preschoolers whose mothers drank two servings of alcohol or more a day during pregnancy. The starting point of this comparison should be explicitly stating the H_0: $\mu = 100$. It is assumed that the average IQ score of preschoolers whose mothers drank was the same as the population. Hence, the H_0 is the no effect hypothesis. Then, the alternative hypothesis, H_1, needs to be stated as the opposite of H_0. If the researchers are interested in whether the sample mean is the same or different from the population mean, the alternative hypothesis is H_1: $\mu \neq 100$. The unequal sign includes both greater and less than so it does not have a specific direction. Therefore, it is a nondirectional hypothesis. Nondirectional hypothesis testing is only interested in whether or not the sample could come from a population with the hypothesized mean. A nondirectional hypothesis test is also called a **two-tailed test**.

Directional hypothesis testing, on the other hand, is interested in a particular direction, so the alternative hypothesis could be expressed as either higher or lower than the population parameter, H_1: $\mu > 100$ or H_1: $\mu < 100$. A directional hypothesis test is also called a **one-tailed test**. Under a one-tailed test, the alternative hypothesis is stated with a specific direction. Which of these tests to use depends on whether the problem statement contains clear direction with key words such as "higher," "increasing," "improving," "lower," "decreasing," or "deteriorating." When the problem statement contains these key words for a particular direction, one-tailed tests are appropriate. To complete the example of the impact of maternal drinking on preschoolers'

Two-tailed test: H_0: $\mu = 100$

H_1: $\mu \neq 100$

Right-tailed test: H_0: $\mu = 100$

H_1: $\mu > 100$

Left-tailed test: H_0: $\mu = 100$

H_1: $\mu < 100$

This list illustrates that H_0 is the "no effect" hypothesis. It states that the average IQ score of preschoolers whose mothers drank during pregnancy is the same as the population mean. Please note that the "=" sign is always in H_0, the null hypothesis.

The alternative hypothesis, H_1, is the hypothesis that researchers turn to when H_0 is rejected. It states that there is some kind of difference between the average IQ scores of preschoolers whose mothers drank while pregnant and the population mean. The nature of the difference depends on whether the test is a two-tailed test, a right-tailed test, or a left-tailed test. H_1 in a two-tailed test states that the average IQ score of preschoolers whose mothers drank differed from the population mean. H_1 in a right-tailed test states that the average IQ score of preschoolers whose mothers drank was higher than the population mean. H_1 in a left-tailed test states that the average IQ score of preschoolers whose mothers drank was lower than the population mean. Researchers can choose the best pair among this list of all possible pairs of hypotheses according to their purposes. Based on well-documented studies on potential harm of alcohol on fetal development, the logical choice in this study is the left-tailed test with H_1: $\mu < 100$.

You might have seen other textbooks list one-tailed tests as H_0: $\mu \leq 100$ and H_1: $\mu > 100$ in a right-tailed test or H_0: $\mu \geq 100$ and H_1: $\mu < 100$ in a left-tailed test, so that all possible outcomes are included in the pair of hypotheses. These kinds of expressions are regarded as outdated. In this book, for the sake of simplicity and clarity, all null hypotheses remain with "=" sign only. For example, a swimming coach is testing the effectiveness of a particular training routine. The purpose of the training routine is to drop swimmers' times and make them swim faster. Therefore, the hypothesis test is set up as a left-tailed test. If the swimmers' times do not change, then we have to conclude that the training routine has no effect (i.e., we fail to reject H_0). Obviously, if the swimmers' times get slower due to the training routine, clearly not the effect intended by the coach, we have to conclude that the training routine has no effect.

Step 2: Identify the Rejection Zone for the Hypothesis Test

As discussed in the previous section, social scientists collectively agree that Type I errors need to be kept under control to protect the general public from falsely claimed effects. The probability of committing a Type I error is α which is the key to establish the standard to evaluate the strength of the scientific evidence for the hypothesis test. Different α levels may be assigned according to the nature of the research. In a research project with strong theoretical reasoning behind it, or in a project with many similar studies being conducted repeatedly and resulted in consistent findings, a strict standard may apply, such as $\alpha = .01$. In contrast, in a research project of an exploratory nature, without much theoretical reasoning to back it up, a weak standard may suffice, such as $\alpha = .10$. Usually for research in social sciences, the default level is $\alpha = .05$.

The α level helps identify the **rejection zone**. The rejection zone is bounded by the critical value of a statistic. Once the rejection zone is set, decision rules can be clearly stated. If any calculated test statistic falls in the rejection zone, the decision is to reject H_0. Otherwise, if the test statistic does not fall in the rejection zone, we fail to reject H_0. The term *fail to reject H_0* is a double negative and seems cumbersome. I will explain in more detail why "fail to reject H_0" is preferred to "accept H_0" in Step 4, which is to make the correct conclusion.

Rejection zone is determined by two factors: the α level and the specified pair of hypotheses (e.g., either directional hypotheses or nondirectional hypotheses). Let's continue the previous IQ example to identify the rejection zones assuming $\alpha = .05$. First, we use $\alpha = .05$ to identify the rejection zone for a two-tailed test or a nondirectional test.

Two-tailed test: $H_0: \mu = 100$

$H_1: \mu \neq 100$

Rejection zone: $|Z| > Z_{\alpha/2}$

The rejection zone is specified by the critical value of Z. In a two-tailed test, the significance level, α, is evenly divided into two tails, and the critical value of Z is specified as $Z_{\alpha/2}$. Assuming $\alpha = .05$, the probability of each tail is $\alpha/2 = .05/2 = .025$. We can easily identify the left tail $P(Z < -z) = .025$ from the Z Table; the z value is -1.96. Based on the symmetrical nature of the curve, the right tail $P(Z > z) = .025$; the z value is 1.96. Therefore, the best way to express the rejection zone for a two-tailed test is $|Z| > 1.96$, as shown in Figure 7.1.

FIGURE 7.1 ● Rejection Zone for $|Z| > 1.96$

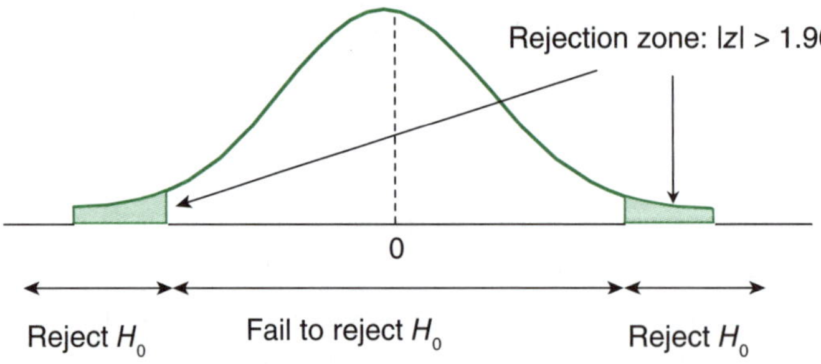

Next, we use $\alpha = .05$ to identify the rejection zone for a right-tailed test.

Right-tailed test: $H_0: \mu = 100$

$H_1: \mu > 100$

Rejection zone: $Z > Z_\alpha$

In a right-tailed test, the α is solely in the right tail. Assuming $\alpha = .05$, the probability of the right tail is .05. According to the Z table, when the tail probability is .05, the critical value of Z is between 1.64 and 1.65. I prefer to use the more conservative (i.e., a higher standard and more difficult to reach) standard, so I pick the value 1.65. Therefore, the best way to express the rejection zone for a right-tailed test is $Z > 1.65$, as shown in Figure 7.2.

FIGURE 7.2 ● Rejection Zone for $Z > 1.65$

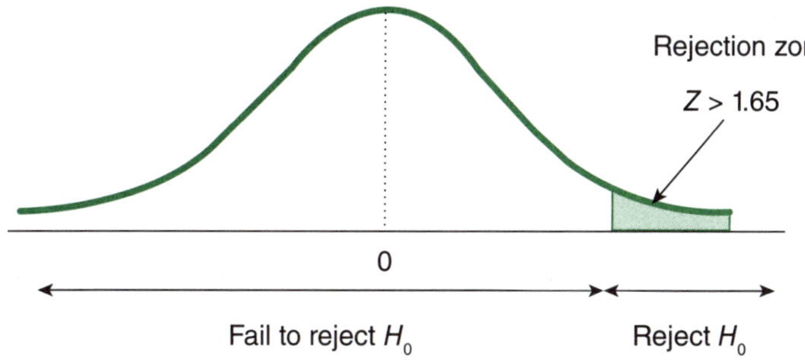

Next, we use $\alpha = .05$ to identify the rejection zone for a left-tailed test.

Left-tailed test: $H_0: \mu = 100$

$H_1: \mu < 100$

Rejection zone: $Z < -Z_\alpha$

The rejection zone for a left tailed test is $Z < -1.65$, as shown in Figure 7.3.

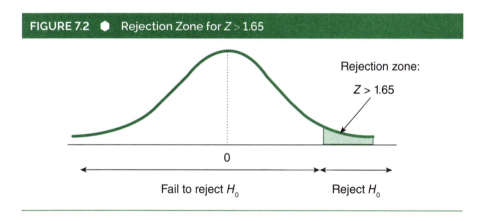

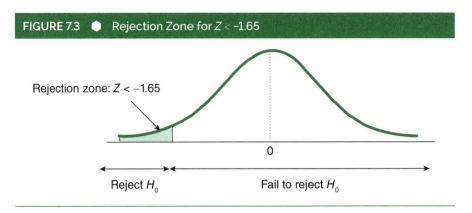

FIGURE 7.3 ● Rejection Zone for $Z < -1.65$

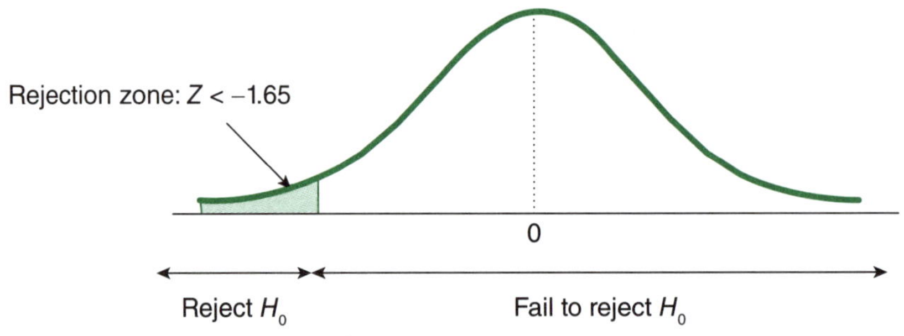

Step 3: Calculate the Test Statistic

When we use a sample mean, $\overline{X}$, to compare with the population mean, μ, with the population standard deviation, σ provided in the problem statement, we use the Z formula in Chapter 6 to calculate the test statistic. The calculated Z value is based on the data collected by the researchers. The calculated Z value shows the strength of the evidence in the research.

$$Z = \frac{\overline{X} - \mu}{\frac{\sigma}{\sqrt{n}}}$$

We will learn other statistics in the following chapters to deal with different kinds of comparisons. The basic principle of a test statistic is to calculate the difference between a sample statistic and a hypothesized population parameter, and then divide it by the standard error of the statistic.

$$\text{Test statistic} = \frac{\text{Sample statistic} - \text{Population parameter}}{\text{Standard error of the statistic}}$$

After the four-step hypothesis procedure has been introduced, a few examples will illustrate the process in its entirety.

Step 4: Make the Correct Conclusion

After identifying the rejection zone in Step 2 and calculating the test statistic in Step 3, the next step is to compare the calculated value with the rejection zone. If the calculated value falls in the rejection zone, we reject H_0. If the calculated value does not fall in the rejection zone, we fail to reject H_0. In statistics, we prefer to use precise language. The decision made by researchers after conducting a simple hypothesis test is to reject H_0 or fail to reject H_0, and such decisions are **mutually exclusive and collectively exhaustive**. This means that the two decisions (1) "reject H_0" and (2) "fail to reject H_0" could not be true at the same time, and they cover all possible outcomes.

Let's discuss the difference between "accept H_0" and "fail to reject H_0." The analogy of the legal system helps illustrate the distinction. In the legal system, there are two possible outcomes from a jury decision: (1) "guilty" or (2) "not guilty." "Not guilty" is also a double negative phrase but there is no movement to change "not guilty" to a simple phrase "innocent" anytime soon. To be clear, "not guilty" is not the same as "innocent" because "not guilty" simply means there is not enough evidence to convince all jurors that the defendant has committed the crime beyond a reasonable doubt. "Innocent" is an affirmative statement that the defendant ***did not*** commit the crime and "innocent" belongs in a subcategory of the "not guilty." However, "innocent" and "not guilty" are not 100% synonymous. In some cases, a highly suspicious defendant may be found "not guilty" because a key piece of evidence was thrown out by the judge, a key eyewitness went missing, the police mishandled the evidence, the murder weapon was never found, or a brilliantly convincing alternative story was argued by the defense lawyers. These were just some examples of why "not guilty" was not the same as "innocent."

Now let's apply the same principle to hypothesis test in scientific research. "Fail to reject H_0" simply means that the evidence is not strong enough to support the claimed effect. "Accept H_0" is an affirmative statement that there is *no effect*. "Accept H_0" belongs in a subcategory of the "fail to reject H_0."

Here is a real-life example to illustrate why this distinction matters. I attended a research presentation made by a job candidate seeking a tenure-track position at a university. The candidate presented his research to the faculty members in the department where the hiring decision was going to be made. The candidate's research topic was on spatial orientation. According to the previous literature, spatial orientation is an ability that consistently shows significant gender differences. After his presentation, I asked, "Did you find any gender difference in your study on spatial orientation?" He answered, "There were no gender differences in my study." The answer would have been much better if he said, "There was not enough evidence to support the claim that gender differences existed in my study." "There were no gender differences in my study" is the "accept H_0" answer. "There was not enough evidence to support the claim that gender differences exist" is the "fail to reject H_0" answer. The fail to reject H_0 answer is logically and technically superior to the accept H_0 answer. The job candidate's research was not focused on gender differences, and the research design was not conducive to show whether they existed or not.

There simply was not enough evidence on gender differences in his study. Being able to answer questions correctly during a job interview might have a big impact on the hiring decision.

Let's move on to a few examples of how the four-step hypothesis test works in action.

Examples of the Four-Step Hypothesis Test in Action

The four-step hypothesis test is a standard scientific approach to test the strength of the empirical evidence on a claim about a population. Such a four-step hypothesis test procedure will be useful not only in this chapter but also in all of the later chapters. In other words, it is definitely worth the effort to learn how to conduct hypothesis tests accurately. Learning by doing is the best strategy, so let's go through a couple of examples to illustrate the four-step process.

EXAMPLE 7.1

A standardized IQ test is known to be normally distributed and has $\mu = 100$ and $\sigma = 10$. There is a common belief that students in wealthier school districts tend to have higher IQ scores. Researchers randomly selected a sample of 25 students from a wealthy school district. The average IQ of this sample was $\bar{X} = 103$. Conduct a test to verify if the average IQ score for students from this wealthy district is higher than 100, assuming $\alpha = .05$.

We use the four-step hypothesis testing process to answer this question.

Step 1: State the pair of hypotheses.

Based on the problem statement "If the average IQ score for students from this wealthy district is higher than 100," the problem indicates a directional hypothesis. The key word "higher" indicates a right-tailed test. The alternative hypothesis, H_1, is $\mu > 100$. Then H_0 covers the equal sign, $\mu = 100$. It is important to point out that the research interest was to compare the average IQ score of students from this wealthy district to the general population mean. The research interest was not limited to the mean IQ of 25 students in this sample. Thus, the hypotheses must be phrased to describe the "research population," which refers to all students in this wealthy district.

H_0: $\mu = 100$

H_1: $\mu > 100$

Step 2: Identify the rejection zone for the hypothesis test.

We know this is a right-tailed test from Step 1. When the tail probability is .05, according to the Z table, the Z value is 1.65. The rejection zone for the right-tailed test is $Z > 1.65$. It is clear and useful to draw a simple normal distribution curve with the rejection zone shaded as shown in Figure 7.4.

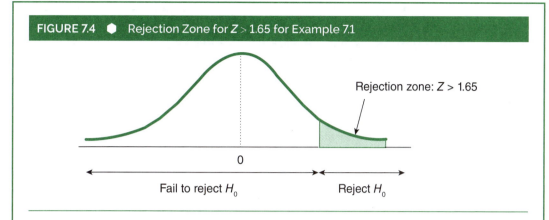

FIGURE 7.4 ● Rejection Zone for $Z > 1.65$ for Example 7.1

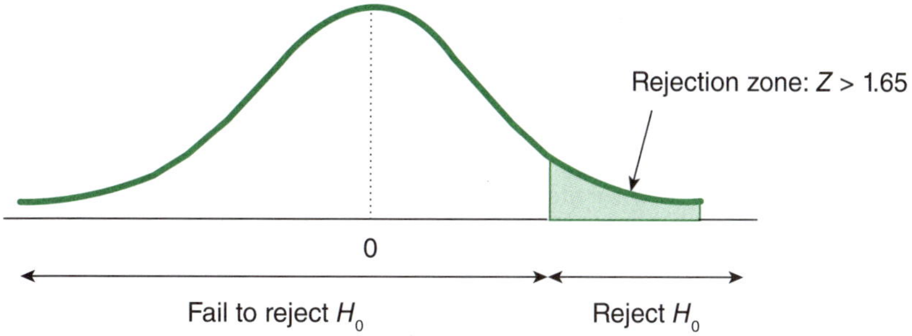

Step 3: Calculate the test statistics.

The problem statement asks for a comparison between students in this wealthy district and the general population mean with a known σ. Therefore, the Z test is the correct statistic to accomplish this purpose. It is easy to list all the numbers stated in the problem statement, so as to separate out what we know and what we need to figure out.

What we know:

$\mu = 100$

$\sigma = 10$

$\bar{X} = 103$

$n = 25$

We need to figure out Z. Next, insert all the numbers in the Z-test formula.

$$Z = \frac{\bar{X} - \mu}{\frac{\sigma}{\sqrt{n}}}$$

$$Z = \frac{103 - 100}{10/\sqrt{25}} = \frac{3}{2} = 1.5$$

Step 4: Make the correct conclusion.

Compare the calculated $Z = 1.5$ with the rejection zone $Z > 1.65$. The calculated Z does not fall in the rejection zone. Therefore, we fail to reject H_0. The evidence was not strong enough to support the claim that the average IQ score of students from this wealthy district was higher than 100 at the $\alpha = .05$ level.

Statistics are widely available in different situations. Besides IQ scores or standardized test scores, life expectancy is a commonly available statistic locally, nationally, and internationally. Let's use life expectancy as an example to conduct a hypothesis test in the next example.

EXAMPLE 7.2

Hawaiians are known to have an active lifestyle, enjoying the fresh air and natural beauty the islands have to offer. The life expectancy of the U.S. population across all 50 states is $\mu = 78.62$ years with a standard deviation $\sigma = 16.51$ years. A randomly selected sample of 100 Hawaiians shows an average life expectancy $\bar{X} = 82.52$. Did Hawaiians' average life expectancy significantly differ from 78.62, assuming $\alpha = .01$?

We apply the four-step hypothesis testing process to answer this question.

Step 1: State the pair of hypotheses.

Life span depends on heredity, individual action, and environment. Some people live a very long life, while others might not be so lucky. The standard deviation, $\sigma = 16.51$, reflects such individual differences. The problem statement asks, "Did Hawaiians' average life expectancy significantly differ from 78.62?" Such a question did not specify any direction of the difference. Therefore, a nondirectional hypothesis is the correct procedure to use. The pair of hypothese is stated below.

H_0: $\mu = 78.62$

H_1: $\mu \neq 78.62$

Step 2: Identify the rejection zones for the hypothesis test.

From Step 1, we know that we need to conduct a two-tailed test. The problem statement specifies $\alpha = .01$. When the tail probability is $\alpha/2 = .01/2 = .005$, according to the Z table, the left tail is $P(Z < -z) = .005$, and the z value is between −2.57 and −2.58. I prefer to use the more conservative approach, so I pick the value of −2.58. Based on the symmetrical nature of the curve, the right tail is $P(Z > z) = .005$, and the z value is 2.58. Therefore, the best way to express the rejection zones for a two-tailed test is $|Z| > 2.58$, as shown in Figure 7.5.

FIGURE 7.5 The Rejection Zones for $|Z| > 2.58$

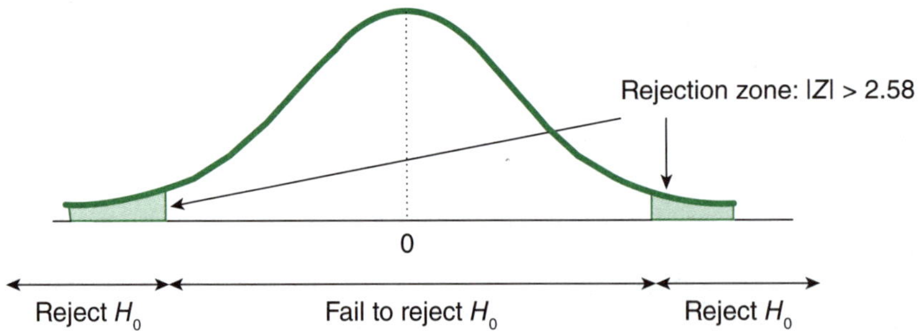

Step 3: Calculate the statistic.

The problem statement asked for a comparison between the average life span of people in Hawaii and the population mean. The population standard deviation, σ, was provided; therefore, the Z test is the correct statistic to accomplish this purpose. Let's list all the numbers stated in the problem statement and separate out what we know from what we need to figure out.

Let's list everything we know from the problem statement.

$\mu = 78.62$

$\sigma = 16.51$

$\bar{X} = 82.52$

$n = 100$

We need to figure out the Z. Next, insert all the numbers in the Z-test formula.

$$Z = \frac{\bar{X} - \mu}{\frac{\sigma}{\sqrt{n}}}$$

$$Z = \frac{82.52 - 78.62}{16.51/\sqrt{100}} = \frac{3.9}{1.65} = 2.36$$

Step 4: Make the correct conclusion.

Compare the calculated $Z = 2.36$ with the rejection zone, $|Z| > 2.58$. The calculated Z is not within the rejection zone. Therefore, we failed to reject H_0. The evidence was not strong enough to support the claim that Hawaiians' average life expectancy was significantly different from 78.62.

According to the problem, $\alpha = .01$, which is a very strict standard for a hypothesis test. If we relax the standard to the customary level of $\alpha = .05$, the rejection zone would become $|Z| > 1.96$. With a calculated Z value = 2.36, the outcome of a hypothesis test at $\alpha = .05$ would be to reject H_0. It is important to know that the αlevel is determined before the statistic is calculated. Researchers decide in advance what significance level is appropriate for the research purpose.

After going through a couple of examples, it is clear that the four-step hypothesis testing procedure is a standardized scientific process to test the strength of the empirical evidence on a claim about the population. Let's recap that process.

Step 1: Explicitly state the pair of hypotheses. This means to declare the purpose of the hypothesis test and identify whether it is a directional test or a nondirectional test.

Step 2: Identify the rejection zone for the hypothesis test. Based on the predetermined α level, the rejection zone can be specified by the critical value of Z. The rejection zone serves as a criterion for decision making on the outcomes of the hypothesis test.

Step 3: Calculate the test statistic. Many different statistical tests will be discussed throughout this book. So far, we have only covered the Z test.

Step 4: Make the correct conclusion. The research conclusion is made by comparing the calculated value of the test statistic and the rejection zone. If the calculated test statistic is within the rejection zone, we reject H_0. It means that the evidence is strong enough to claim an effect. If the calculated test statistic is not within the rejection zone, we fail to reject H_0. It means that the evidence is not strong enough to support the claim of an effect.

POP QUIZ

4. What values are needed to figure out the rejection zones for Z tests?
   a. The α level
   b. One-tailed test or two-tailed test
   c. Sample size
   d. Both (a) and (b)
   e. All of the above

5. For a two-tailed Z test with $\alpha = .10$, the critical values that set the boundaries for the rejection zones are
   a. $Z > 1.65$
   b. $Z > 1.96$
   c. $|Z| > 1.65$
   d. $|Z| > 1.96$

FURTHER DISCUSSIONS ON DIRECTIONAL VERSUS NONDIRECTIONAL HYPOTHESIS TESTS

Directional hypothesis tests versus nondirectional hypothesis tests is a topic that needs to be discussed further in detail. The most noticeable difference between one-tailed tests and two-tailed tests is the rejection zone. Under a one-tailed test, the probability of committing a Type I error, α, is only on one side of the distribution curve. When conducting a two-tailed test, α needs to be evenly divided into two tails, $\alpha/2$. Assuming $\alpha = .05$, let's examine the rejection zones for these situations.

Right-tailed test:

Rejection zone: $Z > Z_\alpha$

$Z > 1.65$

Left-tailed test:

Rejection zone: $Z < -Z_\alpha$

$Z < -1.65$

Two-tailed test:

Rejection zone: $|Z| > Z_{\alpha/2}$

$|Z| > 1.96$

The critical value is the value that sets the boundary for the rejection zone. Assuming the same α level, the critical value of Z is higher in a two-tailed test than in a one-tailed test. A higher critical value of Z means that the strength of the evidence needs to be stronger in a two-tailed test than in a one-tailed test to be able to reject H_0. I will illustrate this point by using an example.

EXAMPLE 7.3

Hawaiians are known to have an active lifestyle, enjoying the fresh air and natural beauty the islands have to offer. The life expectancy of the U.S. population across all 50 states is $\mu = 78.62$ years with a standard deviation $\sigma = 16.51$ years. A randomly selected sample of 100 Hawaiians showed an average life expectancy $\bar{X} = 82.52$ years. Was Hawaiians' average life expectancy significantly higher than 78.62 years, assuming $\alpha = .01$?

If you thought this example looked familiar, you were correct. This problem used the same numbers as in Example 7.2. The only difference was the way the question was phrased at the end. It indicated a directional hypothesis test. Let's apply the four-step hypothesis testing procedure to answer this question.

Step 1: State the pair of hypotheses.

According to the problem, "Was Hawaiians' average life expectancy significantly higher than 78.62 years?" The question contained a clear direction indicated by the key word "higher." Therefore, a directional hypothesis, in this case, a right-tailed test, is the correct approach. The pair of hypotheses is stated below.

H_0: $\mu = 78.62$

H_1: $\mu > 78.62$

Step 2: Identify the rejection zone for the hypothesis test.

From Step 1, we knew that we needed to conduct a right-tailed test. The problem statement specified $\alpha = .01$. When the tail probability was $\alpha = .01$, according to the Z table, the critical value of Z is 2.33. The rejection zone is $Z > 2.33$, as shown in Figure 7.6.

(Continued)

(Continued)

FIGURE 7.6 ⬢ The Rejection Zone for $Z > 2.33$

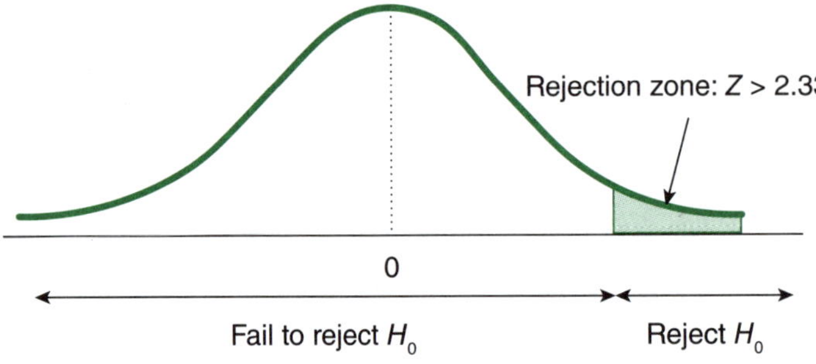

Step 3: Calculate the test statistic.

The statement asks to compare the Hawaiians' average life expectancy with the population mean with a known σ, so the *Z* test is the correct statistic to accomplish this purpose. It is easy to list all the numbers in the problem statement, so as to separate out what we know and what we need to figure out.

What we know:

$\mu = 78.62$

$\sigma = 16.51$

$\bar{X} = 82.52$

$n = 100$

We need to figure out the *Z*. Next, insert all the numbers in the *Z*-test formula.

$$Z = \frac{\bar{X} - \mu}{\frac{\sigma}{\sqrt{n}}}$$

$$Z = \frac{82.52 - 78.62}{16.51/\sqrt{100}} = \frac{3.9}{1.65} = 2.36$$

Step 4: Make the correct conclusion.

Compare the calculated $Z = 2.36$ with the rejection zone, $Z > 2.33$. The calculated *Z* fell in the rejection zone. Therefore, we rejected H_0. Hawaiians' life expectancy was significantly higher than 78.62 years.

You probably noticed the different conclusions reached in Examples 7.2 and 7.3 using the same numbers. In Example 7.2, we conducted a two-tailed test but failed to reject H_0. However, in Example 7.3, we conducted a one-tailed test and rejected H_0.

The choice of conducting a one-tailed test or a two-tailed test should not be based on which one is easier to reject H_0 to produce a statistically significant result. When the research topic is fairly new or exploratory or there is no particular theoretical reasoning to suggest a directional test, two-tailed tests are the preferred standard operating procedures. One-tailed tests are usually reserved for research topics that have been studied repeatedly and consistent results have been obtained, or there are theoretical arguments strongly suggesting directional outcomes.

POP QUIZ

6. When exploratory research is conducted without extensive prior research or theoretical reasoning to suggest a particular direction of relationship, the preferred hypothesis tests should be __________.
   a. two-tailed tests
   b. one-tailed tests
   c. whichever is easier to produce significant results
   d. randomly selected

EXERCISE PROBLEMS

1. According to the ACT Profile Report—National, the ACT composite scores are normally distributed with $\mu = 21.0$ and $\sigma = 5.4$. Many charter schools received funding from the state government without providing accountability measures as required by the public schools. A sample of 36 high school seniors was randomly selected from charter schools. Their average ACT composite score was 19.8. Was the average ACT composite score of seniors from charter schools lower than 21.0, assuming $\alpha = .10$?

2. The expenditure per pupil for K–12 public schools in the United States has a normal distribution with a mean $\mu = \$12,611.84$ and a standard deviation $\sigma = \$4,148.46$. Researchers randomly selected a group of 50 high performing public schools, and their average expenditure per pupil was \$13,526.91. Was the average expenditure per pupil from high performing public schools higher than \$12,611.84, assuming $\alpha = .05$?

3. A car manufacturer claims that a new model Q has an average fuel efficiency $\mu = 35$ miles per gallon on the highway, and a standard deviation $\sigma = 5$. The American Automobile Association randomly selected 40 new Qs and measured their fuel efficiency at 33 miles per gallon. Was the car manufacturer truthful in advertising Q's fuel efficiency as being 35 miles per gallon, assuming $\alpha = .10$?

Solutions

1. The problem statement is whether the average ACT composite score of seniors from charter schools was lower than 21.0. "Lower" is a key word for direction; therefore, a left-tailed test is appropriate.

   Step 1: State the pair of hypotheses.

   Left-tailed test: H_0: $\mu = 21.0$

   H_1: $\mu < 21.0$

   Step 2: Identify the rejection zone.

   For a left-tailed test with $\alpha = .10$, the critical value of Z is 1.28. The rejection zone is $Z < -1.28$, as shown in Figure 7.7.

FIGURE 7.7 ● Rejection Zone for $Z < -1.28$

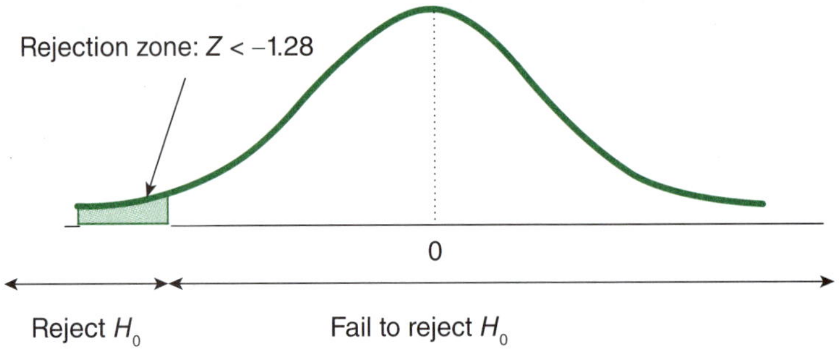

   Step 3: Calculate the test statistic.

   When comparing the average ACT score of students from charter schools to a population mean with a known σ, the use of a Z statistic is the correct approach.

   List all the numbers given in the problem statement.

   $\mu = 21.0$

   $\sigma = 5.4$

   $\bar{X} = 19.8$

   $n = 36$

$$Z = \frac{\bar{X} - \mu}{\sigma / \sqrt{n}} = \frac{19.8 - 21.0}{5.4 / \sqrt{36}} = \frac{-1.2}{0.9} = -1.33$$

Step 4: Make a conclusion.

Compare the calculated Z value = –1.33 from Step 3 and the rejection zone $Z < -1.28$ from Step 2. The calculated Z is within the rejection zone, so we reject H_0. The evidence is strong enough to support the claim that the average ACT of seniors from charter schools was lower than 21.0 at $\alpha = .10$.

2. According to the problem statement, "Was the average expenditure per pupil from high performing public schools higher than $12,611.84?" This indicates clear direction, so a one-tailed test (in particular, a right-tailed test) is the correct approach.

   Step 1: State the pair of hypotheses.

   Right-tailed test: H_0: μ = $12,611.84

   H_1: μ > $12,611.84

   Step 2: Identify the rejection zone.

   For a right-tailed test with $\alpha = .05$, the critical value of Z is 1.65. The rejection zone is shown in Figure 7.8.

FIGURE 7.8 ● Rejection Zone for $Z > 1.65$ for Exercise Problem 2

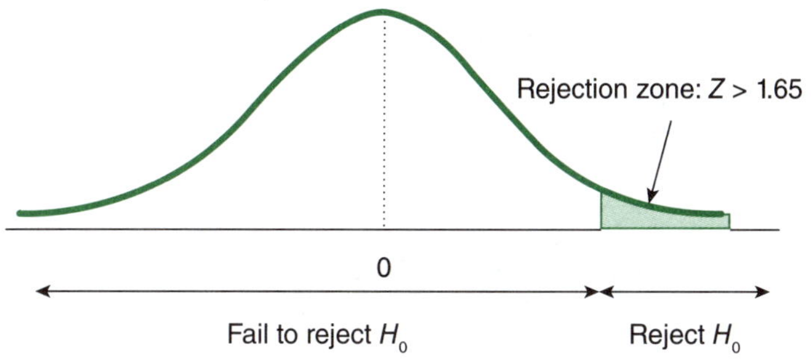

Step 3: Calculate the test statistic.

When comparing the average expenditure per pupil from high performing public schools to a population mean with a known σ, the use of a Z statistic is the correct approach.

List all the numbers given in the problem statement.

$\mu = 12611.84$

$\sigma = 4148.46$

$\bar{X} = 13526.91$

$n = 50$

$$Z = \frac{\bar{X} - \mu}{\sigma / \sqrt{n}} = \frac{13526.91 - 12611.84}{4148.46 / \sqrt{50}} = \frac{915.07}{586.68} = 1.56$$

Step 4: Make the correct conclusion.

Compare the calculated Z value = 1.56 from Step 3 and the rejection zone $Z > 1.65$ from Step 2. The calculated Z does not fall in the rejection zone, so we fail to reject H_0. The evidence is not strong enough to support the claim that the average expenditure per pupil from high performing public schools was higher than $12,611.84.

3. The problem statement asks about the truth of an advertisement. Therefore, the H_0 is that we assume the company is telling the truth unless the evidence is strong enough to prove otherwise. A two-tailed test is appropriate in this case.

   Step 1: State the pair of hypotheses.

   Two-tailed test: H_0: $\mu = 35$

   H_1: $\mu \neq 35$

   Step 2: Identify the rejection zone.

   For a two-tailed test with $\alpha = .10$, the critical value of Z is 1.65. The rejection zone, $|Z| > 1.65$, is shown in Figure 7.9.

FIGURE 7.9 ● Rejection Zone for |Z|> 1.65 for Exercise Problem 3

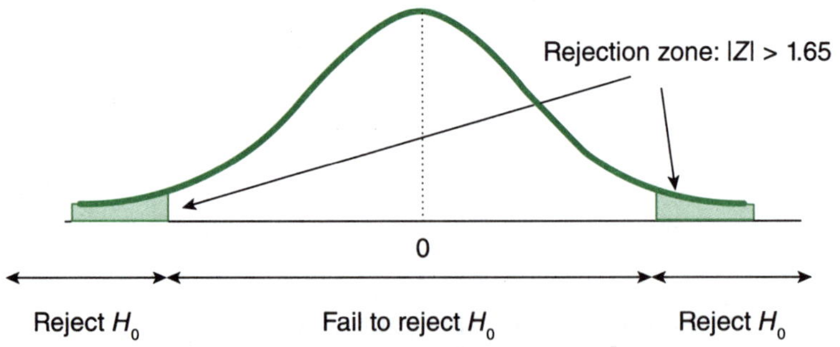

Step 3: Calculate the test statistic.

When comparing the average fuel efficiency with the population mean with a known σ, the use of a Z statistic is the correct approach.

List all the numbers given in the problem statement.

$\mu = 35$

$\sigma = 5$

$\bar{X} = 33$

$n = 40$

$$Z = \frac{\bar{X} - \mu}{\sigma / \sqrt{n}} = \frac{33 - 35}{5 / \sqrt{40}} = \frac{-2}{0.79} = -2.53$$

Step 4: Make a conclusion.

Compare the calculated Z value = –2.53 from Step 3 and the rejection zone, $|Z| > 1.65$ from Step 2. The calculated Z fall within the rejection zone, so we reject H_0. The evidence was strong enough to support the claim that the company was not truthful in its advertisement.

What You Learned

Hypothesis testing involves problem-solving skills. At the outset, one must understand the problem statement to determine if any key word indicates a clear direction of the test. If so, a one-tailed test is the correct approach. If not, a two-tailed test is the correct solution. Due to the limited statistical topics covered so far, only Z test is used to illustrate hypothesis testing. All examples presented in Chapter 7 deploy the four-step hypothesis test procedure.

Step 1: State the pair of hypotheses.

Two-tailed test: H_0: $\mu = xx$ (*xx* is the value specified as the population mean in question)

H_1: $\mu \neq xx$

Right-tailed test: H_0: $\mu = xx$

H_1: $\mu > xx$

Left-tailed test: H_0: $\mu = xx$

H_1: $\mu < xx$

Step 2: Identify the rejection zone.

The rejection zone is the standard we use to evaluate the strength of the evidence based on the data collected in a sample.

Two-tailed test rejection zone: $|Z| > Z_{\alpha/2}$

Right-tailed test rejection zone: $Z > Z_{\alpha}$

Left-tailed test rejection zone: $Z < -Z_{\alpha}$

Step 3: Calculate the statistic.

Comparing a sample mean and a population mean with a known σ, the Z statistic is the correct approach. According to the problem statement, gather the values of the sample mean, sample size, population mean, and population standard deviation and place them in the rightful spots in the Z formula.

$$Z = \frac{\bar{X} - \mu}{\frac{\sigma}{\sqrt{n}}}$$

Step 4: Make the correct conclusion.

Compare the calculated Z value from Step 3 and the rejection zone from Step 2. If the calculated Z falls in the rejection zone, we reject H_0. If the calculated Z does not fall in the rejection zone, we fail to reject H_0. The four-step hypothesis testing procedure will also be applicable in the following chapters with some minor adjustments when needed.

Key Words

Alternative hypothesis, H_1: The hypothesis that researchers turn to when the null hypothesis is rejected. The alternative hypothesis is also called the research hypothesis. 180

Effect size: The effect size is a standardized measure of the difference between the sample statistic and the hypothesized population parameter in units of standard deviation. 184

Hypothesis testing: Hypothesis testing is the standardized process to test the strength of the evidence for a claim about a population. 185

Mutually exclusive and collectively exhaustive: Mutually exclusive and collectively exhaustive means that events have no overlap—if one is true, the others cannot be true—and they cover all possible outcomes. 191

Null hypothesis, H_0: The null hypothesis is the no effect hypothesis. It states the effect that the researchers are trying to establish does not exist. 180

One-tailed test (directional test): A one-tailed test is a directional hypothesis test, which is usually conducted when the research topic suggests a consistent directional relationship with the backing of either theoretical reasoning or repeated empirical evidence. 186

Power: Power is defined as the sensitivity to detect an effect when an effect actually exists. The power of a statistical test is the probability of correctly rejecting H_0 when H_0 is false. Power is positively correlated with all of the three factors: (1) sample size, (2) α level, and (3) the effect size. 183

Rejection zone: The rejection zone is bounded by the critical value of a statistic. It is the standard used to judge the strength of the empirical evidence. When the calculated test statistics fall in the rejection zone, the correct decision is to reject H_0. 188

Two-tailed test (nondirectional test): A two-tailed test is a nondirectional hypothesis test, which is usually done when the research topic is exploratory or there are no reasons to expect consistent directional relationships from either theoretical reasoning or previous studies. 186

Type I error: A Type I error is a mistake of rejecting H_0 when H_0 is true. In other words, a Type I error is to falsely claim an effect when the effect actually does not exist. The symbol α represents the probability of committing a Type I error. 182

Type II error: A Type II error is a mistake of failing to reject H_0 when H_0 is false. In other words, a Type II error is failing to claim an effect when the effect actually exists. The symbol β represents the probability of committing a Type II error. 182

Learning Assessment

Multiple Choice: Circle the Best Answer to Every Question

1. The statistical power refers to the sensitivity to
   a. reject the null hypothesis (H_0) when H_0 is actually true
   b. reject the null hypothesis (H_0) when H_0 is actually false
   c. fail to reject the null hypothesis (H_0) when H_0 is actually true
   d. fail to reject the null hypothesis (H_0) when H_0 is actually false
2. A Type II error refers to
   a. claim a treatment effect when the effect actually exists
   b. claim a treatment effect when the effect actually does not exist
   c. fail to claim a treatment effect when the effect actually exists
   d. fail to claim a treatment effect when the effect does not exist

3. A normal adult population is known to have a mean $\mu = 65$ on the WPA cognitive ability test. A higher score means better cognitive ability. A researcher studied the detrimental effects of chemotherapy on brain function, so she randomly selected a group of 16 patients who received chemo last month to measure their WPA score. What is the alternative hypothesis for this study?
   a. H_1: $\mu \geq 65$
   b. H_1: $\mu \leq 65$
   c. H_1: $\mu > 65$
   d. H_1: $\mu < 65$
4. In a one-tailed test, H_0: $\mu = 120$, H_1: $\mu < 120$, the rejection zone is
   a. on the right tail of the distribution
   b. on the left tail of the distribution
   c. in the middle of the distribution
   d. evenly split into two tails
5. Putting innocent people in jail is analogous to falsely claiming a research effect when the effect does not exist. Both are __________ errors.
   a. Type I
   b. Type II
   c. Type III
   d. Type IV
6. In a one-tailed test, H_0: $\mu = 10$, H_1: $\mu > 10$, the rejection zone is
   a. on the right tail of the distribution
   b. on the left tail of the distribution
   c. in the middle of the distribution
   d. evenly split into two tails
7. Type I error means that a researcher has
   a. claimed a treatment effect when the effect actually exists
   b. claimed a treatment effect when the effect actually does not exist
   c. failed to claim a treatment effect when the effect actually exists
   d. failed to claim a treatment effect when the effect does not exist
8. For a two-tailed test with $\alpha = .01$, the critical values that set the boundaries for the rejection zones are
   a. $Z > 1.96$
   b. $Z > 2.58$
   c. $|Z| > 1.96$
   d. $|Z| > 2.58$

9. The starting point of a hypothesis test is the null hypothesis, which
   a. states that the effect does not exist
   b. is denoted as H_1
   c. is always stated in terms of sample statistics
   d. states that the effect exists

10. A Type I error means that a researcher has
    a. rejected the null hypothesis (H_0) when H_0 is actually true
    b. rejected the null hypothesis (H_0) when H_0 is actually false
    c. accepted the null hypothesis (H_0) when H_0 is actually true
    d. accepted the null hypothesis (H_0) when H_0 is actually false

11. A researcher is conducting a study to evaluate a program that claims to increase short-term memory capacity. The short-term memory capacity score is normally distributed with $\mu = 7$. Both theoretical reasoning and previous empirical research have shown a strong positive effect of this kind of program. Which of the following is the correct statement of the alternative hypothesis H_1?
    a. $\mu \neq 7$
    b. $\mu = 7$
    c. $\mu > 7$
    d. $\mu < 7$

Free Response Questions

12. Assume that anxiety scores as measured by an anxiety assessment inventory are normally distributed with $\mu = 20.5$ and $\sigma = 4.3$ A sample of 45 patients who were undergoing treatment for anxiety was randomly selected, and their mean anxiety score $\overline{X} = 22$. Was the average anxiety score for patients who are under treatment significantly different from 20.5? Assume $\alpha = .10$.

13. Adult male's body mass index (BMI) is normally distributed in the United States with a mean $\mu = 25.6$ and a standard deviation $\sigma = 3.9$. Researchers randomly selected $n = 42$ men from the population. What is the standard error of the mean for the BMI in this sample?

14. National data on student loans indicate that the student loan amount has a mean $\mu = \$31,172$. The distribution of student loans is approximately normal with a standard deviation $\sigma = \$25,643$. A random sample of 35 students from a private university reported a mean student debt load of \$38,750. Is the average student loan from this private university higher than \$31,172? Assume $\alpha = .05$.

15. According to the National Association of Builders, the average single-family home size in the United States is normally distributed with a mean $\mu = 2,392$ square feet and a standard deviation $\sigma = 760$ square feet. Thirty-six single-family houses were randomly selected from Cleveland, and the mean size was 2,025 square feet. Was the average square footage of houses in Cleveland smaller than 2,392 square feet? Assume $\alpha = .05$.

Answers to Pop Quiz Questions

1. a
2. d
3. d
4. d
5. c
6. a

8

ONE-SAMPLE *t* TEST AND DEPENDENT-SAMPLE *t* TEST

Learning Objectives

After reading and studying this chapter, you should be able to do the following:

- Explain how a *t* test differs from a *Z* test
- Describe the purpose of a one-sample *t* test, and how it is conducted
- Name the three pieces of information needed to identify the critical *t* value that set the boundary of rejection zone in a *t* test
- Describe what counterbalance is in conducting a dependent-sample *t* test and explain why it is important
- Describe the four-step hypothesis procedure for one-sample *t* test and dependent-sample *t* test
- Conduct one-sample *t* test and dependent-sample *t* test in Excel
- Calculate and interpret confidence intervals
- Calculate and interpret effect sizes

WHAT YOU KNOW AND WHAT IS NEW

In Chapter 7, you learned to conduct a hypothesis test using the *Z* test, $Z = \frac{\bar{X} - \mu}{\sigma / \sqrt{n}}$, to compare a sample mean with the hypothesized population mean when σ is known. The general statistical test formula is expressed as (Sample statistic – Population parameter)/(Standard error of the statistic). In *Z* tests, the population standard deviation, σ, is always readily available in the problem statement to allow the calculation of standard error of the mean, $\sigma / \sqrt{n}$. However, in reality, σ is mostly unknown in a population. Instead of giving up comparing sample means with population means altogether, statisticians figured out how to modify the process to come up with a close-enough process to estimate the significance of the difference between a sample mean and a population mean.

The slightly modified tests are called one-sample *t* tests. In general, one-sample *t* tests maintain the main features of the *Z* test to perform comparisons between sample means and population means. Frankly, *t* tests bring us one step closer to the reality where we can test the significance of the difference between a sample mean and its corresponding population mean when σ is unknown. Since σ is unknown, the sample standard deviation, *s*, is used as the best estimate of σ. The basic formula for a *t* test can be expressed as *t* = (Sample statistic – Population parameter)/(Estimated standard error of the statistic).

Dependent-sample *t* tests are very similar to the one-sample *t* tests. Dependent-sample *t* tests are used to calculate the difference in the same sample before and after a treatment or the difference in a variable between two samples with an explicit one-on-one paired relationship. Once the differences are calculated, you will need to calculate the mean, standard deviation, and standard error of the differences to conduct a hypothesis test for dependent samples. Because there is only one set of differences, the process of dependent-sample *t* test is essentially identical as one-sample *t* test.

THE UNKNOWN σ AND CONDUCTING THE ONE-SAMPLE *t* TEST

Up to this point in this book, in all the examples and homework problems when a sample mean is compared with the population mean, the population standard deviation, σ, has been readily provided. However, in reality, σ, in most cases, is unknown. To work around this problem, *s* is used as the best estimate of σ. The sample variance, s^2, is an unbiased estimate of the population variance, σ^2. The sample variance formula was first discussed in Chapter 3. Let's repeat it here again.

$$s^2 = \frac{\Sigma(X - \bar{X})^2}{df} = \frac{SS}{df} = \frac{SS}{n-1}$$

The degrees of freedom for sample variance are $n - 1$ because the sample mean, $\bar{X}$, is used to estimate the population mean in the process of calculating the variance. This restricts the number of values that are free to vary. Accordingly, 1 *df* is lost. Hence, the degrees of freedom for sample variance are $n - 1$. When s is used to estimate the unknown σ, the estimated standard error of the mean $s_{\bar{X}}$ is calculated simply by replacing σ with s in the standard error of the mean formula, $\sigma_{\bar{X}} = \sigma / \sqrt{n}$, and you get $s_{\bar{X}} = s / \sqrt{n}$. This estimated standard error of the mean is the denominator of the one-sample *t*-test formula. The good news is that the numerator of the one-sample *t* test is exactly the same as the numerator of the *Z* test. Putting the numerator and the denominator together, you get the one-sample *t*-test formula

$$t = \frac{\bar{X} - \mu}{s_{\bar{X}}} = \frac{\bar{X} - \mu}{\frac{s}{\sqrt{n}}}$$

with $df = n - 1$. The estimated standard error of the mean is an important distinction between *Z* tests and one-sample *t* tests.

THE *t* DISTRIBUTION WITH A SPECIFIC CURVE FOR EVERY DEGREE OF FREEDOM

When conducting a hypothesis test for a *Z* test, the calculated *Z* needs to be compared with the critical *Z* value that sets the boundary of the rejection zone, and this critical *Z* value can be found in the *Z* Table as stated in Chapters 5, 6, and 7. The same principle applies to conducting a *t* test. The calculated *t* needs to be compared with the critical *t* value that sets the boundary of the rejection zone. The difference is that there is only one *Z* curve, but there is a specific *t* curve for every degree of freedom. The critical value of *t* that sets the boundary of the rejection zone varies for every degree of freedom, and the variations are noticeable, especially for small sample sizes, $n < 30$.

Figure 8.1 shows five different *t* curves with five different degrees of freedom. The curve with the highest peak represents the standard normal distribution curve, which is the *Z* curve or a *t* curve with $df = \infty$; the other curves with lower peaks represent the *t* curves with $df = 5, 3, 2$, or 1. The curve with the lowest peak represents the *t* curve with $df = 1$, the curve with the second lowest peak represents the *t* curve with $df = 2$, and the curve with the third lowest peak represents the *t* curve with $df = 3$.

These *t* curves are especially developed to conduct comparisons between sample means and population means, particularly when the sample size is small, $n < 30$, and the

FIGURE 8.1 ● *t* Curves With Five Different Degrees of Freedom

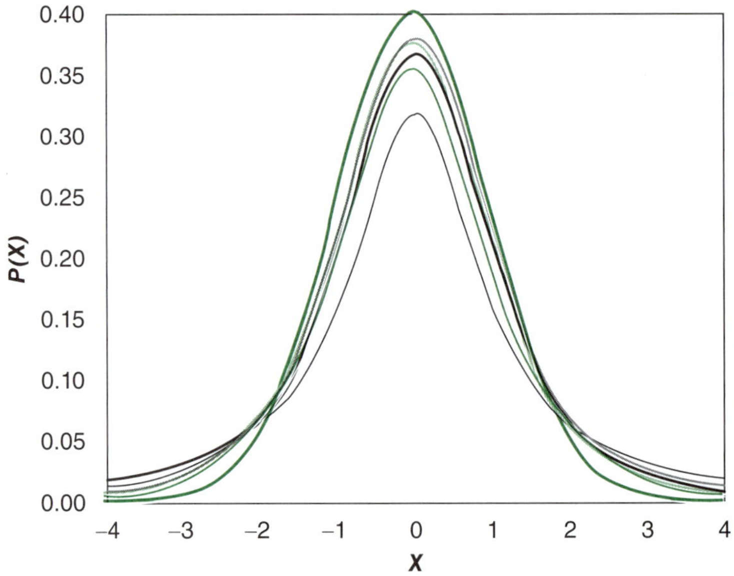

Note: ©IkamusumeFan (2013). This image is not altered and is licensed under the Creative Commons Attribution-Share Alike 3.0 Unported license.

population standard deviation, σ, is unknown. The slight modifications in *t* tests as compared with *Z* tests are listed below.

1. The population standard deviation, σ, is unknown. The sample standard deviation, *s*, is used as the best estimate of σ.
2. It is important to use the degrees of freedom, $df = n - 1$, to calculate the sample variance and sample standard deviation. The degrees of freedom for the one-sample *t* test are $n - 1$.
3. There is a unique *t*-distribution curve for every degree of freedom. It is important especially for small sample sizes, $n < 30$. The critical value that sets the boundary of the rejection zone varies for every degree of freedom.

Here is the formula for a **one-sample *t* test**, according to these modifications.

$$t = \frac{\bar{X} - \mu}{s_{\bar{X}}} = \frac{\bar{X} - \mu}{\frac{s}{\sqrt{n}}}, \text{ with } df = n - 1$$

where

$\bar{X}$ = the sample mean

μ = the population mean

n = the sample size

s = the sample standard deviation $= \sqrt{\frac{\Sigma(X - \bar{X})^2}{df}} = \sqrt{\frac{SS}{df}} = \sqrt{\frac{SS}{n-1}}$

$s_{\bar{X}}$ = the estimated standard error of the mean $= \frac{s}{\sqrt{n}}$

If you put the *t*-test formula and *Z*-test formula next to each other, you can see that the only difference is that σ in the *Z* test is replaced by *s* in the *t* test.

$$Z = \frac{\bar{X} - \mu}{\frac{}{\sqrt{n}}} \qquad t = \frac{\bar{X} - \mu}{\frac{}{\sqrt{n}}}$$

The critical values for *t* tests can be found in the *t* Distribution Table or ***t* Table** in Appendix B. The *t* Table provides critical values of *t* that set the boundaries of rejection zones for *t* tests given three pieces of information: (1) one-tailed or two-tailed tests, (2) the *df* of the test, and (3) the α level. The numbers in the top row of the *t* Distribution Table specify various α levels for one-tailed tests and the numbers in the second row specify various α levels for two-tailed tests. The numbers in the first column identify the degrees of freedom (*df*) for the *t* test, and all the numbers inside the *t* Table are critical values that set the boundaries of rejection zones. The shape of *t* curves is symmetric, as in the *Z* distribution, which means that when you cut the *t* curve in half along the peak (i.e., the location of mean, median, and mode), the left side is a mirror image of the right side. In small sample sizes, $n < 30$, *t* curves tend to have lower peaks and higher tails than the *Z* curve. As the sample sizes get bigger and bigger (especially, $n > 30$), *t* curves approximate the *Z* curve, and the differences between the two become less and less noticeable. It is appropriate to conduct *t* tests when the population is normally distributed or when sample size is large (i.e., $n > 30$). As stated in the central limit theorem, either of these two conditions allows the sampling distribution of the means to be normally distributed. Normal distribution is required to appropriately apply *t* tests. Table 8.1 shows a portion of the *t* Table. We will go over examples to illustrate how to identify critical *t* values that set the boundaries of rejection zones.

TABLE 8.1 ● A Portion of the t Table

	One-Tailed α Level							
	.25	.20	.15	.10	.05	.025	.01	.005
	Two-Tailed α Level							
df	.50	.40	.30	.20	.10	.05	.02	.01
1	1	1.376	1.963	3.078	6.314	12.71	31.82	63.66
2	0.816	1.061	1.386	1.886	2.92	4.303	6.965	9.925
3	0.765	0.978	1.25	1.638	2.353	3.182	4.541	5.841
4	0.741	0.941	1.19	1.533	2.132	2.776	3.747	4.604
5	0.727	0.92	1.156	1.476	2.015	2.571	3.365	4.032
6	0.718	0.906	1.134	1.44	1.943	2.447	3.143	3.707
7	0.711	0.896	1.119	1.415	1.895	2.365	2.998	3.499
8	0.706	0.889	1.108	1.397	1.86	2.306	2.896	3.355
9	0.703	0.883	1.1	1.383	1.833	2.262	2.821	3.25
10	0.7	0.879	1.093	1.372	1.812	2.228	2.764	3.169
11	0.697	0.876	1.088	1.363	1.796	2.201	2.718	3.106
12	0.695	0.873	1.083	1.356	1.782	2.179	2.681	3.055

EXAMPLE 8.1

Find the critical t value that sets the boundary of the rejection zone in each subquestion.

a. A two-tailed test with $\alpha = .01$ and $df = 8$

b. A right-tailed test with $\alpha = .10$ and $df = 11$

c. A left-tailed test with $\alpha = .05$ and $df = 5$

a. You might remember that you have to calculate $\alpha/2$ when conducting a two-tailed test using the Z test, because the Z Table only lists one-tailed probability. However, you don't have to do that in the t test. Because the t Table lists both one-tailed α and two-tailed α, you simply need to identify the column with the correct α level. Please make a note that one-tailed α is the same column as two-tailed 2α, such as one-tailed $\alpha = .05$ is the same as two-tailed $\alpha = .10$.

In the second row, find the vertical column corresponding to two-tailed $\alpha = .01$, and then find $df = 8$ in the first column; the value in the intersection is 3.355. The rejection zone for this two-tailed test is expressed as $|t| > 3.355$.

b. In the top row, find the vertical column corresponding to one-tailed $\alpha = .10$, and then find $df = 11$ in the first column; the value in the intersection is 1.363. The rejection zone for this right-tailed test is $t > 1.363$.

c. There are only positive *t* values reported in the *t* Table. Due to the symmetrical characteristic of the *t* distribution, negative *t* values can be identified through the *t* Table by adding the negative sign to the value reported in the *t* Table. In the top row, find the vertical column corresponding to one-tailed $\alpha = .05$, and then find $df = 5$ in the first column. The value in the intersection is 2.015. The rejection zone for this left-tailed test is $t < -2.015$.

Alternatively, you can use the **T.INV(probability,df)** function in Excel to find the critical values of *t*. This function is designed to identify the critical value based on the probability and *df* provided $P(t < t_c)$. Please note that the **T.INV** function is coded to identify the probability to the left of the critical value. I will demonstrate how to use **T.INV** function to identify, two-tailed, right-tailed, and left-tailed critical values of *t* tests.

a. You need to divide the α into two in a two-tailed test, $\alpha/2 = .005$ and $df = 8$. Select a blank cell in Excel, type "=**T.INV(.005,8)**", then hit **Enter**. You see the critical value for the left tail as –3.35539. Therefore, due to symmetry of *t* curves, the rejection zone for two-tailed is $|t| > 3.355$.

b. For a right-tailed *t* with $\alpha = .10$, the probability for the left tail is $1 - \alpha = .90$. Select a blank cell in Excel, type "=**T.INV(.90,11)**", then hit **Enter**. You see the critical value as 1.36343. The rejection zone for the right-tailed test is $t > 1.363$.

c. For a left-tailed test, select a blank cell in Excel, type "=**T.INV(.05,5)**", then hit **Enter**. You see the critical value for the left tail as –2.01505. The rejection zone is $t < -2.015$.

THE FOUR-STEP HYPOTHESIS TEST IN ONE-SAMPLE *T* TESTS

After learning the *t*-test formula and finding critical values in the *t* Table, let's conduct hypothesis tests in one-sample *t* tests. The four-step hypothesis testing procedure, as discussed in Chapter 7, is still applicable in conducting *t* tests. Here are the four-step hypothesis test procedure in the one-sample *t* tests.

Step 1: Explicitly state the pair of hypotheses.

Two-tailed test: H_0: $\mu = xx$ (xx is the value specified as the population mean in question)

H_1: $\mu \neq xx$

(Continued)

(Continued)

Right-tailed test:

H_0: $\mu = xx$

H_1: $\mu > xx$

Left-tailed test:

H_0: $\mu = xx$

H_1: $\mu < xx$

Step 2: Identify the rejection zone.

Rejection zone for a two-tailed test: $|t| > t_{\alpha/2}$, with $df = n - 1$

Rejection zone for a right-tailed test: $t > t_{\alpha}$, with $df = n - 1$

Rejection zone for a left-tailed test: $t < -t_{\alpha}$, with $df = n - 1$

Step 3: Calculate the test statistic.

$$t = \frac{\bar{X} - \mu}{s_{\bar{X}}} = \frac{\bar{X} - \mu}{\frac{s}{\sqrt{n}}}, \text{ with } df = n - 1$$

Step 4: Make the correct conclusion.

Compare the calculated t value in Step 3 to the rejection zone in step 2. If the calculated t falls within the rejection zone, we reject H_0. If the calculated t is not within the rejection zone, we fail to reject H_0.

EXAMPLE 8.2

ACT composite scores are normally distributed with a $\mu = 21$. A sample of 25 students who participated in experiential learning program was randomly selected from a local high school and their ACT score distribution was a $\bar{X} = 23$ and $s = 5.2$. Do the data support the claim that the average ACT composite score of students who participated in experiential learning program from this high school was higher than 21, using $\alpha = .05$?

The population standard deviation, σ, is unknown in this problem statement. It is not possible to conduct a Z test; therefore, a t test is an appropriate procedure. We use the four-step hypothesis testing process to conduct a t test.

Step 1: State the pair of hypotheses.

Based on the problem statement, "the average ACT score of students who participated in experiential learning program from this high school was higher than 21," thus the problem indicates a directional hypothesis. The key word "higher" indicates a right-tailed test. It is easy to start with

H_1: the alternative hypothesis in which the researchers are interested is $\mu > 21$. Then H_0 covers the equal sign, $\mu = 21$.

H_0: $\mu = 21$

H_1: $\mu > 21$

Step 2: Identify the rejection zone for the hypothesis test.

We need three pieces of information to identify the critical *t* value that sets the boundary of the rejection zone: (1) one-tailed test or two-tailed test, (2) the *df*, and (3) α level. We know that we are conducting a right-tailed test, $\alpha = .05$ and $df = n - 1 = 25 - 1 = 24$ based on the problem statement. According to the *t* Distribution Table, the critical *t* value is 1.711. The rejection zone for the right-tailed test is $t > 1.711$, as shown in Figure 8.2.

FIGURE 8.2 ● Using the *t* Table to Identify Rejection Zone $t > 1.711$

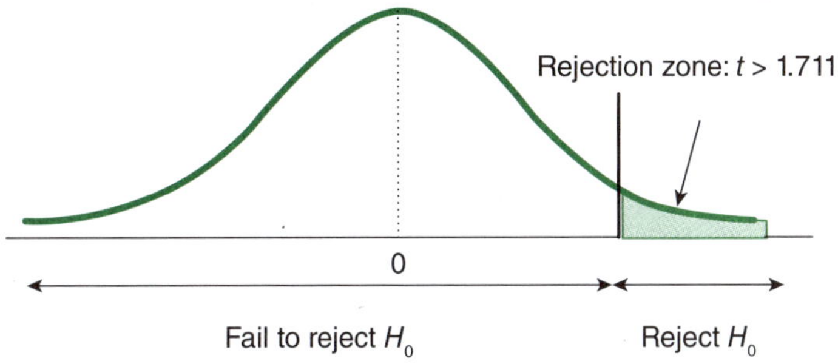

Step 3: Calculate the test statistic.

The problem statement asks for a comparison between the average ACT score of students who participated in experiential learning program in this high school and the population mean. Therefore, the *t* test is the correct statistic to accomplish this purpose. Let's list all the numbers stated in the problem statement, so as to separate out what we know and what we need to figure out.

What we know:

$\mu = 21$

$s = 5.2$

$\bar{X} = 23$

$n = 25$

We need to figure out *t*. Next, insert all the numbers in the *t*-test formula.

$$t = \frac{\bar{X} - \mu}{\frac{s}{\sqrt{n}}}$$

$$t = \frac{23 - 21}{\frac{5.2}{\sqrt{25}}} = \frac{2}{1.04} = 1.923$$

(Continued)

(Continued)

Step 4: Make the correct conclusion.

Compare the calculated $t = 1.923$ with the rejection zone $t > 1.711$. The calculated t falls in the rejection zone. Therefore, we reject H_0. At $\alpha = .05$, the evidence is strong enough to support the claim that the average ACT score of students who participated in experiential learning program from this high school was higher than 21.

Besides standardized exams, such approaches can also be applied in medical tests, especially using a group of patients with a particular medical condition compared with a known standard (i.e., a population mean) as shown in the next example.

EXAMPLE 8.3

HDL stands for "high-density lipoprotein" and the levels of HDL are normally distributed with a $\mu = 50$ mg/dl. A sample of 16 heart attack patients was randomly selected from a large hospital, and their HDL levels were measured with $\bar{X} = 40$ mg/dl and $s = 9$ mg/dl. Did the data support the claim that the average HDL of heart attack patients was lower than 50 mg/dl, using $\alpha = .01$?

The population standard deviation, σ, is unknown; therefore, a t test is an appropriate statistical procedure to answer this problem.

Step 1: State the pair of hypotheses.

The problem statement asks if "the average HDL of heart attack patients was lower than 50 mg/dl." "Lower" is a key word for a left-tailed test.

Left-tailed test: H_0: $\mu = 50$ mg/dl

H_1: $\mu < 50$ mg/dl

Step 2: Identify the rejection zone.

A left-tailed test with $\alpha = .01$ and $df = n - 1 = 16 - 1 = 15$.

Rejection zone: $t < -2.602$ (as shown in Figure 8.3).

FIGURE 8.3 ● Using the *t* Distribution Table to Identify Rejection Zone $t < -2.602$

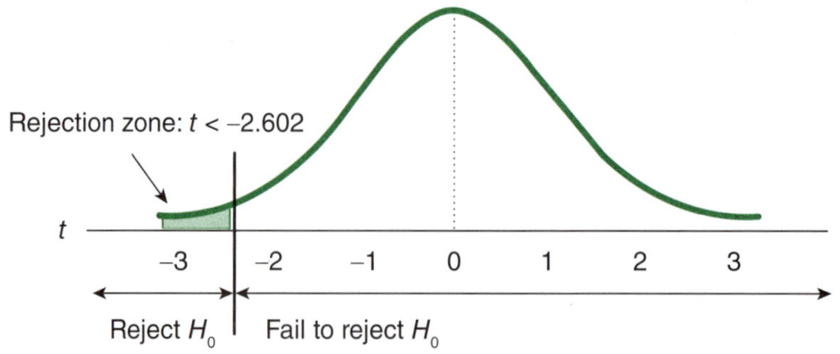

Step 3: Calculate the test statistic.

List all the numbers given in the problem statement.

$\mu = 50$ mg/dl

$s = 9$ mg/dl

$\bar{X} = 40$ mg/dl

$n = 16$

We need to figure out t. Next, insert all the numbers in the t-test formula.

$$t = \frac{\bar{X} - \mu}{\frac{s}{\sqrt{n}}}$$

$$t = \frac{40 - 50}{\frac{9}{\sqrt{16}}} = \frac{-10}{2.25} = -4.444$$

Step 4: Make the correct conclusion.

Compare the calculated $t = -4.444$ with the rejection zone $t < -2.602$. The calculated t falls in the rejection zone. Therefore, we reject H_0. At $\alpha = .01$, the evidence was strong enough to support the claim that the average HDL of heart attack patients was lower than 50 mg/dl.

The t tests allow us to compare a sample mean $\bar{X}$ with a hypothesized population mean without requiring σ. The sample standard deviation, s, is used as the best estimate of the σ. Both $\bar{X}$ and s can be calculated directly from reported individual scores in a sample. Of course, they require the formulas that were covered in Chapter 3. You'd better refresh your memory on calculating sample mean and sample standard deviation before attempting to solve such problems.

A Trip Down the Memory Lane

The sample mean and standard deviation are basic descriptive statistics that remain important regardless of what statistics you do. When given all values from a sample, you are able to calculate the sample mean by adding all the values and dividing by the sample size, $\bar{X} = \Sigma X / n$. Calculating sample standard deviation starts with $SS = \Sigma(X - \bar{X})^2$, and then $s = \sqrt{SS/df}$.

EXAMPLE 8.4

Reaction time is defined as the time taken to complete a task. Depending on the complexity of the task, the average time to perform a task differs. The reaction time for seeing a green light and pushing a button is normally distributed with a $\mu = 220$ milliseconds. A group of nine student athletes was randomly selected, and their reaction times were measured at 273, 289, 138, 121, 285, 252, 129, 120, and 130. Did student athletes' reaction times differ from 220 milliseconds, using $\alpha = .10$?

The population standard deviation, σ, is unknown; therefore, we use the four-step hypothesis testing process to conduct a *t* test to answer this problem.

Step 1: State the pair of hypotheses.

The problem statement asks, "Did student athletes' reaction times differ from 220 milliseconds?" This statement did not contain any information regarding direction. Therefore, a nondirectional hypothesis is the correct approach.

H_0: $\mu = 220$

H_1: $\mu \neq 220$

Step 2: Identify the rejection zone for the hypothesis test.

From Step 1, we know that we are conducting a two-tailed test. The problem statement specifies $\alpha = .10$, and $df = n - 1 = 9 - 1 = 8$. According to the *t* Table, the critical *t* value is 1.86. The rejection zone is $|t| > 1.86$, as shaded in Figure 8.4.

FIGURE 8.4 ● Using the *t* Distribution Table to Identify Rejection Zone $|t| > 1.86$

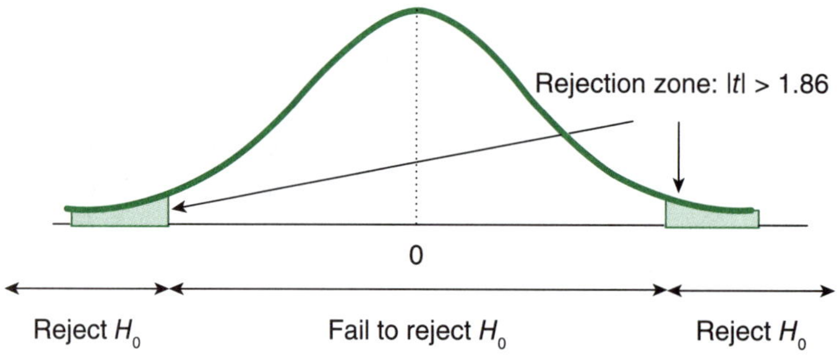

Step 3: Calculate the statistic.

What we know:

$\mu = 220$ milliseconds

$n = 9$

When given all measures in a sample, you can calculate $\bar{X}$ and s needed for conducting a t test. You need to refresh your memory on what you learned in Chapter 3.

$$\bar{X} = \frac{\Sigma X}{n}$$

$$SS = \Sigma(X - \bar{X})^2$$

$$s = \sqrt{\frac{\Sigma(X - \bar{X})^2}{df}} = \sqrt{\frac{SS}{df}} = \sqrt{\frac{SS}{n-1}}$$

Let's put all the measures in a table and demonstrate the step-by-step process in Table 8.2.

First, we need to figure out the sample mean, $\bar{X}$. Then we construct the second column $X - \bar{X}$, and the third column $(X - \bar{X})^2$, and add all the numbers in the third column to obtain the sum of squares, SS.

$$\bar{X} = \Sigma X / n = 1737 / 9 = 193$$

$$SS = \Sigma(X - \bar{X})^2 = 49164$$

$$s = \sqrt{\frac{\Sigma(X - \bar{X})^2}{df}} = \sqrt{\frac{SS}{df}} = \sqrt{\frac{SS}{(n-1)}} = \sqrt{\frac{49164}{8}} = \sqrt{6145.5} = 78.4$$

Next, insert all the numbers in the t-test formula.

$$t = \frac{\bar{X} - \mu}{\frac{s}{\sqrt{n}}}$$

$$t = \frac{193 - 220}{\frac{78.4}{\sqrt{9}}} = \frac{-27}{26.1} = -1.034$$

TABLE 8.2 ● Nine Student Athletes' Reaction Times

X	$X - \bar{X}$	$(X - \bar{X})^2$
273	80	6,400
289	96	9,216
138	–55	3,025
121	–72	5,184
285	92	8,464
252	59	3,481
129	–64	4,096
120	–73	5,329
130	–63	3,969
1,737		49,164

Step 4: Make the correct conclusion.

Compare the calculated $t = -1.034$ with the rejection zone $|t| > 1.86$. The calculated t did not fall in the rejection zone. Therefore, we failed to reject H_0. At $\alpha = .10$, the evidence was not strong enough to support the claim that the average reaction time of student athletes differed from 220 milliseconds.

After going through several examples on conducting t tests, you should get a sense that t tests are more realistic than Z tests because they don't require the population standard deviation, σ. Using the sample standard deviation, s, as the best estimate of σ allows scientists to compare a particular sample mean and a hypothesized population mean to obtain useful information. However, if σ is known or provided, there is no need to estimate its value, so a Z test should apply in such a situation.

It is worth repeating that, for every degree of freedom, there is a specific t curve. That is the reason why it needs three values—(1) one-tailed test versus two-tailed test, (2) α level, and (3) *df*—to identify the critical values of t that set off the boundaries of rejection zones.

POP QUIZ

1. Which one of the following pieces of information is *not* needed to identify the critical values of t that set off the boundaries of the rejection zones?
   a. The significance level, α
   b. One-tailed test versus two-tailed test
   c. The degrees of freedom for the t test
   d. The population standard deviation, σ
2. When the problem statement provides the population mean, μ, the sample mean, $\bar{X}$, the population standard deviation, σ, the sample standard deviation, s, and the sample size, n, to compare the sample mean with the population mean, which one of the following is the correct approach to conduct hypothesis tests?
   a. $Z = \frac{X - \mu}{\sigma}$
   b. $Z = \frac{X - \bar{X}}{s}$
   c. $Z = \frac{\bar{X} - \mu}{\frac{\sigma}{\sqrt{n}}}$
   d. $t = \frac{\bar{X} - \mu}{\frac{s}{\sqrt{n}}}$

CONFIDENCE INTERVALS IN ONE-SAMPLE *t* TESTS

Two of the most common types of statistical inferences are hypothesis tests and confidence intervals. You have learned the standard procedure to conduct a hypothesis test for a claim about a population. Now it is time to introduce **confidence interval** (CI). A CI is defined as an interval of values calculated from sample data and such an interval is likely to include the population parameter. CIs provide additional information that researchers' can rely on to draw conclusions. The calculated CIs can convey the quality of the estimation or statistical inference. As an inferential technique, the CI does not require a hypothesis about the population mean and does not have to be used in conjunction with a hypothesis. A CI can be appropriately applied when the sampling distribution of the means is normally distributed. Constructing a CI involves two steps: first, obtaining a point estimate, and second, calculating an interval estimate.

Point Estimate

A **point estimate** is a single value from the sample statistics to estimate a population parameter. For example, to estimate a population mean, a sample mean is the best value for that purpose. When a school district is asked to provide a single value to represent its 12th graders' SAT scores, the best estimate is to randomly select a group of 12th graders in the district to calculate their mean SAT score. The point estimate is the sample mean, $\bar{X}$.

However, this single value does not provide the quality of the estimate. It may be close to or far away from the population parameter. Therefore, statisticians extend the point estimate to an interval estimate.

Interval Estimate

An **interval estimate** refers to a range of values used to estimate a population parameter. The width of the interval estimate depends on the confidence level. A **confidence level** is $(1 - \alpha)$, which is the complement to α, the significance level. When $\alpha = .05$ is used as the default value, a 95% CI is the default width. The confidence level is customarily expressed as a percentage, so it is expressed as $(1 - \alpha)100\%$. CIs are calculated as the point estimate $\bar{X}$ ± the margin of error, *E*. The margin of error is calculated as the critical value of *t* for a two-tailed test, $t_{\alpha/2}$, multiplying the estimated standard error of the mean, $s/\sqrt{n}$.

$$E = t_{\alpha/2}\left(\frac{s}{\sqrt{n}}\right)$$

The symbol $t_{\alpha/2}$ denotes the critical value of *t* for a two-tailed test with a significance level = α. The $\alpha/2$ is simply a *subscript*, not a math operation. Since the margin of error is calculated by the multiplication of $t_{\alpha/2}$ and the estimated standard error of the mean, as α decreases, the confidence level $(1 - \alpha)$ increases, and the CI widens. Therefore, a 99% CI is wider than a 95% CI and a 95% CI is wider than a 90% CI when the estimated standard error of the mean remains the same.

The CIs are calculated as the point estimate ± the margin of error. In estimating a CI for a population mean, the CI centers at $\bar{X}$. The upper limit of the CI is calculated by adding the margin of error, $\bar{X} + t_{\alpha/2}(s/\sqrt{n})$, and the lower limit of the CI is calculated by subtracting the margin of error, $\bar{X} - t_{\alpha/2}(s/\sqrt{n})$. As you can see, CI always centers at the point estimate, then pushes it to the right by adding the margin of error to reach the upper limit and pushes it to the left by subtracting the margin of error to reach the lower limit. That's the reason why CI always uses the two-tailed $t_{\alpha/2}$. Therefore, putting the point estimate and the margin of error together creates the CI for μ.

$$\text{CI for } \mu = \bar{X} \pm t_{\alpha/2}\left(\frac{s}{\sqrt{n}}\right)$$

The rounding rule for constructing a CI is to report the lower and upper limits with one more decimal place than reported in the sample mean. Let's demonstrate the process by using an example.

EXAMPLE 8.5

Let's use the reaction time example stated in Example 8.4. The reaction time for seeing a green light and pushing a button is normally distributed with a $\mu = 220$ milliseconds. A group of nine student athletes was randomly selected, and their reaction times were measured at 273, 289, 138, 121, 285, 252, 129, 120, and 130.

a. Calculate and interpret the 90% CI for the student athletes' reaction time.
b. Compare the 90% CI with the result from the two-tailed hypothesis test with $\alpha = .10$.

a. When given all measures in a sample, you have to calculate $\bar{X}$ and s before constructing a 90% CI. We can use the numbers calculated in Example 8.4 to construct the 90% CI.

$$\bar{X} = \Sigma X / n = 1737 / 9 = 193$$

$$SS = \Sigma(X - \bar{X})^2 = 49164$$

$$s = \sqrt{\frac{\Sigma(X-\bar{X})^2}{df}} = \sqrt{\frac{SS}{df}} = \sqrt{\frac{SS}{(n-1)}} = \sqrt{\frac{49164}{8}} = 78.4$$

These are the same values as calculated in Table 8.2. A little repetition is good for learning statistics.

The point estimate of μ is $\bar{X} = 193$.

The margin of error is $t_{\alpha/2}\left(\frac{s}{\sqrt{n}}\right)$.

Let's break the calculation down to $t_{\alpha/2}$ and $\frac{s}{\sqrt{n}}$.

With a 90% CI, it means $(1 - \alpha) = .90$; therefore, $\alpha = .10$.

The critical value of t in a two-tailed test with $\alpha = .10$ is the same as a one-tailed test with $\alpha = .05$. The symbol $t_{\alpha/2}$ denotes the critical value of t for a two-tailed test with significance level $= \alpha$. The critical value of t for a two-tailed test, $\alpha = .10$ and $df = 8$, is 1.86.

$$\frac{s}{\sqrt{n}} = \frac{78.4}{\sqrt{9}} = 26.13$$

Now you can calculate the margin of error, $t_{\alpha/2}\frac{s}{\sqrt{n}} = 1.86\frac{78.4}{\sqrt{9}} = 48.6$

$$90\%\ \text{CI of } \mu = \bar{X} \pm t_{\alpha/2}\frac{s}{\sqrt{n}} = 193 \pm 48.6$$

The lower limit and the upper limit of the 90% CI are reported in brackets [144.4, 241.6] with the lower limit on the left and the upper limit on the right, inside the brackets. The 90% CI [144.4, 241.6] is constructed from the sample of nine student athletes to make an inference about the population mean for all student athletes. The correct way to interpret the 90% CI is that we are 90% confident that the population mean for all student athletes is included in this interval. When the hypothesized value of the H_0 (e.g., $\mu = 220$) is included in the brackets [144.4, 241.6], we conclude that there are no significant differences between student athletes' reaction time and the general population's.

b. The result from the two-tailed hypothesis test with $\alpha = .10$ as reported in Example 8.4 is that the calculated t ($t = -1.034$) did not fall in the rejection zone ($|t| > 1.86$). Therefore, we failed to reject H_0. At $\alpha = .10$, the evidence was not strong enough to support the claim that student athletes' average reaction time differed from 220 milliseconds. The outcome from a 90% CI is consistent with the result from the two-tailed hypothesis test with $\alpha = .10$.

In general, conducting a two-tailed hypothesis test with significance level $= \alpha$ generates the same conclusion as constructing a $(1 - \alpha)100\%$ CI. The range of values of the CI provides additional information for interpretation. When the hypothesized population mean under H_0 is within the CI, there is no significant difference between the sample mean and the population mean. When the population mean is not within the CI, there is a significant difference between the sample mean and the population mean.

Let's go through another example to provide more practice.

EXAMPLE 8.6

The licensed drivers' ages in a suburb are normally distributed with a $\mu = 48$. A sample of nine drivers was randomly selected because of their moving violation citations. Their ages were 17, 28, 33, 45, 65, 21, 19, 38, and 40.

(Continued)

(Continued)

a. What was the 95% CI for the average age of drivers who received moving violation citations?

b. Interpret the 95% CI.

a. Use the following formulas to calculate the $\bar{X}$ and s.

$$\bar{X} = \Sigma X / n$$

$$SS = \Sigma(X - \bar{X})^2$$

$$s = \sqrt{\frac{\Sigma(X - \bar{X})^2}{df}} = \sqrt{\frac{SS}{df}} = \sqrt{\frac{SS}{(n-1)}}$$

TABLE 8.3 ● Nine Drivers' Ages

X	$X - \bar{X}$	$(X - \bar{X})^2$
17	−17	289
28	−6	36
33	−1	1
45	11	121
65	31	961
21	−13	169
19	−15	225
38	4	16
40	6	36
306		1,854

Let's put all the measures in a table and demonstrate the step-by-step process in Table 8.3.

We calculate $(X - \bar{X})$ in the second column, and $(X - \bar{X})^2$ in the third column, and add all the numbers in the third column to obtain *SS*.

First, we need to figure out the sample mean $\bar{X} = \Sigma X/n = 306/9 = 34$. The point estimate of μ is $\bar{X} = 34$.

$$SS = \Sigma(X - \bar{X})^2 = 1854$$

$$s = \sqrt{\frac{\Sigma(X - \bar{X})^2}{df}} = \sqrt{\frac{SS}{df}} = \sqrt{\frac{SS}{(n-1)}} = \sqrt{\frac{1854}{8}} = \sqrt{231.75} = 15.2$$

The critical *t* value for a two-tailed test, $\alpha = .05$, $df = 8$ is $t_{\alpha/2} = 2.306$

The margin of error is $t_{\alpha/2}\frac{s}{\sqrt{n}} = 2.306\frac{15.2}{\sqrt{9}} = 11.7$

95% CI for $\mu = \bar{X} \pm t_{\alpha/2}\frac{s}{\sqrt{n}} = 34 \pm 11.7 = [22.3, 45.7]$

b. The 95% CI [22.3, 45.7] for the average age of drivers who were cited for moving violations did not include the population mean under H_0, $\mu = 48$. There was a significant difference between the average age of drivers who received moving violations and the population mean for all licensed drivers in this suburb. The average age of drivers who received moving violations was younger than the average age of all licensed drivers.

POP QUIZ

3. When the 95% CIs do *not* include the hypothesized population parameter under H_0, the correct interpretation is that
   a. there is significant difference between the sample mean and the population mean at $\alpha = .05$.
   b. there is no significant difference between the sample mean and the population mean at $\alpha = .05$.
   c. we could not reach any conclusion with 95% CIs.
   d. the evidence is not strong enough to reject H_0.
4. The 90% CIs of *t* tests provide the same conclusions as
   a. right-tailed *t* tests with $\alpha = .10$.
   b. two-tailed *t* tests with $\alpha = .10$.
   c. left-tailed *t* tests with $\alpha = .10$.
   d. two-tailed *t* tests with $\alpha = .05$.

INTRODUCING DEPENDENT-SAMPLE *t* TESTS

The **dependent-sample *t* test** is designed for the comparisons of two interval or ratio measures from the same sample, such as measuring the score on a science test before and after a group of students go through an experimental teaching method. The differences of the before and after test scores demonstrate the effectiveness of this experimental teaching method. Since the same group of students is measured repeatedly, this is the reason why dependent-sample t tests are also labeled as "within-subject" design or "repeated measures." In some rare cases, dependent-sample t tests can be applied to different samples only when the two samples have an explicit one-on-one paired relationship, such as job satisfaction ratings obtained from identical twins or expressed voting preferences from spouses. When dependent-sample t tests involve two separate samples, they are labeled as "paired-sample t tests" or "matched-sample t tests." The paired nature of the samples allows direct calculation of the differences between them; thus, the standard deviation of the differences and standard error of the differences can also be derived. Due to the fact that there is only one set of differences, the dependent-sample t tests follow exactly the same procedure as one-sample t tests.

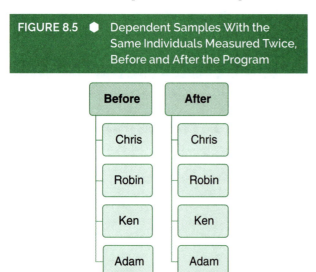

FIGURE 8.5 ● Dependent Samples With the Same Individuals Measured Twice, Before and After the Program

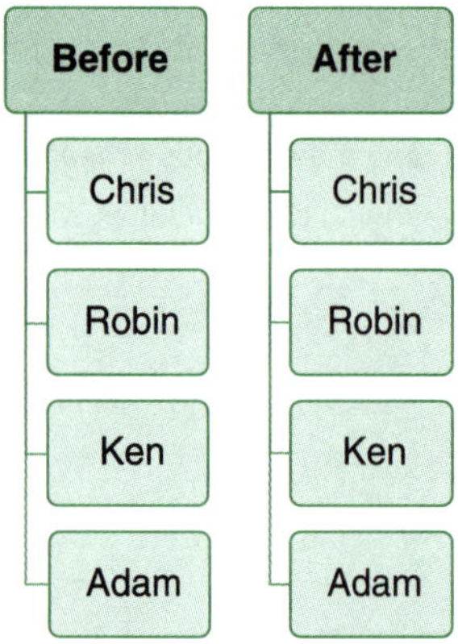

The advantage of using dependent samples is to remove the potentially confounding effect of individual differences because the measurements are taken from the same individuals. For example, randomly selecting four volunteers to participate in a diet program. These four participants' weights are measured before the program, and again after the program. The effect of the diet program is shown by comparing the participants' weight before and after the program, as shown in Figure 8.5. Measuring the weight difference before and after the diet program removes the confounding factor of individual differences and increases the statistics power. However, it is likely that fatigue, memory, and/or practice effects may influence the results when participants are measured repeatedly in some situations. Research participants might not always be available to be measured repeatedly. Attrition of research participants in a study using repeated measures can decrease the size of usable observations (i.e., sample size).

POP QUIZ

5. Which one of the following research studies calls for a dependent-sample t test?
   a. Comparing the psychological stress of married men versus single men
   b. Comparing the happiness of pet owners versus non-pet owners
   c. Comparing cholesterol levels for the same group of patients before and after they take a new drug for 6 months
   d. Comparing the financial savviness of home owners versus renters

CALCULATIONS AND THE HYPOTHESIS TESTING PROCESS IN DEPENDENT-SAMPLE *T* TESTS

The dependent-sample t tests start with calculating the difference scores between the before and after measurements to reflect the treatment effect (i.e., an experimental teaching method, a diet program, or a newly developed vaccine). We are going to go over the calculations and the hypothesis test process for the dependent-sample t tests in the following section.

Mean, Standard Deviation, and Standard Error of the Differences

The explicit one-on-one paired relationship of two measures either from the same sample or from two matched samples allows the direct calculation of the differences, D. For example, we calculate $D = X_{\text{after}} - X_{\text{before}}$, where X_{after} stands for the measure after the treatment and X_{before} stands for the measure before the treatment. It is logical to use the $X_{\text{after}} - X_{\text{before}}$ to show the treatment effect. When you add up every difference divided by the sample size, you get the mean difference.

$$\text{Mean difference} = \bar{D} = \frac{\Sigma D}{n}$$

Actually, when you treat the differences, D, as a variable X, the mean, sum of squares (SS), standard deviation, and standard error formulas are exactly the same as the formulas discussed in Chapter 3.

$$\text{Sum of squares of the differences} = SS_{\text{D}} = \Sigma(D - \bar{D})^2$$

$$\text{Standard deviation of the differences} = s_{\text{D}} = \sqrt{\frac{SS_{\text{D}}}{df}} = \sqrt{\frac{\Sigma(D - \bar{D})^2}{n-1}}$$

$$\text{Standard error of the differences} = s_e = \frac{s_{\text{D}}}{\sqrt{n}}$$

At this point, it should be clear that the same formulas used to calculate the mean, standard deviation, and standard error of a variable apply to the calculations of the mean, standard deviation, and standard error of the differences in a dependent-sample t test.

Hypothesis Testing for Dependent-Sample *t* Tests

The purpose of a hypothesis test in dependent samples is to investigate whether there are significant differences between the measures before and after the treatment. The focus of the test is the differences. The procedure for conducting a hypothesis test for dependent samples is similar to the four-step hypothesis testing process in one-sample t tests, because there is only one set of differences.

The four-step hypothesis test for dependent-sample t tests:

Step 1: Explicitly state the pair of hypotheses.

There are two-tailed tests, right-tailed tests, and left-tailed tests for dependent-sample t tests. Based on the problem statement, you need to judge whether a directional hypothesis test or a nondirectional hypothesis test is appropriate. Then choose one of the following three options accordingly.

Two-tailed test: $H_0: \mu_D = 0$

$H_1: \mu_D \neq 0$

Right-tailed test: $H_0: \mu_D = 0$

$H_1: \mu_D > 0$

Left-tailed test: $H_0: \mu_D = 0$

$H_1: \mu_D < 0$

Step 2: Identify the rejection zone.

You need three pieces of information to identify the critical values that set the boundaries of the rejection zone: a one-tailed or a two-tailed test, the *df*, and the α level.

Rejection zone for a two-tailed test: $|t| > t_{\alpha/2}$, with $df = n - 1$

Rejection zone for a right-tailed test: $t > t_{\alpha}$, with $df = n - 1$

Rejection zone for a left-tailed test: $t < -t_{\alpha}$, with $df = n - 1$

Step 3: Calculate the test statistic.

$$\text{The general principle of a } t \text{ test} = \frac{\text{Sample statistic} - \text{Population parameter}}{\text{Estimated standard error of the statistic}}$$

In dependent samples,

$$t = \frac{\bar{D} - \mu_D}{s_e}$$

Because $\mu_D = 0$ under the null hypothesis, the t formula can be further simplified to

$$t = \frac{\bar{D}}{s_e}$$

Step 4: Make the correct conclusion.

If the calculated *t* from Step 3 is within the rejection zone, we reject H_0. If the calculated *t* from Step 3 is not within the rejection zone, we fail to reject H_0.

Let's use an example to go through the four-step hypothesis test procedure.

EXAMPLE 8.7

Researchers conducted a study to investigate the effect of a diet program. Table 8.4 shows the weights of the four participants—Chris, Robin, Ken, and Adam—before the diet program and after the diet program. Conduct a hypothesis test at $\alpha = .10$ to investigate whether there was an effect of the diet program.

TABLE 8.4 ● Four Participants' Weights Before and After a Diet Program

Participant	Before	After
Chris	140	131
Robin	187	190
Ken	179	162
Adam	210	185

Step 1: Explicitly state the pair of hypotheses.

There is no key word to indicate a direction in the problem statement, so a two-tailed hypothesis test is appropriate.

$H_0: \mu_D = 0$

$H_1: \mu_D \neq 0$

Step 2: Identify the rejection zone.

Based on the problem statement, we use a two-tailed test, $\alpha = .10$, and $df = n - 1 = 4 - 1 = 3$. The critical value in the *t* Table is 2.353, so the rejection zone for a two-tailed test is $|t| > 2.353$.

Step 3: Calculate the test statistic.

The calculation for mean, standard deviation, and standard error of the differences is shown step-by-step in Table 8.4a, where the difference score $D = X_{after} - X_{before}$.

TABLE 8.4a ● Calculation of Mean, Standard Deviation, and Standard Error of the Differences for the Diet Program Participants

Participant	Before	After	D	$(D - \bar{D})$	$(D - \bar{D})^2$
Chris	140	131	−9	$-9 - (-12) = 3$	9
Robin	187	190	3	$3 - (-12) = 15$	225
Ken	179	162	−17	$-17 - (-12) = -5$	25
Adam	210	185	−25	$-25 - (-12) = -13$	169
			−48		428

(Continued)

(Continued)

$$\bar{D} = \frac{\Sigma D}{n} = \frac{-48}{4} = -12$$

$$SS_D = \Sigma(D - \bar{D})^2 = 9 + 225 + 25 + 169 = 428$$

$$s_D = \sqrt{\frac{\Sigma(D - \bar{D})^2}{n-1}} = \sqrt{\frac{428}{3}} = \sqrt{142.67} = 11.94$$

$$s_e = s_D / \sqrt{n} = 11.94 / \sqrt{4} = 5.97$$

$$t = \frac{\bar{D}}{s_e} = \frac{-12}{5.97} = -2.010$$

Step 4: Make the correct conclusion.

The calculated $t = -2.010$ is not within the rejection zone. We thus failed to reject H_0 at $\alpha = .10$ level. The evidence was not strong enough to support the claim that the diet program had an effect on weight. The researchers should consider increasing the sample size in the future to evaluate the effectiveness of their diet program. Small sample size produced large standard error of the mean difference to make it difficult to reject H_0.

Whenever the problem statements ask whether a program has an effect without specifying a direction, such as a positive effect or a negative effect, you have to use a nondirectional hypothesis test (i.e., a two-tailed test). When the problem statements contain key words for a specific direction—either a positive effect, improvement, increasing scores, better performance, or a negative effect, detrimental effect, decreasing scores, or worse performance—you have to use a directional hypothesis test (i.e., a right-tailed test or a left-tailed test, accordingly). You have to understand exactly what the problem is asking to be able to conduct the correct hypothesis test.

EXAMPLE 8.8

Researchers had long suspected that texting has a detrimental effect on driving behavior. To avoid the unnecessary high risk of actually driving on the roads, driving behavior was measured via a driving simulator. Mistakes such as failing to stay within the lane, driving at least 10 miles above or below the speed limit, and failing to use the turn signal were automatically recorded by the simulator. To avoid individual differences in driving behavior, participants were tested twice in different conditions: "driving while texting" and "driving without texting." There were nine participants in the study. Their driving mistakes are reported in Table 8.5. Did the data support the claim that texting had a detrimental effect on driving behavior, using $\alpha = .05$?

You may follow the same formulas as stated in Example 8.7 to solve this problem or you can learn how to use Excel to conduct dependent-sample *t* tests. A solid understanding of the formulas helps you conduct tests using Excel correctly. Excel is especially useful when dealing with large samples with hundreds of cases.

The problem statement indicated participants were tested twice in different conditions, so the dependent-sample *t* test is the correct test. I am going to show you how to use Excel to conduct a dependent-sample *t* test.

1. Enter Table 8.5 data in Excel first, then click on the **Data** tab, then click on the **Data Analysis** on the upper right side. In the **Analysis Tools** Window, select "**t-Test Paired Two Sample for Means**", then click **OK** as shown in Figure 8.6.

TABLE 8.5 ● Driving Mistakes When Driving While Texting and Driving Without Texting

Texting	No Texting
35	23
46	39
23	12
33	10
29	14
41	29
31	17
22	13
37	23

FIGURE 8.6 ● Screenshot of the t-Test: Paired Two Sample for Means in Excel for Example 8.8

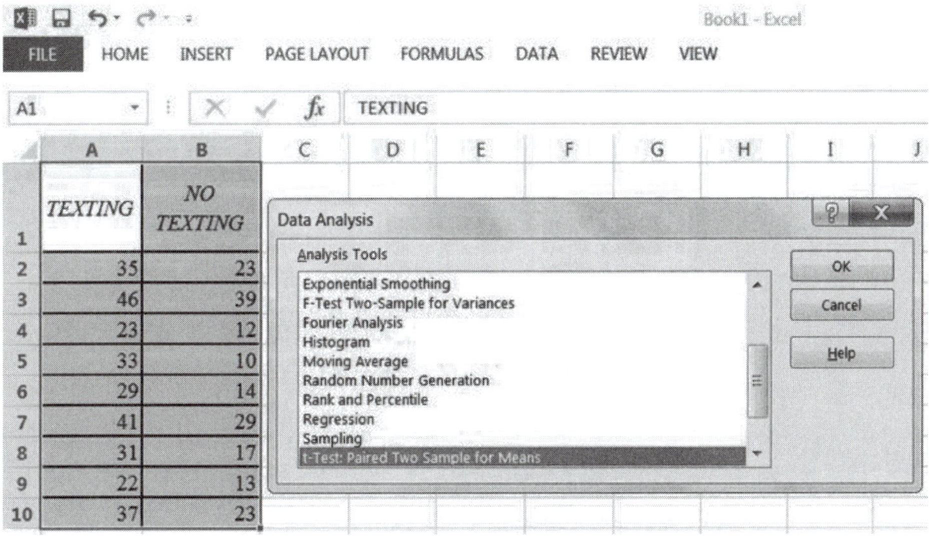

(Continued)

(Continued)

2. A window pops up to ask about the data range for the two groups. The beauty of conducting statistical analyses using Excel is that you are able to analyze large samples as easy as analyze a handful of observations. Highlight the data range (from first data point to the last data point) for texting condition and place them in the **Variable 1** then highlight the data range for no texting condition and place them in the **Variable 2** as shown in Figure 8.7. Since the variable labels (as shown in row 1) are included in the input, make sure to check the box for **Labels**. Doing so ensure the results will be properly labeled. The difference is automatically calculated as **(Variable 1 – Variable 2)** in Excel. Then click **OK**.

FIGURE 8.7 ● Screenshot of Variable Set Up for Example 8.8

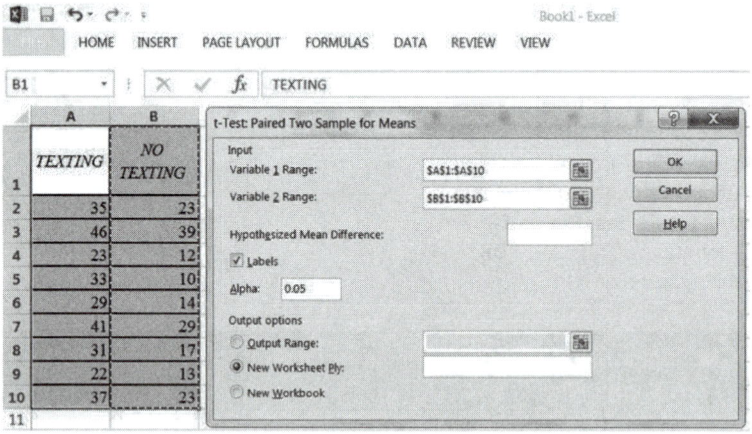

3. The results of the dependent-sample *t* test appear in a separate worksheet in Excel as shown in Figure 8.8.

FIGURE 8.8 ● Screenshot of the Results in Example 8.8

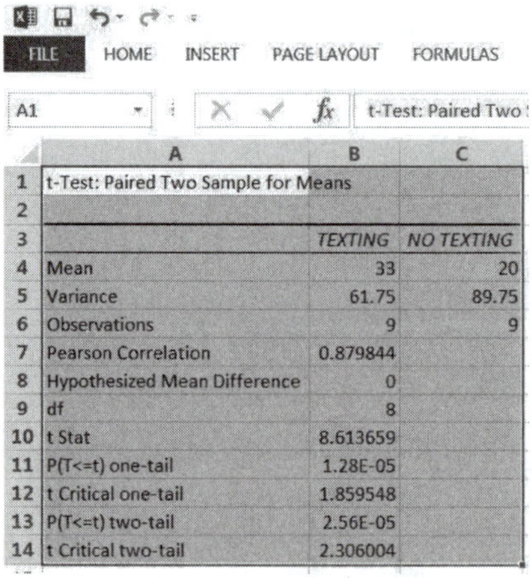

When statistics results are computed by software packages, users don't have to compare the calculated test statistic to a table full of critical values that set the boundaries of rejection zones. The statistics software packages automatically calculate a *p* (probability) value associated with the calculated test statistic. This is important for you to know and understand whenever you use a statistical software package: the *p value associated with a test statistic* is defined as the probability of obtaining the magnitude of the calculated test statistic assuming H_0 is true. The calculated *p* value is compared with a predetermined significance level, α (i.e., .05 or .01). If the calculated *p* value is less than α, it means that obtaining such a magnitude of the calculated test statistic is highly unlikely under the assumption that H_0 is true. Thus, it is logical for researchers to reject H_0 when the calculated *p* value is less than α. On the other hand, when the calculated *p* value is larger or equal to α, the researchers fail to reject H_0.

Let's apply the four-step hypothesis test for dependent sample using the Excel results.

Step 1: The problem statement asks "if texting had a detrimental effect on driving." *Detrimental* is a key word to indicate a directional hypothesis. If texting was detrimental to driving, the number of mistakes would be higher under the texting condition. The difference was calculated as the driving mistakes under texting minus those under no texting. The "detrimental" effect meant that larger number of mistakes were made under driving while texting than driving without texting; therefore, the differences were positive, thus a **right-tailed** test was the correct approach.

Difference = (Mistakes under texting – Mistakes under no texting)

H_0: $\mu_D = 0$

H_1: $\mu_D > 0$

Step 2: Identify the rejection zone.

Based on the problem statement, you use a right-tailed test, $\alpha = .05$, and $df = n - 1 = 9 - 1 = 8$. The critical value in the *t* Table is 1.860, so the rejection zone for the right-tailed test is $t > 1.86$. Or use the *p* value associated with the *t* test, the rejection zone is a one-tailed $p < .05$.

Step 3: Calculate the test statistic.

The calculated $t = 8.164$ as shown in Figure 8.8.

Step 4: Make the correct conclusion.

The result shows that the calculated $t = 8.614$ and the critical value of the right-tailed *t* is 1.860. We reject H_0. Or you can use the calculated one-tailed *p* value = .0000128 to compare with $\alpha = .05$, the result is highly significant because the *p* value associated with the calculated *t* is much smaller than $\alpha = .05$. You see the $P(T <= t)$ one tail is 1.28E-05 in Figure 8.8. This is a scientific notation that means 1.28×10^{-5}. So either using the critical *t* value or the calculated *p* value, the evidence was strong enough to support that texting had a detrimental effect on driving behavior.

As mentioned previously, the result of a dependent-sample *t* test is not confounded by individual differences, because the same individuals are used in different conditions. However, fatigue, learning, memory, practice, or other time-related factors are likely to influence the results when participants are measured repeatedly. One procedure that might improve the design of a study using dependent samples is to **counterbalance** the order of presentations. Counterbalance is a cautionary procedure in conducting repeated measures of the same sample to make sure that the order of presentation is balanced out. In a research study where participants are exposed to both Conditions A and B, half of the research participants should receive Condition A first and the other half receive Condition B first. For example, in this case, a good way to counterbalance is to randomly assign half of the participants to the "driving while texting" condition first and the other half to the "driving without texting" first. The purpose of counterbalance is to make sure that research results outcomes are strictly due to different conditions instead of effects due to other confounding factors such as fatigue, learning, memory, or practice. Counterbalance is necessary to enhance research's internal validity.

Confidence Interval of the Differences

The CI for the dependent sample mean difference applies the same principle of calculating a CI from a one-sample t test: CI = Point estimate ± Margin of error.

In conducting dependent-sample t tests, the procedure is modified to

$$(1-\alpha)100\% \text{ CI for } \mu_D = \bar{D} \pm t_{\alpha/2}\, s_e$$

Let's use the numbers in Example 8.8 to practice the CI formula for the dependent-sample t test. It also allows us to compare the widths of CI with different levels of α.

EXAMPLE 8.9

Continue to use the same data in Example 8.8 to calculate a 95% CI for μ_D under two driving conditions. Also, construct a 90% CI for μ_D, and comment on the width of the CI under different confidence levels.

Although Excel does not provide direct options for conducting CI, we can still use the basic results reported in the Figure 8.8. We found that $n = 9$, $\bar{D} = 33 - 20 = 13$, and $t = 8.614$. However, we still need to figure out s_e for the differences to calculate the CI. Based on the calculated t formula and the result, we can solve for the s_e.

$$t = \frac{\bar{D}}{s_e}$$

$$8.614 = \frac{13}{s_e}$$

$$s_e = \frac{13}{8.614} = 1.51$$

$(1 - \alpha)100\% = 95\%$, $\alpha = .05$, and $df = n - 1 = 9 - 1 = 8$; the two-tailed critical value of $t_{\alpha/2} = 2.306$.

$$95\%\ \text{CI for } \mu_D = \bar{D} \pm t_{\frac{\alpha}{2}}\, s_e = 13 \pm 2.306(1.51) = 13 \pm 3.48 = [9.52, 16.49]$$

The 95% CI indicated that we were 95% confident that the differences between driving while texting and driving without texting were between 9.52 and 16.48. Zero was not included in the 95% CI; therefore, we concluded that the difference between these two conditions was significant.

Construct a 90% CI for μ_D using the same sample statistics.

$(1 - \alpha)100\% = 90\%$, $\alpha = .10$, and $df = n - 1 = 9 - 1 = 8$; the two-tailed critical value of $t_{\alpha/2} = 1.86$.

$$90\%\ \text{CI for } \mu_D = \bar{D} \pm t_{\frac{\alpha}{2}}\, s_e = 13 \pm 1.86(1.51) = 13 \pm 2.81 = [10.19, 15.81]$$

The 95% CI was wider than the 90% CI because it contained a wider margin of errors with a lower low and higher high in the interval.

Effect Size for the Dependent-Sample *t* Tests

The dependent-sample *t* test is a procedure utilized by social scientists when they investigate differences between two conditions measured by the same group of participants or differences measured by two samples with explicit one-on-one paired relationship. However, when large sample sizes are deployed in a study, even a trivial difference between two means can be statistically significant. Indeed, a statistically significant result might not be scientifically or practically meaningful. Therefore, the sixth edition of the *Publication Manual of the American Psychological Association* (2010) states,

> For the reader to appreciate the magnitude or importance of a study's findings, it is almost always necessary to include some measure of effect size in the Results section. Whenever possible, provide a confidence interval for each effect size reported to indicate the precision of estimation of the effect size. (p. 34)

An effect size is defined as a standardized measure of the strength of a test statistic. A general formula to calculate an effect size $= \dfrac{\text{Mean Difference}}{\text{Standard deviation of the differences}}$. The specific formula to calculate the effect size for a dependent-sample *t* test is Cohen's *d*.

$$\text{Cohen's } d = \frac{\bar{D}}{s_D}$$

The effect size expresses the standardized difference in units of standard deviation of the differences. When the effect size equals 1, the interpretation is that the mean difference is 1 standard deviation apart from each other. The general interpretation of the magnitude of effect size is that 0.2 is small, 0.5 is medium, and 0.8 is large (Cohen, 1988). In addition, effect size calculation can provide powerful insights in meta-analyses. A meta-analysis is the analysis of analyses. It is the process of synthesizing research results by using rigorous scientific methods to retrieve, select, and statistically analyze the results from previous studies on the same topic. When multiple studies with the same independent and dependent variables are collected, analyzed, and statistically integrated under a meta-analysis, it offers meaningful generalizations or can examine potential confounding factors for inconsistent results. Effect sizes are necessary tools to allow such statistical integration across different studies.

Let's look at an example for calculating effect size in a dependent-sample *t* test.

EXAMPLE 8.10

Continue to use data in Example 8.8, calculate the effect size for the difference in driving mistakes under a texting condition versus under a no-texting condition. Use the results from Figure 8.8, $n = 9$, $\bar{D} = 13$, and $s_e = 1.51$ to solve this problem.

The effect size formula for a dependent-sample *t* test: Cohen's $d = \frac{\bar{D}}{s_D}$

We can solve s_D, based on the formula $s_e = \frac{s_D}{\sqrt{n}}$

$$1.51 = \frac{s_D}{\sqrt{9}}$$

$$s_D = 1.51 \times \sqrt{9} = 4.53$$

$$\text{Cohen's } d = \frac{\bar{D}}{s_D} = \frac{13}{4.53} = 2.87$$

The effect size for the difference in driving mistakes between driving under texting versus driving without texting is 2.87. The effect size is very large.

In summary, we have covered the four-step hypothesis testing procedure, CIs, and effect sizes for dependent-sample *t* tests. Although the process starts with two variables, the explicit paired relationship of these variables allows direct calculation of the differences between them. The focus of the dependent-sample *t* test is on the differences. Then you apply the same formulas you learned in Chapter 3 to calculate the mean, standard deviation, and standard error of the differences. Or use the Excel *t* Test: Paired Two Sample for Means function. The general formula for CI is $(1-\alpha)100\%$ CI = Point estimate ± Margin of error. The CI for $\mu_D = \bar{D} \pm t_{\alpha/2} s_e$. The effect size is Cohen's $d = \frac{\bar{D}}{s_D}$.

POP QUIZ

6. The alternative hypothesis for a right-tailed test of a dependent-sample t test is expressed as
   a. $\mu_1 - \mu_2 = 0$.
   b. $\mu_D = 0$.
   c. $\mu_1 - \mu_2 > 0$.
   d. $\mu_D > 0$.

7. The correct formula to calculate the effect size for dependent-sample t tests is
   a. Cohen's $d = \frac{\bar{D}}{s_D}$.
   b. Cohen's $d = \frac{\bar{X}_1 - \bar{X}_2}{s_p}$.
   c. Glass's $\Delta = \frac{\bar{X}_1 - \bar{X}_2}{s_{control}}$.
   d. Glass's $\Delta = \frac{\bar{X}_1 - \bar{X}_2}{s_{larger}}$.

EXERCISE PROBLEMS

1. A car manufacturer advertised the fuel efficiency of one of its popular models as $\mu = 29$ miles per gallon. Fuel efficiency was normally distributed. A sample of nine cars was randomly selected by a third-party nonprofit organization, and the cars' fuel efficiency numbers were measured as 17, 28, 33, 35, 30, 21, 19, 29, and 31. Use $\alpha = .05$ to test whether there was truth in the advertisement of this car manufacturer.

2. Adult male height is normally distributed with $\mu = 70$ inches. A group of 25 college male swimmers was randomly selected with $\bar{X} = 73$ inches and $s = 5.5$ inches.
   a. Use $\alpha = .01$ to conduct a test to verify whether the average height of swimmers was taller than 70 inches.
   b. Construct and interpret the 95% CI for college male swimmers' average height.

3. A pharmaceutical company designed a study to test a new drug's effect on lowering cholesterol. Researchers randomly selected 10 volunteers and measured their cholesterol before they started taking the new drug and after they had taken the new drug for 2 months. The participants' cholesterol levels are reported in

TABLE 8.6 ● Cholesterol Level of Participants Before and After Taking the New Drug

Before	After
240	210
198	183
267	240
211	203
255	245
229	216
236	227
203	204
274	212
242	211

Table 8.6. Assume that the cholesterol levels were normally distributed. Use Excel to answer the following questions.

a. Was the new drug effective in lowering cholesterol, using $\alpha = .01$?
b. What was the effect size for the difference in cholesterol levels before and after taking the new drug?
c. What was the 98% CI for the difference in cholesterol levels before and after taking the new drug?

Solutions

1. The population standard deviation, σ, is unknown; therefore, the four-step hypothesis testing process to conduct a one-sample t test is the correct approach to answer this problem.

Step 1: State the pair of hypotheses.

The problem statement asks, "Was there truth in the advertisement?" The question does not contain any key word regarding direction. Therefore, a nondirectional hypothesis is the correct approach.

H_0: $\mu = 29$ miles per gallon

H_1: $\mu \neq 29$ miles per gallon

Step 2: Identify the rejection zone for the test.

From Step 1, we know that we are conducting a two-tailed test. The problem statement specifies $\alpha = .05$ and $df = n - 1 = 9 - 1 = 8$. According to the t Table, the critical t value is 2.306. The rejection zone is $|t| > 2.306$, as shaded in Figure 8.9.

FIGURE 8.9 ● Using the t Distribution Table to Identify Rejection Zone $|t| > 2.306$

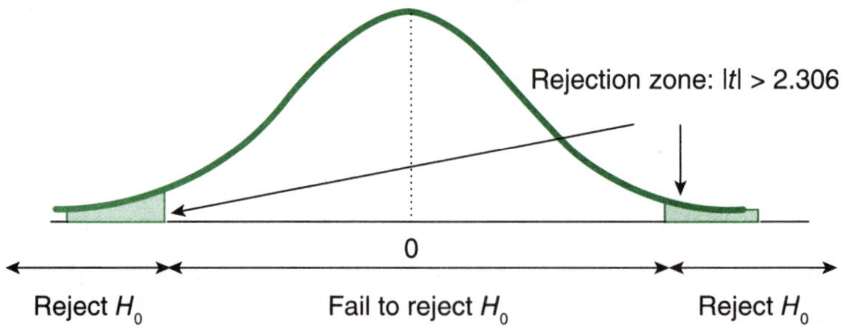

Step 3: Calculate the test statistic.

Let's list all the numbers stated in the problem statement and separate what we knew from what we need to figure out.

What we knew:

$\mu = 29$ miles per gallon

$n = 9$

When given all measures in a sample, you have to calculate $\bar{X}$ and s before conducting a t test.

$$\bar{X} = \Sigma X / n$$

$$SS = \Sigma(X - \bar{X})^2$$

$$s = \sqrt{\frac{\Sigma(X - \bar{X})^2}{df}} = \sqrt{\frac{SS}{df}} = \sqrt{\frac{SS}{n-1}}$$

You may use what you learn in Chapter 3 to calculate these numbers using your calculator. Or you may learn a new way to solve the problem by using Excel's built-in functions. Enter the values of X in Excel, and label **A1** as X; the values go in **A2** to **A10** shown in Figure 8.10. To get $\bar{X}$, simply type in the function "**=AVERAGE(A2:A10)**" in a blank cell, **A12**, to display the sample mean. In this command, **A2** specifies the location of the first value of the sample and **A10** specifies the location of the last value in the sample. Then hit **Enter**. To get the sample standard deviation, s, simply type in "**=STDEV.S(A2:A10)**" in a blank cell, **A13**, to calculate the sample standard deviation. Then hit **Enter**. As shown in Figure 8.10, the sample mean is 27 and the standard deviation is 6.422616. It is rounded to $s = 6.4$.

FIGURE 8.10 ● Screen Shot of Using Excel Built-in Functions to Calculate $\bar{X}$ and s

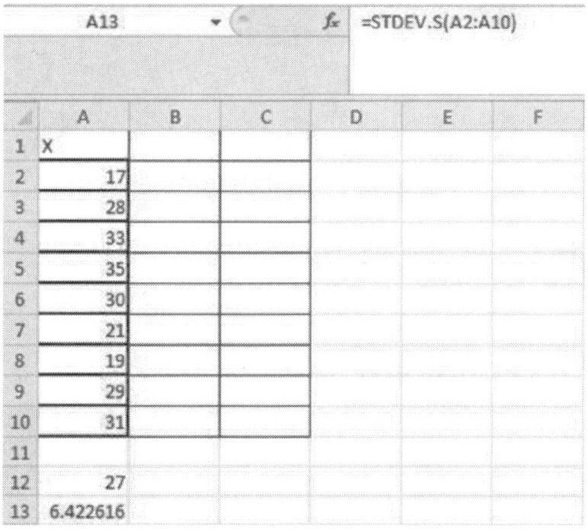

Next, insert all the numbers in the t-test formula.

$$t = \frac{\bar{X} - \mu}{\frac{s}{\sqrt{n}}}$$

$$t = \frac{27 - 29}{\frac{6.4}{\sqrt{9}}} = \frac{-2}{2.13} = -0.939$$

Step 4: Make the correct conclusion.

Compare the calculated $t = -0.939$ with the rejection zone $|t| > 2.306$. The calculated t did not fall in the rejection zone. Therefore, we failed to reject H_0. At $\alpha = .05$, the evidence was not strong enough to support the claim that the car manufacturer failed to meet the truth in advertisement standard. Thus, we accepted that there was truth in car dealer's advertisement. It is worth noting that the hypothesis test procedure started by assuming the car dealer was telling the truth; therefore, H_0: $\mu = 29$. The alternative hypothesis was that the car dealer was not telling the truth; therefore, H_1: $\mu \neq 29$. It is very important to be able to explicitly state the pair of hypotheses according to the purpose of the research.

2. The population standard deviation, σ, is unknown in this problem statement. We use the four-step hypothesis testing process to conduct a t test to answer this problem.

   a. Use $\alpha = .01$ to conduct a hypothesis test.

Step 1: State the pair of hypotheses.

Based on the problem statement, "the average height of swimmers was taller than 70 inches" indicates a directional hypothesis. The key word "taller" indicates a right-tailed test.

H_0: $\mu = 70$

H_1: $\mu > 70$

Step 2: Identify the rejection zone for the hypothesis test.

We need three pieces of information to identify the critical value that set the boundary of the rejection zone. This is a right-tailed test, $\alpha = .01$ and $df = n - 1 = 25 - 1 = 24$ based on the problem statement. According to the t Distribution Table, the critical t value is 2.492. The rejection zone for the right-tailed test is $t > 2.492$, as shown in Figure 8.11.

FIGURE 8.11 ● Using the *t* Distribution Table to Identify the Rejection Zone $t > 2.492$

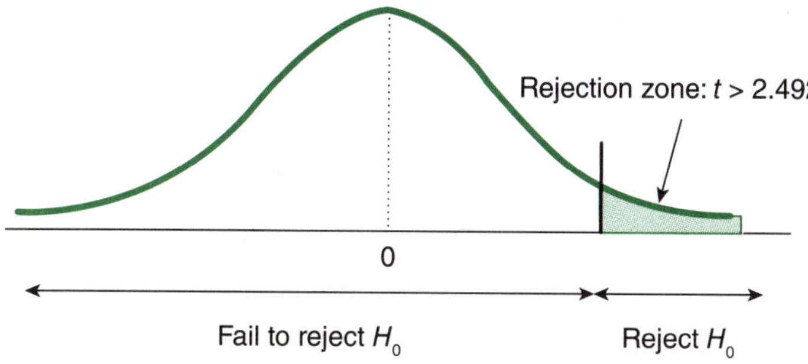

Step 3: Calculate the statistics.

Let's list all the numbers that we knew from the problem statement.

$\mu = 70$ inches

$s = 5.5$ inches

$\bar{X} = 73$ inches

$n = 25$

We need to figure out *t*. Next, insert all the numbers in the *t*-test formula.

$$t = \frac{\bar{X} - \mu}{\frac{s}{\sqrt{n}}}$$

$$t = \frac{73 - 70}{\frac{5.5}{\sqrt{25}}} = \frac{3}{1.1} = 2.727$$

Step 4: Make the correct conclusion.

Compare the calculated $t = 2.727$ with the rejection zone $t > 2.492$. The calculated *t* was in the rejection zone. Therefore, we rejected H_0. The evidence was strong enough to support the claim that the average height of swimmers was taller than 70 inches.

b. Construct and interpret the 95% CI for college male swimmers' average height.

The formula for 95% CI for μ is $\bar{X} \pm t_{\alpha/2} \frac{s}{\sqrt{n}}$.

With α = .05, df = 24, two-tailed critical value of t is 2.064, $\bar{X} = 73$ inches, and $\frac{s}{\sqrt{n}} = 1.1$,

The 95% CI for μ is 73 ± 2.064 (1.1) = 73 ± 2.270 = [70.730, 75.270]. The value under the H_0, μ = 70 is not included in the 95% CI. The average swimmer's height was significantly different from 70 inches at α = .05 level.

3. After entering Table 8.6 data in Excel, click on the **Data** tab, then click on the **Data Analysis** on the upper right side. In the **Analysis Tools Window**, select "**t-Test Paired Two Sample for Means**", then click **OK** as shown in the screenshot in Figure 8.12. Follow the same step-by-step instructions as stated in Example 8.9. There are two particular alerts: (1) the treatment effect is defined as the difference of patient's cholesterol after taking the drug minus that before taking the drug. Since the Excel automatically calculate the difference by using (**Variable 1 – Variable 2**), we have to set up the **After** as **Variable 1** with the data range **$B$1:$B$11** and **Before** as **Variable 2** with the data range **$A$1:$A$11** as shown in Figure 8.12, (2) the problem statement specified **Alpha** = 0.01 which differs from the default 0.05, so change the α level to **0.01**. Then click **OK**.

FIGURE 8.12 ● Screenshot of Setting Up Before and After Cholesterol Scores in Excel

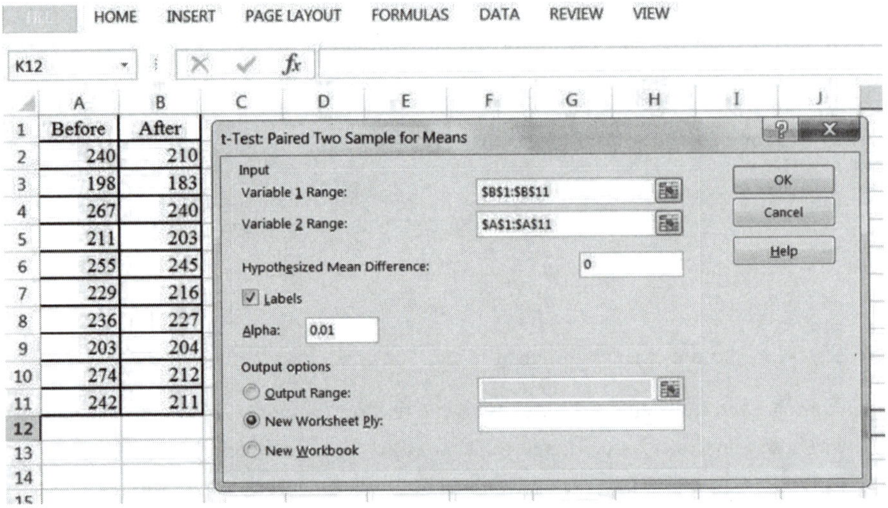

The calculation results are shown in Figure 8.13. Notice that both one tailed and two tailed critical values of t are shown only in positive values. In a left-tailed test, users need to add the negative sign on their own.

FIGURE 8.13 ● Screenshot of the Cholesterol Scores Results

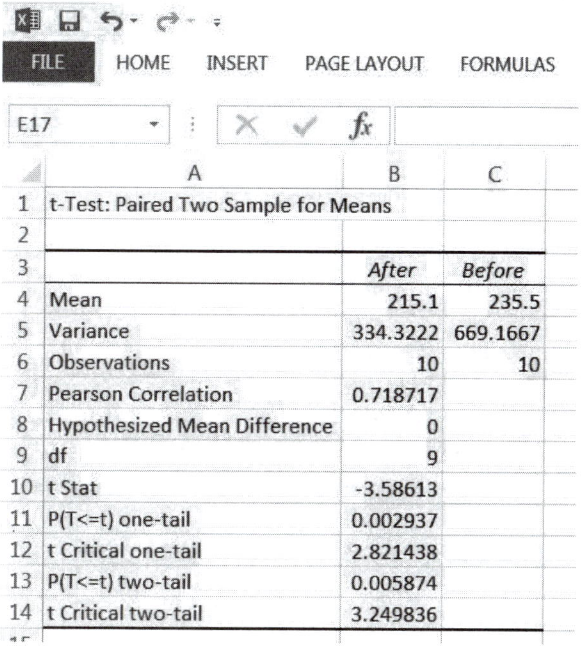

	A	B	C
1	t-Test: Paired Two Sample for Means		
2			
3		*After*	*Before*
4	Mean	215.1	235.5
5	Variance	334.3222	669.1667
6	Observations	10	10
7	Pearson Correlation	0.718717	
8	Hypothesized Mean Difference	0	
9	df	9	
10	t Stat	-3.58613	
11	P(T<=t) one-tail	0.002937	
12	t Critical one-tail	2.821438	
13	P(T<=t) two-tail	0.005874	
14	t Critical two-tail	3.249836	

Let's use the Excel results to answer the following three questions.

a. Was the new drug effective in lowering cholesterol, using $\alpha = .01$?

b. What was the effect size for the difference in cholesterol levels before and after taking the new drug?

c. What was the 98% CI for the difference in cholesterol levels before and after taking the new drug?

Step 1: The problem statement asks "if the new drug was effective in lowering cholesterol." *Lowering* is a key word to indicate a directional hypothesis. The treatment effect is defined as the cholesterol level after taking the drug minus that before taking the drug. The "lowering" effect is likely to make the difference negative so a left-tailed test is the correct approach.

Difference = (cholesterol after drug – cholesterol before drug)

H_0: $\mu_D = 0$

H_1: $\mu_D < 0$

Step 2: Identify the rejection zone.

Based on the problem statement, $\alpha = .01$, and $df = n - 1 = 10 - 1 = 9$, the critical value in the t Table is 2.821, so the rejection zone for the left-tailed test is $t < -2.821$. Please note in Figure 8.13 the outcome produced by Excel, the **t Critical one-tail** was listed as 2.821438. Users need to make the necessary adjustment for the left-tailed test by putting the negative sign to the critical value.

Step 3: Calculate the test statistic.

The calculated $t = -3.586$ as shown in the result table.

Step 4: Make the correct conclusion.

The result shows that the calculated $t = -3.586$ is within the rejection zone. We reject H_0. Or you can use the calculated one-tailed p value = .002937 to compare with $\alpha = .01$, the result is highly significant. So either using the critical t value or the calculated p value, we concluded that the new drug was effective in lowering the patients' cholesterol levels.

Based on the results in Figure 8.13, we find $t = -3.586$ and the $\bar{D} = 215.1 - 235.5 = -20.4$.

$$t = \frac{\bar{D}}{s_e} = \frac{\bar{D}}{\frac{s_D}{\sqrt{n}}}$$

$$-3.586 = \frac{-20.4}{\frac{s_D}{\sqrt{10}}}$$

$$s_D = 17.989$$

$$\text{Cohen's } d = \frac{\bar{D}}{s_D} = \frac{-20.4}{17.989} = -1.134$$

This new drug had a large effect size in lowering patients' cholesterol levels.

The critical value of t for 98% CI is $t = 2.821$, $s_e = \frac{s_D}{\sqrt{10}} = \frac{17.989}{\sqrt{10}} = 5.689$

98% CI for $\mu_D = \bar{D} \pm t_{\alpha/2} s_e = -20.4 \pm 2.821(5.689) = -20.4 \pm 16.049 = [-36.449, -4.351]$

The value of H_0: was not included in this 98% CI, therefore, we concluded that there was a significant difference of patients' cholesterol levels before and after taking this new drug.

What You Learned

You have learned how to conduct a proper test between a sample mean and a hypothesized population mean when the population standard deviation, σ, is unknown. The *t* test is appropriate when the sampling distribution of the means is normally distributed. You have also learned that there is a specific *t*-distribution curve for every degree of freedom. You need three pieces of information to identify the critical *t* values that set the boundaries of rejection zones: (1) one-tailed or two-tailed test, (2) the *df*, and (3) the α level.

The four-step hypothesis testing procedure is applicable in conducting a one-sample *t* test. The *t* formula is

$$t = \frac{\bar{X} - \mu}{s_{\bar{X}}} = \frac{\bar{X} - \mu}{\frac{s}{\sqrt{n}}}, \text{ with } df = n = 1$$

The formula for CI for μ is

$$\text{CI for } \mu = \bar{X} \pm t_{\alpha/2}\left(\frac{s}{\sqrt{n}}\right)$$

Dependent-sample *t* tests usually involve the same group of research participants being measured twice—before and after the treatment. On some rare occasions, dependent-sample *t* tests are applied to two closely matched or paired samples. The before–after or the unique paired relationship allows the calculation of differences, thus, the mean, standard deviation, and standard error of the differences. The four-step hypothesis testing procedure is applicable in conducting a dependent-sample *t* test. The *t* formula is

$$t = \frac{\bar{D}}{s_e} = \frac{\bar{D}}{\frac{s_D}{\sqrt{n}}}, \text{ with } df = n - 1$$

The formula for constructing a CI for μ_D is

$$(1-\alpha)\,100\%\text{ CI for } \mu_D = \bar{D} \pm t_{\alpha/2} s_e$$

The formula for calculating the effect size for a dependent-sample *t* test:

$$\text{Cohen's } d = \frac{\bar{D}}{s_D}$$

Key Words

Confidence interval: A confidence interval is defined as an interval of values calculated from sample data to estimate the value of a population parameter. 222

Confidence level: Confidence level is expressed as (1 – α), which is the complement to α, the significance level. 223

Counterbalance: Counterbalance is a cautionary procedure in conducting repeated measures of the same sample to make sure that the order of presentation of different conditions is balanced out so the order does not confound the research results. 236

Dependent-sample *t* tests: Dependent-sample *t* tests are used to test the differences between two measures of one variable from the same sample before and after a treatment or the difference between a variable from two samples with an explicit one-on-one paired relationship. 227

Interval estimate: An interval estimate is also called the margin of error and is calculated as the multiplication between a two-tailed critical *t* value, $t_{\alpha/2}$ and the standard error of the statistics. 223

One-sample *t* test: A one-sample *t* test is designed to conduct a test between a sample mean and a hypothesized population mean when σ is unknown, 212

$$t = \frac{\overline{X} - \mu}{s_{\overline{X}}} = \frac{\overline{X} - \mu}{\frac{s}{\sqrt{n}}}, \text{ with } df = n - 1$$

Point estimate: A point estimate is to use a single value from the sample statistics to estimate a population parameter. 223

***t* Table:** The *t* Distribution Table provides critical values of *t* that set the boundaries of rejection zones for *t* tests given three pieces of information: (1) one-tailed or two-tailed test, (2) the *df*, and (3) the α level. 213

Learning Assessment

Multiple Choice: Circle the Best Answer for Every Question

1. The difference between using a *Z* test and using a *t* test depends on
   a. whether μ is given in the problem statement
   b. whether σ is given in the problem statement
   c. whether $\overline{X}$ is given in the problem statement
   d. whether *n* is given in the problem statement
2. For a two-tailed *t* test with $\alpha = .01$, $df = 20$, what is the critical *t* value that set the boundary for the rejection zone?
   a. $t > 1.96$
   b. $t > 2.58$
   c. $|t| > 1.96$
   d. $|t| > 2.845$
3. For a left-tailed *t* test with $\alpha = .01$, $n = 15$, what is the rejection zone for this test?
   a. $t < -2.602$
   b. $t < -2.624$
   c. $t > 2.602$
   d. $t > 2.624$

4. For a right-tailed *t* test with $\alpha = .01$, $n = 5$, what is the rejection zone for this test?
   a. $t < -3.747$
   b. $t < -3.365$
   c. $t > 3.747$
   d. $t > 3.365$

5. Which one of the following scenarios is *not* suitable to conduct a *t* test?
   a. The population is normally distributed and $n > 30$
   b. The population is *not* normally distributed, and sample size is $n > 30$
   c. The population is normally distributed and $n < 30$
   d. The population is *not* normally distributed and $n < 30$

6. To test whether a sample mean differs from a hypothesized population mean, with $\alpha = .01$, what confidence level should you use if you want to use a confidence interval to get the same result?
   a. 99.5%
   b. 99%
   c. 98%
   d. 95%

7. A sample is selected from a population with $\mu = 80$, and a treatment is administered to the sample. If the sample standard deviation is 9, which set of sample characteristics is most likely to lead to a decision to reject H_0?
   a. $\bar{X} = 81$ for a sample size $n = 10$
   b. $\bar{X} = 81$ for a sample size $n = 100$
   c. $\bar{X} = 83$ for a sample size $n = 10$
   d. $\bar{X} = 83$ for a sample size $n = 100$

8. When *n* is small ($n \le 30$), the *t* distribution
   a. is identical to the normal *Z* distribution
   b. has a lower peak and higher tails than the normal *Z* distribution
   c. has a higher peak and lower tails than the normal *Z* distribution
   d. is a skewed nonsymmetrical distribution

9. Which one of the following studies calls for a dependent-sample *t* test?
   a. A group of 30 students' quiz grades before and after reading the book chapter
   b. A comparable pay study compares salaries between male and female workers on the same position in the same company
   c. A comparison of learning outcomes from two classes of sixth graders is made, one class using lectures and the other using student-centered inquiry
   d. A comparison of job satisfaction is made between married workers and unmarried workers

10. In a dependent-sample *t* test, researchers want to have 10 people's scores on both "driving while texting" and "driving without texting" conditions. How many research participants do the researchers need?

    a. $n = 10$
    b. $n = 20$
    c. $n = 30$
    d. $n = 40$

11. A psychologist investigated the impact of study-abroad programs on students' openness to experience. A group of students were measured on their openness to experience before and after they have studied abroad. Which statistical procedure is appropriate for this study?

    a. Independent-samples *t* test with equal variances assumed
    b. Independent-samples *t* test with equal variances not assumed
    c. One-sample *t* test
    d. Dependent-sample *t* test

Free Response Questions

12. Nationwide, the average total amount of student loan debt is normally distributed with a $\mu = \$37,423$. A sample of 36 students was randomly selected from a public university. Their student loans had $\overline{X} = \$35,812$ and $s = \$9,436$. Conduct a proper statistical procedure to test whether the average student loan in this public university was lower than \$37,423, using $\alpha = .10$.

13. SAT verbal scores are normally distributed with a $\mu = 500$. A local high school instituted a new program to engage students in reading. A sample of nine students from this high school was randomly selected following their participation in this reading program, and their SAT verbal scores were reported: 600, 760, 550, 505, 660, 540, 480, 535, and 455.

    a. Conduct a proper statistical procedure to test whether the new program had an effect on SAT verbal scores, using $\alpha = .05$.
    b. Construct a 95% CI for the average SAT verbal score of students in this high school.
    c. Compare the results from 13a and 13b.

14. A researcher investigated if a relaxation training was effective in reducing the number of headaches for chronic headache sufferers. For a week prior to training, each participant recorded the number of headaches suffered. The participants received the relaxation training. A week after the training, the participants recorded the number of headaches suffered again. The differences in the number of headaches were normally distributed. The numbers are given in Table 8.7. Use Excel to answer this question.

a. Did the data support the claim that the relaxation training was effective in reducing the number of headaches, using $\alpha = .01$? Make sure you include all the necessary steps in the hypothesis test.

b. What was the effect size for the difference in the number of headaches before and after the relaxation training?

c. What was the 98% CI for the difference in the number of headaches before and after the relaxation training?

15. A group of researchers investigated if a diet/nutrition seminar had an effect on prediabetes patients' blood sugar level. Nine participants were randomly selected and given a diet/nutrition seminar. Their blood sugar levels were taken twice; first time 2 weeks before and second time 2 weeks after the seminar. Assume that the differences in the before and after measures were normally distributed. The blood sugar measures before and after the seminar were reported in Table 8.8. You may choose to answer the question by following the formulas or use Excel to answer this question.

a. Did the data support the claim that the seminar had a significant effect on the blood sugar levels, using $\alpha = .10$? Make sure you include all the necessary steps in the right order for your statistical test.

b. What was the effect size for the difference in blood sugar measures before and after the seminar?

c. What was the 80% CI for the difference in blood sugar measures before and after the seminar?

TABLE 8.7 ● Number of Headaches Reported by Participants Before and After Relaxation Training

Before Training	After Training
15	9
10	8
16	12
11	8
17	11
20	12
13	13
5	6
8	9

TABLE 8.8 ● Measures of Blood Sugar Before and After the Seminar

Before	After
110	105
106	104
126	117
117	119
122	114
114	117
108	108
107	100
118	109

Answers to Pop Quiz Questions

1. d
2. c
3. a
4. b
5. c
6. d
7. a

9

INDEPENDENT-SAMPLES *t* TESTS

Learning Objectives

After reading and studying this chapter, you should be able to do the following:

- Define the independent-samples t test and explain when it can be properly used
- Describe the decision rules for equal variances assumed versus equal variances not assumed
- Explain the purpose for the test of the equality of variances
- Describe the five-step hypothesis procedure for independent-samples t tests
- Calculate and interpret confidence intervals
- Describe the roles of Cohen's d and Glass's delta (Δ) in calculating the effect sizes of the independent-samples t tests

WHAT YOU KNOW AND WHAT IS NEW

You have learned the general t-test formula,

$$t = \frac{\text{Sample statistic} - \text{Population parameter}}{\text{Estimated standard error of the statistic}}$$

in particular, you now know when comparing one sample mean with the population mean under σ is unknown situation, the one-sample t test is the appropriate procedure:

$$t = \frac{\bar{X} - \mu}{s_{\bar{X}}} = \frac{\bar{X} - \mu}{\frac{s}{\sqrt{n}}}$$

The degrees of freedom associated with the one-sample t test is $df = n - 1$. The same principle is applicable to the independent-samples t tests, with minimal modifications. When two samples are independent, random samples selected from normally distributed populations, and we are interested in using the sample mean differences to make inferences about the population mean differences, the independent-samples t test is the appropriate option. The sample statistic of comparing two sample means is expressed as $(\bar{X}_1 - \bar{X}_2)$, and the difference between two population means is $(\mu_1 - \mu_2)$. The complicated part is actually in the denominator of the independent-samples t test. Learning to calculate the denominator (i.e., the estimated standard error of the mean difference) is a major task in Chapter 9. You should pay close attention to it.

Two important assumptions need to be considered before using the independent-samples t test. First, the observations from each sample are independent. Second, at least one of the two following conditions is met: (1) the two sample sizes are large ($n > 30$) or (2) both samples are randomly selected from normally distributed populations.

Since these two samples are randomly selected from two separate populations, it is likely that we face one of two possible situations: (1) the two populations have equal variances or (2) the two populations have unequal variances. Different sets of formulas are required in these two situations. They will be introduced in separate sections of this chapter.

INTRODUCING INDEPENDENT SAMPLES AND THE DECISION RULES ON EQUAL VARIANCES

Let's start by defining **independent samples**. Independent samples refer to samples selected from different populations where the values from one population are not related or linked with the values from another population. In the scenario of two independent samples, the samples contain different individuals, and there is no overlap between samples. No individuals can belong in both samples at the same time. Independent samples are also referred to as "between-subject design." For example, researchers are interested in studying married women's psychological stress versus unmarried women's.

Any female participant in this study can belong to only one of the two groups: "married" versus "unmarried." There is no overlap between these two groups since no participant can be both "married" and "unmarried" at the same time. An independent-samples *t* test is appropriate to conduct the comparison of psychological stress between married and unmarried women.

Let's examine the evolution of *t* formulas as we go through the process.

The general *t* formula:

$$t = \frac{\text{Sample statistic} - \text{Population parameter}}{\text{Estimated standard error of the statistic}}$$

The one-sample *t*-test formula that you learned in Chapter 8:

$$t = \frac{\bar{X} - \mu}{s_{\bar{X}}}$$

The independent-samples *t*-test formula that you will learn in this chapter:

$$t = \frac{(\bar{X}_1 - \bar{X}_2) - (\mu_1 - \mu_2)}{s_{(\bar{X}_1 - \bar{X}_2)}}$$

As indicated in the independent-samples *t*-test formula, the numerator is simply obtained by calculating the difference between the sample means, $(\bar{X}_1 - \bar{X}_2)$, and comparing it with the theoretical difference between the population means, $(\mu_1 - \mu_2)$, as stated in the H_0. The H_0 is the no-effect hypothesis and is always expressed with the equals sign in both directional and nondirectional hypothesis tests:

$$H_0\text{: } \mu_1 - \mu_2 = 0$$

Once $\mu_1 - \mu_2 = 0$ gets into the independent-samples *t*-test formula, it can be further simplified:

$$t = \frac{(\bar{X}_1 - \bar{X}_2)}{s_{(\bar{X}_1 - \bar{X}_2)}}$$

As you can see, after the simplification, the independent-samples *t*-test formula no longer contains the population means, μ_1 or μ_2. Population means or standard deviation are not easily found when conducting scientific research. The independent-samples *t*-test formulas are fully functional without assuming or pretending to know the population parameters.

The only thing new is to calculate the estimated standard error of the mean difference, as expressed in the denominator of the *t* formula. In a situation where the population variance, σ^2, is unknown, the sample variance, s^2, is used as the unbiased estimate of σ^2. The estimated standard error of the mean difference has to be logically derived from the two sample variances.

When selecting two samples from two separate populations, it is likely to result in one of two possible situations. The first situation is that two population variances are similar or equivalent to each other. Thus, the similar variances should be pooled together to create a pooled variance. Then the pooled variance is used to calculate the estimated standard error of the mean difference. The second situation occurs when the population variances are not similar to each other. In this situation, dissimilar variances should be treated separately when calculating the estimated standard error of the mean difference. Distinguishing different formulas and applying the correct set of formulas to solve problems are critically important in conducting independent-samples *t* tests. Therefore, the four-step hypothesis testing procedure needs to be modified to a five-step hypothesis testing procedure to include the decision on whether to pool or not to pool the variances as the first step.

DECISION RULES FOR EQUAL VARIANCES ASSUMED VERSUS EQUAL VARIANCES NOT ASSUMED

To pool or not to pool the variances depends on whether the two variances are equal or not. Deciding whether two variances are equal or not is an empirical question that needs an empirical answer. A hypothesis test on the equality of variances needs to be conducted to answer this question. Since it is a hypothesis test, the four-step process is applicable here.

Step 1: State the pair of hypotheses for equality of variances.

$$H_0 : \sigma_1^2 = \sigma_2^2$$

$$H_1 : \sigma_1^2 \neq \sigma_2^2$$

This is the first time that we deal with a hypothesis test regarding the variances. Such a hypothesis test on the equality of variances requires a new test, *F* test. Its formula is stated in Step 3.

Step 2: Identify the rejection zone for the *F* test.

The *F* test requires a new table to identify the critical value that set off the boundary of the rejection zone. The critical *F* value is determined by two different degrees of freedom, $F_{(df1,\ df2)}$, as stated in the *F* Table in Appendix C. Regarding the two degrees of freedom, df_1 is the degrees of freedom for the numerator, which identifies the column, and df_2 is the degrees of freedom for the denominator, which identifies the row. The intersection between the column and the row provides the critical value for the *F* test. *F* values have a skewed distribution with a long tail toward the right side (e.g., high end) of the distribution. *F* values are never negative because both the numerator and the denominator are always positive. The rejection zone is defined as calculated *F* greater than the critical value of *F* given df_1 and df_2 as listed in the *F* Table.

$$\text{Rejection zone: Calculated } F > F_{(df_1,\, df_2)}$$

The critical values for *F* can have many pages, but I only provide one page with $\alpha = .05$ in this book. For instructors who want to avoid using probability-based tables to search for the critical values of *F*, you may use the **F.INV.RT** function in Excel for Microsoft 365 (both for PC and Mac, and Excel 2019. The syntax for this function is **F.INV.RT(probability,deg_freedom1,deg_freedom2)**. The probability is the α level for the *F* test, the deg_freedom1 is the numerator degrees of freedom, and the deg_freedom2 is the denominator degrees of freedom. When you type "**=F.INV.RT(α,df$_1$,df$_2$)**" in a blank cell in Excel, you will find the critical value of a **folded *F* test** (right-tailed test). Such a function in Excel can eliminate the need to use the *F* Table in the book. The concept of folded *F* test is explained in Step 3.

In addition, you can also calculate the *p* value associated with a calculated *F* statistic by using the **F.DIST.RT** function. The syntax for this function is **F.DIST.RT(x,deg_freedom1,deg_freedom2)**. The x stands for the calculated *F* statistic. When you type "**=F.DIST.RT(x, df$_1$,df$_2$)**" in a blank cell in Excel, you will find the *p* value associated with the specified folded *F* statistic value with df_1 and df_2. Once the *p* value associated with the *F* value is identified, the rejection zone is defined as $p < \alpha$, where α is the predetermined acceptable level of the risk of committing the Type I error. The default value of α is .05.

Step 3: Calculate the test statistic.

$$F = \frac{s_1^2}{s_2^2} = \frac{\text{Larger variance}}{\text{Smaller variance}}$$

The reason for designating s_1^2 as the larger variance and s_2^2 as the smaller variance is to force the calculated *F* to be greater than 1 in order to create a right-tailed test for the *F* statistic. This avoids the problem of finding the left-tailed *F* value. The *F* Table in most statistics textbooks only reports the right-tailed values. Therefore, in this calculation, the larger variance is placed as the numerator and the smaller variance is placed as the denominator of the *F* test. When two variances are close to each other, the *F* value is close to 1. When two variances are far apart from each other, the *F* value becomes large. When we deliberately force the calculated *F* to be greater than 1, such procedures are labeled as **folded *F* tests**. Therefore, the group with the larger variance is labeled as Group 1, and the group with the smaller variance is labeled as Group 2. Consistent labeling is particularly important when you conduct a directional hypothesis test.

Step 4: Make the correct conclusion.

The decision rule is that when the calculated *F* value is not within the rejection zone, we fail to reject H_0. Or when the *p* value associated with the calculated *F* is greater or equal to α, we fail to reject H_0. Thus, the correct way to conduct an independent-samples *t* test is to assume equal variances: $H_0 : \sigma_1^2 = \sigma_2^2$. On the other hand, when the calculated *F* value is within the rejection zone or the *p* value associated with the *F* is less than α level (i.e., $p < \alpha$), we reject H_0. Thus, the correct way to conduct an independent-samples *t* test is to use the procedure for equal variances not assumed. To answer the question "How do I find out whether two groups have equal variances or not?" let's use an example to go through the test for equality of variances.

EXAMPLE 9.1

Psychological stress is normally distributed for both men and women. A random sample of 16 men's psychological stress produced a standard deviation of 6.42, and a random sample of 18 women's psychological stress generated a standard deviation of 3.73. Do men and women have equal variances in psychological stress?

As you can tell from the problem statement, we don't need to have the sample mean information to conduct the equality of variance test. The four-step hypothesis testing procedure is applicable in testing equality of variances.

Step 1: State the pair of hypotheses for equality of variances.

$$H_0 : \sigma_1^2 = \sigma_2^2$$

$$H_1 : \sigma_1^2 \neq \sigma_2^2$$

Step 2: Identify the rejection zone for the *F* test.

The consistency of labeling the two groups involved in the test is the key to obtain the correct answer. The group with larger variance is Group 1 so $s_1 = 6.42$, and its $df_1 = n_1 - 1 = 16 - 1 = 15$; therefore, the men's group is labeled as Group 1. And $s_2 = 3.73$, and its $df_2 = 18 - 1 = 17$; therefore, the women's group is labeled as Group 2.

The critical value of the $F_{(df_1, df_2)} = F_{(15,17)} = 2.31$. The rejection zone is $F > 2.31$. Or if you choose to use the *p* value associated with the calculated *F* to make the decision, the decision rule is to reject H_0 when the *p* value associated with the calculated *F* is less than .05.

Step 3: Calculate the test statistic.

$$F = \frac{s_1^2}{s_2^2} = \frac{6.42^2}{3.73^2} = \frac{41.2146}{13.9129} = 2.96$$

Step 4: Make the correct conclusion.

The calculated $F = 2.96$ is larger than the critical value of $F_{(15,17)} = 2.31$. Or use the **F.DIST.RT(x, df$_1$, df$_2$)** function in Excel to find the *p* value associated with the calculated *F*. In a blank cell in Excel, type "**=F.DIST.RT(2.96,15,17)**", then hit **Enter**. The answer 0.01713 appears. The *p* value associated with $F = 2.96$ with $df_1 = 15$ and $df_2 = 17$ is .0171, which is less than .05. Therefore, we reject H_0. The evidence is strong enough to reject the null hypothesis; therefore, we conclude that the two variances are not equal. We should choose the formulas for equal variances not assumed.

This *F* statistic allows testing for equality of variances from two groups with different sample sizes. The *F* test is preferable to other similar tests when you have to calculate everything by hand. Most statistics software comes with built-in tests for equality of variances. It is convenient to use statistical packages. However, you still need to understand and interpret the printouts from them. As a reminder, the ***p* value associated with a test statistic** is defined as the probability of obtaining the magnitude of the calculated test statistic assuming H_0 is true. This is very convenient because it does not matter what test statistics are calculated, the decision rules for *p* values remain the same. The calculated *p* value is always compared with a predetermined significance level, α (i.e., .05 or .01). If the calculated *p* value is less than α, it means that such a magnitude of the calculated test statistic is highly unlikely under the assumption that H_0 is true. Thus, it is logical for researchers to reject H_0 when the calculated *p* value is less than α. On the other hand, when the calculated *p* value is larger than or equal to α, researchers fail to reject H_0.

In summary, the *p* value associated with a test statistic can be used as a criterion for a hypothesis test. The decision rules are as follows:

When $p < \alpha$, we reject H_0.

When $p \geq \alpha$, we fail to reject H_0.

The result of a hypothesis test on equality of variances is only the beginning of the independent-samples *t* test. It provides an empirical answer to whether two populations have equal variances so we can choose the correct set of formulas to appropriately conduct the independent-samples *t* test. The formulas for equal variances assumed and the formulas for equal variances not assumed are presented in the following sections.

POP QUIZ

1. The F statistic to test the equality of variances specifies the larger variance as the numerator and the smaller variance as the denominator. Such a process is called a folded F test; therefore, the critical values listed in the F Table are
   a. positive and greater than 1.
   b. left-tailed values.
   c. positive and less than 1.
   d. normally distributed.
2. Which degrees of freedom determine the critical values of an F test when conducting a test for equality of variances?
   a. n
   b. $n - 1$
   c. df_1 for the numerator and df_2 for the denominator
   d. $(r - 1)(c - 1)$, where r is the number of rows and c the number of columns

Equal Variances Assumed

When two populations have equal variances, $\sigma_1^2 = \sigma_2^2$, we need to pool the variances together by using the pooled variance (s_p^2) formula:

$$\text{Pooled variance} = s_p^2 = \frac{SS_1 + SS_2}{df_1 + df_2}$$

where

SS_1 = the sum of squares from Sample 1

SS_2 = the sum of squares from Sample 2

df_1 = the degrees of freedom from Sample 1

df_2 = the degrees of freedom from Sample 2

Then we use the pooled variance to calculate the estimated standard error of the mean difference, $s_{(\bar{X}_1 - \bar{X}_2)}$, which is the denominator of the independent-samples t-test formula.

$$s_{(\bar{X}_1 - \bar{X}_2)} = \sqrt{\frac{s_p^2}{n_1} + \frac{s_p^2}{n_2}}$$

Now, to put everything together, when $\sigma_1^2 = \sigma_2^2$, the independent-samples *t*-test formula is

$$t = \frac{\bar{X}_1 - \bar{X}_2}{s_{(\bar{X}_1 - \bar{X}_2)}}$$

Since each degree of freedom has its own specific *t* curve, we need to figure out the degrees of freedom for the independent-samples *t* test. When equal variances are assumed, we need to pool the variances together, and also pool the degrees of freedom together. The degrees of freedom for the independent-samples *t* test with equal variances assumed are $df = df_1 + df_2$. Now we have demonstrated the formula for the independent-samples *t* test with equal variances assumed, and the *df* associated with the *t* test in order to correctly identify the rejection zone for the test. We are ready to go through the entire hypothesis testing process for the independent-samples *t* test with equal variances assumed.

As discussed earlier, the equality-of-variance test has to be done first to allow you to choose the right set of formulas to conduct the independent-samples *t* test. Therefore, the four-step hypothesis testing procedure must be modified to five steps for independent-samples *t* tests.

Step 1: Conduct the equality of variances test.

Step 2: State the pair of hypotheses regarding group means.

Step 3: Identify the rejection zone for the *t* test.

Step 4: Calculate the *t* statistic.

Step 5: Make the correct conclusion.

We will now use a couple of examples to illustrate this five-step hypothesis testing procedure for the independent-samples *t* tests.

EXAMPLE 9.2

Adult women's weights and adult men's weights are both normally distributed. A sample of 13 men was randomly selected with $\bar{X} = 196.08$ pounds and $s = 38.55$ pounds and a sample of 15 women was randomly selected with $\bar{X} = 147.2$ pounds and $s = 28.27$ pounds. Was there a significant difference between men's and women's average weight, using $\alpha = .05$?

(Continued)

(Continued)

Comparisons between randomly selected men and randomly selected women are good examples of independent-samples *t* tests, because there was no connection between men and women in the samples. Notice that the sizes of these two samples do not need to be the same. First, we need to test the equality of variances.

Step 1: Test the equality of variances.

We apply the four-step procedure to test the equality of variances.

Substep 1: State the pair of hypotheses regarding variances.

$$H_0 : \sigma_1^2 = \sigma_2^2$$

$$H_1 : \sigma_1^2 \neq \sigma_2^2$$

Substep 2: Identify the rejection zone for the *F* test.

The group with the larger variance was Group 1, $s_1 = 38.55$ pounds, and its $df_1 = n_1 - 1 = 13 - 1 = 12$; therefore, the men belonged to Group 1. The group with smaller variance was $s_2 = 28.27$ pounds, and its $df_2 = 15 - 1 = 14$; therefore, the women belonged to Group 2.

We found the critical value of the $F_{(df_1, df_2)} = F_{(12, 14)} = 2.53$ in the *F* Table; therefore, the rejection zone was $F > 2.53$. Or the rejection zone was $p < .05$.

Substep 3: Calculate the *F* statistic.

$$F = \frac{s_1^2}{s_2^2} = \frac{38.55^2}{28.27^2} = \frac{1485.91}{799.46} = 1.86$$

Substep 4: Make the correct conclusion.

The calculated $F = 1.86$ was smaller than the critical value of $F_{(12, 14)} = 2.53$. Or the *p* value associated with $F = 1.86$ was .1338—that is, $p > .05$. Therefore, we failed to reject H_0. The evidence was not strong enough to claim that the two variances were unequal. Therefore, the correct independent-samples *t*-test approach was to assume equal variances.

Step 2: State the pair of hypotheses regarding population means.

The problem statement asked, "Was there a significant difference between men's and women's average weight?" There was no hint of direction. Therefore, a two-tailed test was appropriate.

H_0: $\mu_1 - \mu_2 = 0$

H_1: $\mu_1 - \mu_2 \neq 0$

Step 3: Identify the rejection zones for the *t* test.

According to the rule, $df = df_1 + df_2 = n_1 + n_2 - 2 = 26$, $\alpha = .05$, it is a two-tailed test, and the critical *t* value that sets the boundary of the rejection zone is 2.056. The rejection zone is thus $|t| > 2.056$.

Step 4: Calculate the *t* statistic.

When $\sigma_1^2 = \sigma_2^2$, the variances need to be pooled together. The problem statement provides sample sizes, n_1 and n_2, and sample standard deviations, s_1 and s_2, so we need to figure out SS_1 and SS_2 to calculate the pooled variance, s_p^2. In Chapter 3, we learned that $s^2 = SS / df$; therefore, $SS = s^2 df$.

$$SS_1 = s_1^2 df_1 = 38.55^2(13-1) = 17830.92$$

$$SS_2 = s_2^2 df_2 = 28.27^2(15-1) = 11192.4$$

$$s_p^2 = \frac{SS_1 + SS_2}{df_1 + df_2} = \frac{17830.92 + 11192.4}{12 + 14} = \frac{29023.32}{26} = 1116.28$$

$$s_{(\bar{X}_1 - \bar{X}_2)} = \sqrt{\frac{s_p^2}{n_1} + \frac{s_p^2}{n_2}} = \sqrt{\frac{1116.28}{13} + \frac{1116.28}{15}} = \sqrt{85.87 + 74.42} = \sqrt{160.29} = 12.66$$

$$t = \frac{\bar{X}_1 - \bar{X}_2}{s_{(\bar{X}_1 - \bar{X}_2)}} = \frac{196.08 - 147.2}{12.66} = \frac{48.88}{12.66} = 3.861$$

Step 5: Make the correct conclusion.

The calculated $t = 3.861$ is within the rejection zone. Or you may use one of the **T.DIST.2T(x,df)** function in Excel to find the *p* value associated with the calculated *t*. Select a blank cell in Excel, type "**=T.DIST.2T(3.861,26)**", then hit **Enter**. The answer .00067 appears. The *p* value associated with $t = 3.861$ with $df = 26$ is $p = .00067$—that is, $p < .05$. Therefore, either use critical value of *t* or *p* value associated with the calculated *t*, we reject H_0. The evidence was strong enough to support the claim that there was a significant difference between men's and women's average weights at $\alpha = .05$ level.

The independent-samples *t* tests are more complicated than the *Z* tests or one-sample *t* tests because we have to test for equality of variances first to decide which set of formulas is the appropriate one to use. The more examples you practice, the easier this process becomes. In the next example, I will demonstrate how to solve the same problem using Excel's add-on Data Analysis function.

EXAMPLE 9.3

Body weights are normally distributed for both men and women. The randomly selected 13 men's weight and 15 women's weight are listed in the Excel as shown in Table 9.1. Conduct the proper analysis to examine whether there is a significant difference between men's and women's average weight.

(Continued)

(Continued)

The step-by-step instructions to use Excel **Data Analysis** function are provided below. You probably noticed that Example 9.2 and Example 9.3 are the same question solved by different procedures. Example 9.2 was calculated by hand and Example 9.3 is calculated by Excel. As you go through the process, you will learn that Excel is great at doing number crunching but the user needs to have knowledge about the appropriate process to conduct independent-samples *t* tests to get the correct outcomes.

TABLE 9.1 ● Two Random Samples of Men's and Women's Weight in Excel

	A	B	C
1	Men	Women	
2	262	130	
3	247	125	
4	161	189	
5	148	174	
6	152	102	
7	182	148	
8	203	197	
9	190	113	
10	149	126	
11	222	138	
12	183	147	
13	211	155	
14	239	164	
15		176	
16		124	
17			
18	196.077	147.2	mean
19	38.5475	28.2747	s.d.

1. We labeled two groups based on the magnitude of the variance, so we calculated the group mean and standard deviation in Excel. Here are the instructions to calculate means and standard deviations. In **A18**, type "**=Average(A2:A14)**", then hit **Enter** to obtain the mean for men ($\bar{X} = 196.08$) and in **A19**, type "**=STDEV.S(A2:A14)**", then hit **Enter** to obtain the sample standard deviation for men ($s = 38.55$). Do the same in **B18**, type "**=Average(B2:B16)**", then hit **Enter** to obtain the mean for women ($\bar{X} = 147.2$), and in **B19**, type "**=STDEV.S(B2:B16)**", then hit **Enter** to obtain the sample standard deviation for women ($s = 28.27$). Excel does not label the results, so you have to type in the labels: "mean" in **C18** and "s.d." in **C19** as shown in Table 9.1.
2. Based on the mean and the standard deviation for these two groups, you determine that men is Group 1 and women is Group 2. Next, you need to conduct the test for equality of variances. Click on the **Data** tab, then click on the **Data Analysis** on the upper right side. (If you don't see the **Data Analysis** function, it means that you have not properly downloaded the **Data ToolPak**. You need to follow the instruction provided in Chapter 3 to do so or simply Google "how to download Data ToolPak in Excel"). In the **Analysis Tools** window, select "**F-Test Two-Sample for Variances**", then click **OK** as shown in Figure 9.1.
3. A window pops up to ask about the data range for the two groups. Highlight the data range for men's weight and place them in the **Variable 1** slot because men's weight has larger variance. Highlight the data range for women's weight and place them in the **Variable 2** slot because women's weight has smaller variance as shown in the screenshot in Figure 9.2. Since the variable labels (as shown in row 1) is highlighted, make sure to check the box for **Labels** as shown in Figure 9.2. Then click **OK**.

FIGURE 9.1 ● F-Test Two-Sample for Variances in Data Analysis

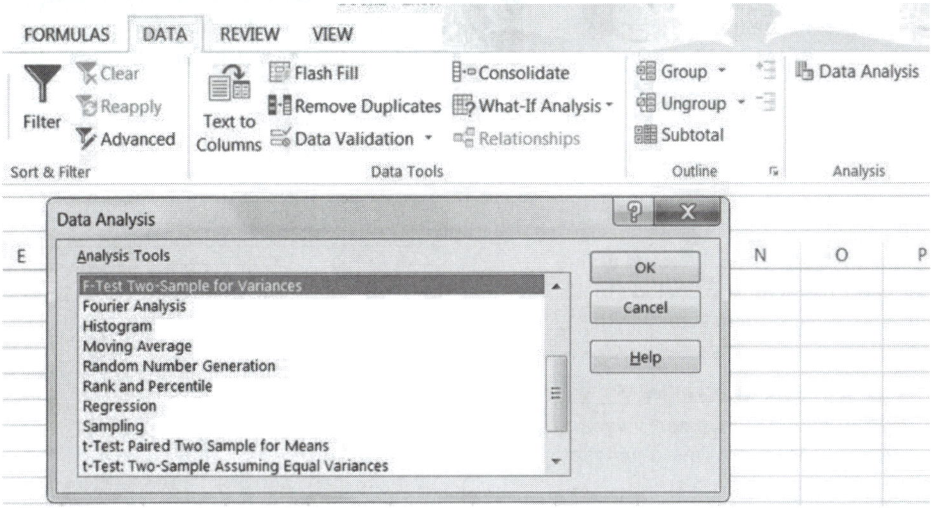

FIGURE 9.2 ● Identify Data Range in F-Test Two-Sample for Variances

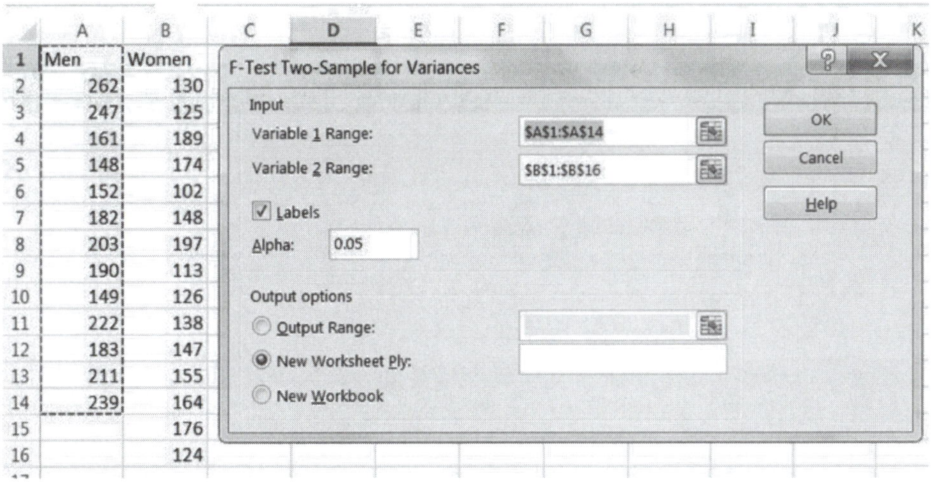

(Continued)

(Continued)

4. The outcome of the equality of variance test appears in a separate worksheet in Excel as shown in Figure 9.3. The result shows that the calculated $F = 1.86$ and the critical value of F is 2.53. We fail to reject H_0. Or you can use the calculated p value $= .134$ to compare with $\alpha = .05$, the result is not significant. Either using the critical F value or the calculated p value, we conclude that equal variances are assumed in this situation.

FIGURE 9.3 ● Results for the F-Test Two-Sample for Variances

	A	B	C
1	F-Test Two-Sample for Variances		
2			
3		*Men*	*Women*
4	Mean	196.0769	147.2
5	Variance	1485.91	799.4571
6	Observations	13	15
7	df	12	14
8	F	1.858649	
9	P(F<=f) one-tail	0.134065	
10	F Critical one-tail	2.534243	

5. Using the results from Figure 9.3, we know that men's and women's weight have equal variances. This knowledge allows us to select the correct procedure in Excel. Click on **Sheet1** located at the bottom of the Excel to go back to the worksheet with the data. Click on the **Data** tab, then click on **Data Analysis**. In the **Analysis Tools** window, select "**t-Test: Two-Sample Assuming Equal Variances**", then click **OK** as shown in Figure 9.4.

FIGURE 9.4 ● t-Test: Two-Sample Assuming Equal Variances in Data Analysis

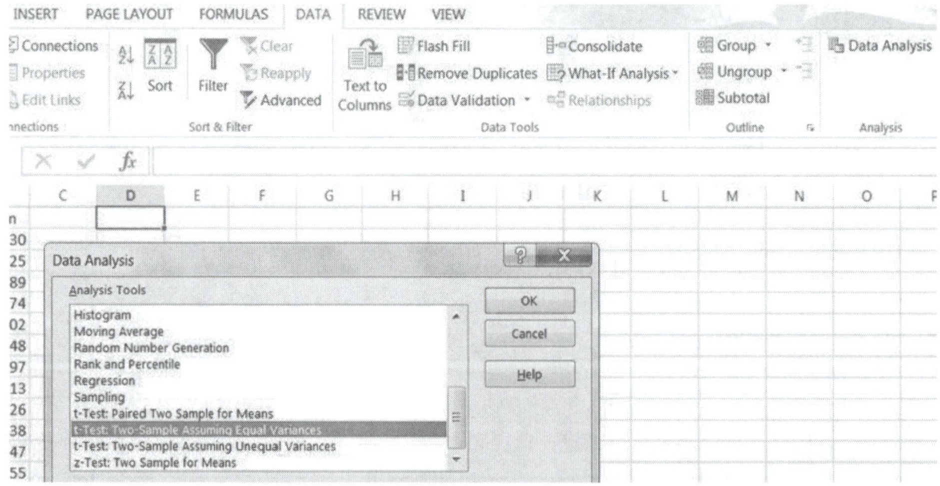

6. Repeat Step 3 to identify the data range needed for this analysis. Keep consistent in labeling the two groups.

7. The outcome of the "**t-Test: Two-Sample Assuming Equal Variances**", appears in a separate worksheet in Excel as shown in Figure 9.5. The result shows that the calculated $t = 3.861$ and the critical value of t with $df = 26$ is 2.056. We reject H_0. Or you can use the calculated two-tailed p value = .000672 to compare with $\alpha = .05$, the result is significant. Either using the critical t value or the calculated p value, we conclude that there is a significant difference between men's and women's average weight at $\alpha = .05$ level.

FIGURE 9.5 ● Results for the *t*-Test: Two-Sample Assuming Equal Variances

A	B	C
t-Test: Two-Sample Assuming Equal Variances		
	Men	*Women*
Mean	196.0769	147.2
Variance	1485.91	799.4571
Observations	13	15
Pooled Variance	1116.282	
Hypothesized Mean Difference	0	
df	26	
t Stat	3.860604	
P(T<=t) one-tail	0.000336	
t Critical one-tail	1.705618	
P(T<=t) two-tail	0.000672	
t Critical two-tail	2.055529	

In summary, an independent-samples *t* test requires the test of equality of variances to be conducted first, so you can select the proper formulas to conduct the *t* test. The hypothesis testing procedure becomes a five-step process. The more you practice the five-step process by reviewing all the examples, exercise problems, and learning assessment questions, the more confident and comfortable you will be with independent-samples *t* tests. The same problem was solved by following the formulas in Example 9.2 then solved by using Excel in Example 9.3. This is to show you that Excel Data Analysis functions do not automatically solve questions for you. Users still have to know what independent-samples *t* tests are and how to solve them correctly. Excel only takes care of the number crunching part. You still need to know which options to use and in what sequence to correctly answer the questions.

When you solve independent-samples t tests by hand, you need to remember that we force the F test for equality of variances to create the right-tailed value only; thus, the larger variance has to be placed in the numerator and is labeled as s_1. The group with the larger variance is labeled as Group 1. Correctly labeling the two groups is very important to conduct a directional hypothesis test using the independent-samples t test.

If the result of the equality-of-variance test is to fail to reject $H_0 : \sigma_1^2 = \sigma_2^2$, we have to assume equal variances to complete the hypothesis test on the group means. Pooled variance needs to be calculated.

$$\text{Pooled variance} = s_p^2 = \frac{SS_1 + SS_2}{df_1 + df_2}$$

Use the pooled variance to calculate the standard error of the mean difference, $s_{(\bar{X}_1 - \bar{X}_2)}$.

$$s_{(\bar{X}_1 - \bar{X}_2)} = \sqrt{\frac{s_p^2}{n_1} + \frac{s_p^2}{n_2}}$$

$s_{(\bar{X}_1 - \bar{X}_2)}$ is the denominator of the t test.

$$t = \frac{\bar{X}_1 - \bar{X}_2}{s_{(\bar{X}_1 - \bar{X}_2)}}$$

The independent-samples t tests have $df = df_1 + df_2 = n_1 + n_2 - 2$.

Next, a different set of formulas will be presented to deal with situations where the two populations do not have equal variances. The good news is that the five-step hypothesis test process is still applicable.

A Trip Down the Memory Lane

Here is an important reminder. You learned that $s^2 = SS / df$ in Chapter 3. This formula involves three values, s, SS, and df, and when given any two of these values in a problem, you will be able to solve the third one. For example, if the problem provides df and s, you can figure out SS. If the problem statement provides df and SS, you can figure out the value of s. If the problem statement provides s and SS, you can figure out the df. Of course, you know that df = n – 1. Although there are many ways in which problems can be stated, as long as you understand the basic principles of the independent-samples t test, you will be able to solve the problems.

POP QUIZ

3. Assume that two samples are randomly selected from two normally distributed populations with equal variances. Sample 1 has $s_1 = 3.4$ and $SS_1 = 104.04$, and Sample 2 has $s_2 = 3$ and $SS_2 = 45$. What is the pooled variance for the two independent samples?

   a. $s_p^2 = 10.65$

   b. $s_p^2 = 13.87$

   c. $s_p^2 = 17.42$

   d. $s_p^2 = 23.29$

Equal Variances Not Assumed

When $\sigma_1^2 \neq \sigma_2^2$, it is inappropriate to pool the variances due to their significant differences. Therefore, different variances must be treated separately, and the estimated standard error of the mean difference is calculated as

$$s_{(\bar{X}_1 - \bar{X}_2)} = \sqrt{\frac{s_1^2}{n_1} + \frac{s_2^2}{n_2}}$$

The formula used to calculate the estimated standard error of the mean difference when equal variances are not assumed is actually simpler than the formula used when equal variances are assumed. However, the degrees of freedom for the independent-samples t test with equal variances not assumed needs to be approximated according to the following formula by Satterthwaite (1946).

$$\text{Satterthwaite's approximated } df = \frac{(w_1 + w_2)^2}{\dfrac{w_1^2}{n_1 - 1} + \dfrac{w_2^2}{n_2 - 1}}$$

where

$$w_1 = \frac{s_1^2}{n_1} \text{ and } w_2 = \frac{s_2^2}{n_2}$$

Such an adjustment results in a smaller df than simply adding df_1 and df_2 together. The smaller df makes the test more conservative because the critical value of t goes higher, and the evidence needs to be stronger to reject H_0. It is obvious that the approximated df is not an integer. When you use the probability-based critical values of *t* tables, the *df* needs to be rounded to an integer. If you round it up to the next integer, it goes against the conservative principle. Therefore, the logical option based on the conservative principle is to round the calculated approximated *df down*. Fortunately, the approximated *df* is provided in most statistics software packages such as Excel, SPSS, and SAS. You really do not need to calculate the Satterthwaite's approximated *df* by hand.

Let's illustrate this new set of formulas by going through the five-step hypothesis testing process for independent-samples *t* tests using a couple of more examples.

EXAMPLE 9.4

ACT scores are normally distributed. A sample of 25 male college freshmen was randomly selected in a public university, and their ACT scores have $\bar{X} = 22.8$ and $s = 3.1$. A sample of 16 female college freshmen was randomly selected from the same university, and their ACT scores have $\bar{X} = 23.5$ and $s = 5.1$. Was the average ACT score for female freshmen higher than that of the male freshmen, using $\alpha = .05$?

The comparison between the randomly selected male and female college freshmen's ACT scores calls for an independent-samples *t* test. Therefore, we need to conduct the five-step hypothesis testing process.

Step 1: Test the equality of variances.

Substep 1: State the pair of hypotheses regarding variances.

$H_0 : \sigma_1^2 = \sigma_2^2$

$H_1 : \sigma_1^2 \neq \sigma_2^2$

Substep 2: Identify the rejection zone for the *F* test.

The larger standard deviation is 5.1 and labeled as s_1 for the female college freshmen, and its $df_1 = n_1 - 1 = 16 - 1 = 15$. Therefore, the female group is labeled as Group 1. The smaller standard deviation is 3.1 and labeled as s_2 for male college freshmen, and its $df_2 = n_2 - 1 = 25 - 1 = 24$. Therefore, the male group is labeled as Group 2.

The critical value of the $F_{(df_1, df_2)} = F_{(15, 24)} = 2.11$.

Substep 3: Calculate the *F* statistic.

$$F = \frac{s_1^2}{s_2^2} = \frac{5.1^2}{3.1^2} = \frac{26.01}{9.61} = 2.71$$

Substep 4: Make the correct conclusion.

The calculated $F = 2.71$ is larger than the critical value of $F_{(15,24)} = 2.11$. Or the *p* value associated with the $F = 2.71$ with $df_1 = 15$ and $df_2 = 24$ is .0143, thus, $p < .05$. Therefore, we reject H_0. The evidence was strong enough to support that the two variances were not equal. Therefore, the correct independent-samples *t*-test approach was equal variances not assumed.

Step 2: State the pair of hypotheses regarding group means.

The problem statement asks, "Was the average ACT score for female freshmen higher than that of male freshmen?" *Higher* is a key word that indicates a direction. In this problem, female freshmen were designated as Group 1 and males as Group 2. Therefore, a right-tailed test is appropriate. It is very important to get the direction of the one-tailed hypothesis test stated exactly as in the problem.

$H_0: \mu_1 - \mu_2 = 0$

$H_1: \mu_1 - \mu_2 > 0$

Step 3: Identify the rejection zone for the *t* test.

$$\text{Satterthwaite's approximated } df = \frac{(w_1 + w_2)^2}{\dfrac{w_1^2}{n_1 - 1} + \dfrac{w_2^2}{n_2 - 1}}$$

where

$$w_1 = \frac{s_1^2}{n_1} = \frac{5.1^2}{16} = \frac{26.01}{16} = 1.63 \text{ and } w_2 = \frac{s_2^2}{n_2} = \frac{3.1^2}{25} = \frac{9.61}{25} = 0.38$$

$$df = \frac{(w_1 + w_2)^2}{\dfrac{w_1^2}{n_1 - 1} + \dfrac{w_2^2}{n_2 - 1}} = \frac{(1.63 + 0.38)^2}{\dfrac{1.63^2}{15} + \dfrac{0.38^2}{24}} = \frac{4.04}{0.183} = 22.08$$

We rounded 22.08 down to 22.

According to the *t* Table, when $df = 22$, $\alpha = .05$, it is a right-tailed test, and the critical *t* value that sets the boundary of the rejection zone is 1.717. The rejection zone for a right-tailed test is $t > 1.717$.

If you use simplified way to pick the smaller of df_1 and df_2 as the *df* for the *t* test, the *df* is 15. The rejection zone for a right-tailed test under $df = 15$, $\alpha = .05$ is $t > 1.753$. *Caution!* Different methods of calculating the *df* when $\sigma_1^2 \neq \sigma_2^2$ will lead to different rejection zones for the *t* test.

(Continued)

(Continued)

Step 4: Calculate the t statistic.

When $\sigma_1^2 \neq \sigma_2^2$, it is inappropriate to pool the variances. So they were treated separately.

$$s_{(\bar{X}_1-\bar{X}_2)} = \sqrt{\frac{s_1^2}{n_1}+\frac{s_2^2}{n_2}} = \sqrt{\frac{26.01}{16}+\frac{9.61}{25}} = \sqrt{1.63+0.38} = \sqrt{2.01} = 1.42$$

$$t = \frac{\bar{X}_1-\bar{X}_2}{s_{(\bar{X}_1-\bar{X}_2)}} = \frac{23.5-22.8}{1.42} = \frac{0.7}{1.42} = 0.493$$

Step 5: Make the correct conclusion.

The calculated $t = 0.493$ is not within the rejection zone. Therefore, we failed to reject H_0. The evidence was not strong enough to support the claim that the average of the female freshmen's ACT scores was higher than that for the male freshmen.

This example illustrates the five-step hypothesis testing process for an independent-samples t test when equal variances are not assumed. This situation requires a complicated approximated *df* formula but actually has a simpler estimated standard error of the mean difference formula as the denominator of the t test. Overall, the process is manageable as long as students are not burdened with pure memorization of formulas.

In the next example, I will demonstrate how to conduct the same question in Example 9.4 but using Excel to solve the problem. Using Excel to solve problems allows you to handle large sample size with hundreds or thousands of cases with the same ease as small sample size. Being able to conduct statistical analyses using Excel is a practical skill that will enhance your ability to apply what you learn in classroom to your work in the future. Please take time to practice these steps of conducting independent-samples t tests using Excel.

EXAMPLE 9.5

ACT scores are normally distributed. A sample of 25 male college freshmen was randomly selected in a public university and a sample of 16 female college freshmen was randomly selected from the same university. Their ACT scores are shown in Table 9.2. Was the average ACT score for female freshmen higher than that of the male freshmen, using $\alpha = .05$?

1. You need to calculate the means and standard deviations from these two groups of raw scores. In **A28**, type "=**Average(A2:A26)**", then hit **Enter** to obtain the average for men's ACT ($\bar{X} = 22.8$), and in **A29**, type "=**STDEV.S(A2:A26)**", then hit **Enter** to obtain the sample standard deviation for men ($s = 3.10$). Do the same in **B28**, type "=**Average(B2:B17)**", then hit **Enter** to obtain the mean for women ($\bar{X} = 23.5$), and in **B29** type "=**STDEV.S(B2:B17)**", then hit **Enter** to obtain the sample standard deviation for women ($s = 5.10$). Excel does not label the results, so you have to type in the labels: "mean" in **C28** and "s.d." in **C29** as shown in Table 9.2.

2. Based on the mean and the standard deviation for these two groups, you determine that women belonged in Group 1 and men belonged in Group 2. Click on the **Data** tab, then click on the **Data Analysis** on the upper right side. In the **Analysis Tools** window, select "**F-Test Two-Sample for Variances**", then click **OK**.

3. A window pops up to ask about the data range for the two groups. Highlight the data range for females' ACT scores **($B$1:$B$17)** and place them in the **Variable 1** because females' ACT scores had larger variance. Highlight the data range for males' ACT scores **($A$1:$A$26)** and place them in the **Variable 2** because males' ACT scores had smaller variance. Since the variable labels (as shown in row 1) is highlighted, make sure to check the box for **Labels** as shown in Figure 9.6. Then click **OK**.

TABLE 9.2 ● Male and Female Freshmen's ACT Scores in Excel

	A	B	C
1	Male ACT	Female ACT	
2	22	31	
3	24	32	
4	25	25	
5	19	29	
6	20	20	
7	21	21	
8	22	23	
9	18	15	
10	26	26	
11	27	27	
12	19	16	
13	19	19	
14	18	18	
15	23	23	
16	24	24	
17	27	27	
18	18		
19	26		
20	27		
21	24		
22	26		
23	23		
24	24		
25	22		
26	26		
27			
28	22.8	23.5	mean
29	3.10	5.10	s.d.

FIGURE 9.6 ● F-Test Two-Sample for Variances of Males' and Females' ACT Scores

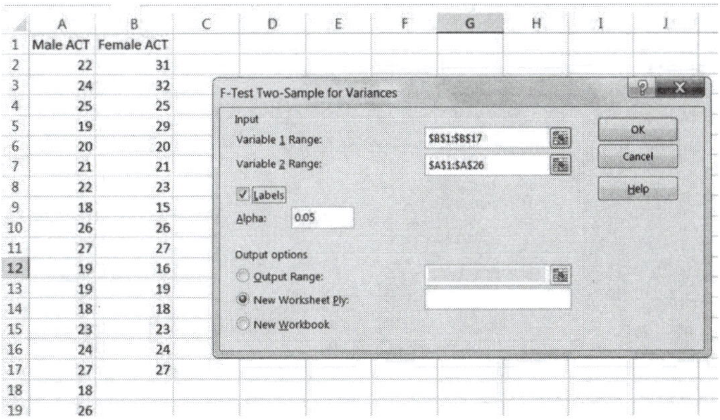

(Continued)

4. The results of the equality of variance test appear in a separate worksheet in Excel as shown in Figure 9.7. The calculated *F* is 2.713 and the critical value of *F* is 2.108. Or you can find the *p* value associated with the calculated *F*, $p = .0142$; thus, $p < .05$. Therefore, either using the critical *F* value or the calculated *p* value, we conclude that equal variances are not assumed in this situation.

FIGURE 9.7 ● The Results for the F-Test Two-Sample for Variances Between Males' and Females' ACT Scores

	A	B	C
1	F-Test Two-Sample for Variances		
2			
3		*Female ACT*	*Male ACT*
4	Mean	23.5	22.8
5	Variance	26	9.583333
6	Observations	16	25
7	df	15	24
8	F	2.713043	
9	P(F<=f) one-tail	0.014214	
10	F Critical one-tail	2.107673	

5. Click on **Sheet1** tab located at the bottom of the Excel to go back to the worksheet where the data are. Click on the **Data** tab, then click on the **Data Analysis**. In the **Analysis Tools** window, select "**t-Test: Two-Sample Assuming Unequal Variances**", then click **OK** as shown in Figure 9.8.

FIGURE 9.8 ● t-Test: Two-Sample Assuming Unequal Variances Between Males' and Females' ACT Scores

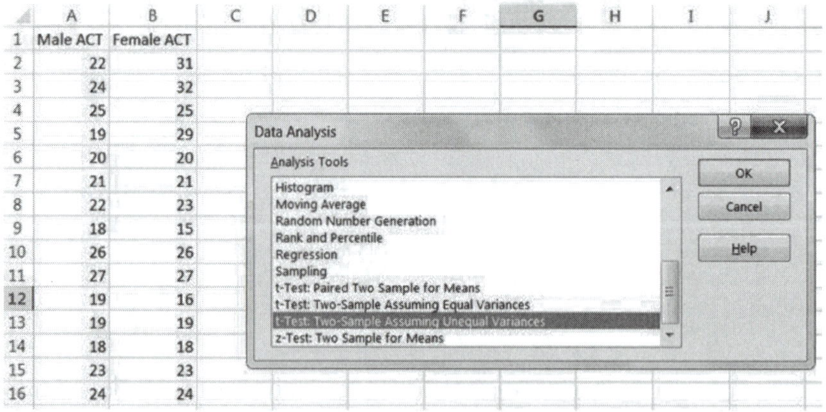

6. Repeat Step 3 to identify the data range needed for this analysis.
7. The results of the "**t-Test: Two-Sample Assuming Unequal Variances**", appear in a separate worksheet in Excel as shown in Figure 9.9. The calculated $t = 0.494$ and the one-tailed critical value of t with $df = 22$ is 1.717. We fail to reject H_0. Or you can use the calculated $p = .313$ to compare with $\alpha = .05$, the result is not significant. Either using the critical t value or the calculated p value associated with the t, we conclude that the difference between men's and women's average ACT scores was not significant at $\alpha = .05$ level. Please note that Excel automatically calculates the *df* according to Satterthwaite's approximated *df* formula and rounded it down, therefore, the $df = 22$.

FIGURE 9.9 ● The Results for the t-Test: Two-Sample Assuming Unequal Variances Between Males' and Females' ACT Scores

	A	B	C
1	t-Test: Two-Sample Assuming Unequal Variances		
2			
3		*Female ACT*	*Male ACT*
4	Mean	23.5	22.8
5	Variance	26	9.583333
6	Observations	16	25
7	Hypothesized Mean Difference	0	
8	df	22	
9	t Stat	0.4939468	
10	P(T<=t) one-tail	0.3131197	
11	t Critical one-tail	1.7171444	
12	P(T<=t) two-tail	0.6262394	
13	t Critical two-tail	2.0738731	

Now you have learned the complete five-step hypothesis testing procedure for the independent-samples *t* tests both by hand and using Excel. Many times, students wonder, "if Excel can do the calculation, why do we still have to learn how to calculate statistical formulas by hand?" The answer became obvious after you compared examples solved by following formulas and calculating by hand versus solved by using Excel. The reason why it is important to learn the correct process and formulas is that have you not learned the five-step hypothesis testing procedure, you would not know to run the "**F-Test Two-Sample for Variances**" first before conducting an independent-samples *t* test. Based on the result of the first test, then select correct test for either "**t-Test: Two-Sample Assuming Equal Variances**", or "**t-Test: Two-Sample Assuming Unequal Variances**". The responsibility to make the right decision is on you because the software will calculate the numbers even if users pick the wrong option. Without adequate understanding of the statistical procedures, there is a 50% chance that users would choose the wrong option. It is clear that equal variances assumed and equal variances not assumed can't be true at the same time. Learning formulas and doing statistical procedures by hand might not be the most efficient way to get to the answers, but it provides a solid foundation to accurately interpret the printouts from statistics software packages.

POP QUIZ

4. Which one of the following characteristics is not required to conduct an independent-samples *t* test?
   a. The observations from the two samples must be independent.
   b. The number of groups needs to be two.
   c. Two samples are randomly selected from two populations.
   d. The two samples have equal variances.
5. When is it appropriate to calculate the pooled variance in conducting an independent-samples *t* test?
   a. When two samples are carefully matched on the group means.
   b. When equal variances are assumed
   c. When two samples are randomly selected from two populations
   d. When equal variances are not assumed

CONFIDENCE INTERVALS OF THE MEAN DIFFERENCES

The confidence interval (CI) is calculated based on the same principle as that introduced in Chapter 8. That is, CI = Point estimate ± Margin of error. The margin of error is the multiplicative product between $t_{\alpha/2}$ and the estimated standard error of the statistic. As discussed in Chapter 8, the $t_{\alpha/2}$ denotes the critical *t* value for a two-tailed α. When putting all together, the $(1-\alpha)100\%$ CI = Point estimate ± $t_{\alpha/2}$(Estimated standard error of the statistic). The higher the confidence level, the wider is the CI.

When conducting independent-samples *t* tests, the CI for $(\mu_1 - \mu_2)$ is expressed as

$$(1-\alpha)100\% \text{ CI for } (\mu_1 - \mu_2) = (\bar{X}_1 - \bar{X}_2) \pm t_{\alpha/2}(s_{(\bar{X}_1 - \bar{X}_2)})$$

When equal variances are assumed, $\sigma_1^2 = \sigma_2^2$, the formula is

$$(1-\alpha)100\% \text{ CI for } (\mu_1 - \mu_2) = (\bar{X}_1 - \bar{X}_2) \pm t_{\alpha/2}\left(\sqrt{\frac{s_p^2}{n_1} + \frac{s_p^2}{n_2}}\right)$$

The lower limit of the (1 – α)100% CI is

$$(\bar{X}_1 - \bar{X}_2) - t_{\alpha/2}\left(\sqrt{\frac{s_p^2}{n_1} + \frac{s_p^2}{n_2}}\right)$$

and the upper limit of the (1 – α)100% CI is

$$(\bar{X}_1 - \bar{X}_2) + t_{\alpha/2}\left(\sqrt{\frac{s_p^2}{n_1} + \frac{s_p^2}{n_2}}\right)$$

When equal variances are not assumed, $\sigma_1^2 \neq \sigma_2^2$, the CI formula is

$$(1-\alpha)100\% \text{ CI for } (\mu_1 - \mu_2) = (\bar{X}_1 - \bar{X}_2) \pm t_{\alpha/2}\left(\sqrt{\frac{s_1^2}{n_1} + \frac{s_2^2}{n_2}}\right)$$

The lower limit of the (1 – α)100% CI is

$$(\bar{X}_1 - \bar{X}_2) - t_{\alpha/2}\left(\sqrt{\frac{s_1^2}{n_1} + \frac{s_2^2}{n_2}}\right)$$

and the upper limit of the (1 – α)100% CI is

$$(\bar{X}_1 - \bar{X}_2) + t_{\alpha/2}\left(\sqrt{\frac{s_1^2}{n_1} + \frac{s_2^2}{n_2}}\right)$$

The interpretation of the CI is very straightforward. If the value of the H_0 is included in the CI, the evidence is not strong enough to support that there is a significant difference between these two group means. If the value of the H_0 is not included in the CI, the evidence is strong enough to support that there is a significant difference between these two group means at the α level. The value of H_0 in the independent-samples t tests is 0. Let's use the numbers in Examples 9.2 and 9.4 to run through these CI calculations and interpretations.

EXAMPLE 9.6

Previously, in Example 9.2, we had 13 men's mean weight, $\bar{X}_1 = 196.08$ pounds, $s_1 = 38.55$ pounds; 15 women's mean weight, $\bar{X}_2 = 147.2$ pounds, $s_2 = 28.27$ pounds; and the estimated standard error of the mean difference, $s_{(\bar{X}_1 - \bar{X}_2)} = 12.66$ pounds. We also established that two groups had equal variances. What is the 95% CI for the difference between men's and women's average weight?

According to the formula, $(1-\alpha)100\%$ CI for $(\mu_1 - \mu_2) = (\bar{X}_1 - \bar{X}_2) \pm t_{\alpha/2}(s_{(\bar{X}_1 - \bar{X}_2)})$, we need to calculate the point estimate and the margin of error.

The point estimate is $(\bar{X}_1 - \bar{X}_2) = (196.08 - 147.2) = 48.88$

The margin of error for 95% CI is the multiplicative product of the two-tailed critical value of t times the estimated standard error of the mean difference. The two-tailed critical value of t under the $df = 26$ and $\alpha = .05$ is 2.056 and the estimated standard error of the mean difference is 12.66, so the error of margin is the multiplication of $2.056 \times 12.66 = 26.029$. Therefore, the low end of the 95% CI is $48.88 - 26.029 = 22.851$ and the high end of the 95% CI is $48.88 + 26.029 = 74.909$. Zero is not included in the 95% CI [22.851, 74.909]; therefore, there is significant difference between men's and women's average weight.

EFFECT SIZES

The independent-samples t test is a procedure utilized by social scientists when they want to study group mean differences among different individuals randomly selected from two different populations. An effect size is defined as a standardized measure of the strength of a test statistic. In independent-samples t tests, effect sizes are calculated as the standardized difference between two sample means in units of standard deviation. Effect size formulas of independent-samples t tests are different under conditions of equal variances assumed versus equal variances not assumed.

When equal variances are assumed, $\sigma_1^2 = \sigma_2^2$, the effect size is calculated as the sample mean difference divided by the estimated pooled standard deviation (Cohen, 1988) and is referred to as **Cohen's *d***.

$$\text{Cohen's } d = \frac{\bar{X}_1 - \bar{X}_2}{s_p}$$

where

$$s_p = \sqrt{\frac{SS_1 + SS_2}{df_1 + df_2}}$$

When equal variances are not assumed, $\sigma_1^2 \neq \sigma_2^2$, it is inappropriate to pool the variances. The effect size is calculated by **Glass's delta (Δ)**. Glass's Δ is defined as the sample mean difference divided by the estimated standard deviation of the control group (Glass et al., 1981; Hedges, 1981). When there is no distinction between the experimental group and the control group, then the larger of s_1 and s_2 should be used as the denominator. This provides a more conservative estimate.

$$\text{Glass's } \Delta = \frac{\bar{X}_1 - \bar{X}_2}{s_{\text{control}}} \text{ or } \frac{\bar{X}_1 - \bar{X}_2}{s_{\text{larger}}}$$

When the effect size equals 1, the interpretation is that two sample means are 1 standard deviation apart from each other. The general interpretation of the magnitude of effect size is that 0.2 is small, 0.5 is medium, and 0.8 is large (Cohen, 1988).

EXAMPLE 9.7

Let's continue working with the numbers calculated from Example 9.2, where 13 men's mean weight was $\bar{X}_1 = 196.08$ pounds, $s_1 = 38.55$ pounds, and 15 women's mean weight, $\bar{X}_2 = 147.2$ pounds, and $s_2 = 28.27$ pounds. The test of equality of variances showed that equal variances assumed between two groups and the pooled variance was $s_p^2 = 1116.28$.

Let's start by choosing the correct formula. Under equal variances assumed situation, the formula is Cohen's $d = \frac{\bar{X}_1 - \bar{X}_2}{s_p}$.

We knew that $s_p^2 = 1116.28$. therefore, $s_p = \sqrt{1116.28} = 33.41$.

$$\text{Cohen's } d = \frac{\bar{X}_1 - \bar{X}_2}{s_p} = \frac{196.08 - 147.2}{33.41} = \frac{48.88}{33.41} = 1.46$$

The effect size of the difference between men's and women's average weight turned out to be large, $d = 1.46$, which means that the average men's weight was 1.46 standard deviations higher than the average women's weight. The difference was significant and practically meaningful.

Next, let's use an example to go through the process of calculating effect size when equal variances are not assumed.

EXAMPLE 9.8

Let's use the comparison of the average BMI between runners and nonrunners. The average BMI of 21 randomly selected nonrunners was 26.89 with an $s = 6.01$, and the average BMI of 20 randomly selected runners was 19.15 with an $s = 3.25$. Calculate the effect size for the group mean difference in this study.

The essential data are as follows:

Nonrunners have larger s than runners, so nonrunners belong in Group 1 and runners belong in Group 2.

Nonrunners: $n_1 = 21$, $\bar{X}_1 = 26.89$, and $s_1 = 6.01$.

Runners: $n_2 = 20$, $\bar{X}_2 = 19.15$, and $s_2 = 3.25$.

The folded F test shows that the calculated $F = \frac{6.01^2}{3.25^2} = 3.42$ is larger than the critical value of $F_{(20,19)} = 2.16$. Or use the **F.DIST.RT(x, df$_1$, df$_2$)** function in Excel to find the p value associated with the calculated F. Select a blank cell in Excel, type "**=F.DIST.RT(3.42,20,19)**", then hit **Enter**. The answer 0.00485 appears. Thus, the p value associated with $F = 3.42$ is .00485. Therefore, we reject H_0. The evidence was strong enough to claim that the two variances were not equal.

When $\sigma_1^2 \neq \sigma_2^2$, it is inappropriate to pool the variances, so the effect size formula should be calculated by Glass's Δ.

$$\text{Glass's } \Delta = \frac{\bar{X}_1 - \bar{X}_2}{s_{\text{control}}}$$

In this problem, we simply use the larger standard deviation between the two groups as the denominator to calculate the effect size.

$$\text{Glass's } \Delta = \frac{\bar{X}_1 - \bar{X}_2}{s_{\text{larger}}} = \frac{26.89 - 19.15}{6.01} = 1.29$$

Effect size = 1.29 is a large effect. It means that the nonrunners' average BMI was 1.29 standard deviations apart from the runners' average BMI. It showed that the effect size of the BMI difference between the runners and nonrunners was large and practically meaningful.

There are two different sets of formulas for conducting an independent-samples t test. The decision on which set of formulas to choose depends on the outcome of the test for the equality of variances. The basic principle is that you only pool the variances together when equal variances are assumed, otherwise you have to treat the variances separately. Such different formulas also apply to effect size calculation. When equal variances are assumed, you use Cohen's $d = (\bar{X}_1 - \bar{X}_2) / s_{\text{pooled}}$ to calculate the effect size. When equal variances are not assumed, you use Glass's $\Delta = (\bar{X}_1 - \bar{X}_2) / s_{\text{control}}$ or use the s_{larger} in the denominator to calculate the effect size.

POP QUIZ

6. Which one of the following formulas is appropriate to calculate the effect size when conducting independent-samples *t* tests with equal variances not assumed?

   a. Cohen's $d = \dfrac{\bar{X}_1 - \bar{X}_2}{s_{pooled}}$

   b. Glass's $\Delta = \dfrac{\bar{X}_1 - \bar{X}_2}{s_{larger}}$

   c. Satterthwaite s approximated $df = \dfrac{(w_1 + w_2)^2}{\dfrac{w_1^2}{n_1 - 1} + \dfrac{w_2^2}{n_2 - 1}}$

   d. $s_{(\bar{X}_1 - \bar{X}_2)} = \sqrt{\dfrac{s_1^2}{n_1} + \dfrac{s_2^2}{n_2}}$

EXERCISE PROBLEMS

1. There are many health benefits of owning a pet. Researchers investigated the effect of owning a pet on loneliness based on a validated loneliness scale in which scores were normally distributed. A group of 30 pet owners was randomly selected and their average loneliness score was $\bar{X} = 49.7$ and $s = 10.5$. A group of 31 non–pet owners was randomly selected and their average loneliness was $\bar{X} = 53.5$ and $s = 11.6$.

   a. Did the data support the claim that owning a pet help people feel less lonely, using $\alpha = .05$?

   b. What is the 95% CI for the difference of loneliness between pet owners and non–pet owners?

   c. What is the effect size for the difference of loneliness between pet owners and non–pet owners?

2. A group of psychologists studied the effect of meditation on anxiety. Participants in the experimental group were given a meditation training before the anxiety assessment. Participants in the control group were given the anxiety assessment

without a meditation training. Each participant was randomly assigned to one of the groups. Participants' anxiety scores were reported in Table 9.3.

a. Conduct a proper statistical procedure to test whether meditation training reduces anxiety level, using $\alpha = .05$. Make sure that you include all the necessary steps in the correct order for your statistical test.

b. Construct a 90% CI for the difference of the average anxiety score between the experimental group and the control group.

TABLE 9.3 ● Participants' Anxiety Scores

Experimental Group	Control Group
$s = 7$	$s = 11$
$\bar{X} = 70$	$\bar{X} = 78$
$SS = 784$	$SS = 1{,}815$

Solutions

1.a.

Step 1: Test the equality of variances.

Substep 1: State the pair of hypotheses regarding variances.

$$H_0 : \sigma_1^2 = \sigma_2^2$$

$$H_1 : \sigma_1^2 \neq \sigma_2^2$$

Substep 2: Identify the criterion for the F test.

For the consistency of labeling the groups, we identified the larger standard deviation, $s_1 = 11.6$, and its $df_1 = n_1 - 1 = 31 - 1 = 30$; so the non–pet owners were labeled as Group 1. We identified the smaller standard deviation, $s_2 = 10.5$, and its $df_2 = 30 - 1 = 29$; so the pet owners are labeled as Group 2.

The critical value of the $F_{(df_1, df_2)} = F_{(30,29)} = 1.85$

Substep 3: Calculate the statistic.

$$F = \frac{s_1^2}{s_2^2} = \frac{11.6^2}{10.5^2} = \frac{134.56}{110.25} = 1.22$$

Substep 4: Make the correct conclusion.

The calculated $F = 1.22$ was smaller than the critical value of $F_{(30,29)} = 1.85$. Or the p value associated with $F = 1.22$ was $p = .2972$. Therefore, we failed to reject H_0. The evidence was not strong enough to reject the null hypothesis. Therefore, the correct independent-samples *t*-test approach was to assume equal variances.

Step 2: State the pair of hypotheses regarding group means.

The problem statement asks, "Did the data support the claim that owning a pet help people feel less lonely?" "Less" is a strong hint for a directional hypothesis. You need to conduct a one-tailed test. If owning a pet help people feel less lonely, pet owners (Group 2) would feel less lonely than non–pet owners (Group 1). Therefore, a right-tailed test is appropriate.

$$H_0: \mu_1 - \mu_2 = 0$$

$$H_1: \mu_1 - \mu_2 > 0$$

Step 3: Identify the rejection zones for the *t* test.

According to the *t* Table for $df = df_1 + df_2 = n_1 + n_2 - 2 = 59$, $\alpha = .05$; it is a right-tailed test, and the closest df in the *t* Table is $df = 60$. Thus, the critical t value that sets the boundary of rejection zone is 1.671. The rejection zone for a two-tailed test is $t > 1.671$.

Step 4: Calculate the *t* statistic.

When $\sigma_1^2 = \sigma_2^2$, the variances need to be pooled together.

$$SS_1 = s_1^2 df_1 = 11.6^2(30) = 4036.8$$

$$SS_2 = s_2^2 df_2 = 10.5^2(29) = 3197.25$$

$$s_p^2 = \frac{SS_1 + SS_2}{df_1 + df_2} = \frac{4036.8 + 3197.25}{30 + 29} = \frac{7234.05}{59} = 122.61$$

$$s_{(\bar{X}_1 - \bar{X}_2)} = \sqrt{\frac{s_p^2}{n_1} + \frac{s_p^2}{n_2}} = \sqrt{\frac{122.61}{31} + \frac{122.61}{30}} = \sqrt{3.955 + 4.087} = \sqrt{8.042} = 2.836$$

$$t = \frac{\bar{X}_1 - \bar{X}_2}{s_{(\bar{X}_1 - \bar{X}_2)}} = \frac{53.5 - 49.7}{2.836} = \frac{3.8}{2.836} = 1.340$$

Step 5: Make the correct conclusion.

The calculated $t = 1.340$ was not within the rejection zone. Or you can use the **T.DIST.RT(x,df)** function in Excel to find the p value associated with the right-tailed *t* test. Select a blank cell in Excel, type "**=T.DIST.RT(1.340,59)**", then hit **Enter**. The answer 0.09269 appears. The p value associated with the right-tailed t = 1.340 was .09269, which was greater than the level. Therefore, we failed to reject H_0. The evidence was not strong enough to support the claim that owning a pet help people feel less lonely.

1.b. You need to find the 95% CI for the difference of loneliness between pet owners and non–pet owners. According to the formula, $(1-\alpha)100\%$ CI for $(\mu_1 - \mu_2) = (\bar{X}_1 - \bar{X}_2) \pm t_{\alpha/2} s_{(\bar{X}_1-\bar{X}_2)}$, we need to calculate the point estimate and the margin of error.

The point estimate is $(\bar{X}_1 - \bar{X}_2) = 53.5 - 49.7 = 3.8$

The margin of error is $t_{\alpha/2} s_{(\bar{X}_1-\bar{X}_2)}$. When df = 59, two-tailed α = .05, the critical value of t = 2.000 and $s_{(\bar{X}_1-\bar{X}_2)} = 2.836$, the margin of error is 5.672.

95% CI for $(\mu_1 - \mu_2)$ was 3.8 ± 5.672. The low end and high end of the 95% CI were [−1.872, 9.472]. Zero was included in the 95% CI; therefore, there was no significant differences between pet owners' and non–pet owners' loneliness scores.

1. c. When $\sigma_1^2 = \sigma_2^2$, the effect size is calculated as Cohen's $d = (\bar{X}_1 - \bar{X}_2) / s_p$. Based on the answer from 1.a, $s_p = \sqrt{122.61} = 11.07$.

$$\text{Effect size } d = \frac{\bar{X}_1 - \bar{X}_2}{s_p} = \frac{53.5 - 49.7}{11.07} = \frac{3.8}{11.07} = 0.34$$

The effect size of the loneliness scores between pet owners and non–pet owners was 0.34 standard deviation apart from each other. The effect was small.

2. a.

Step 1: Test the equality of variances.

Substep 1: State the pair of hypotheses regarding variances.

$$H_0 : \sigma_1^2 = \sigma_2^2$$

$$H_1 : \sigma_1^2 \neq \sigma_2^2$$

Substep 2: Identify the rejection zone for the F test.

Neither df nor n was explicitly stated in the problem, so you need to use the formula of $SS = s^2 \times df$ to find out the df for both groups. When two values are provided in this formula, you can figure out the third one. Let's figure out one group at a time. For the consistency of labeling the groups, the larger standard deviation was $s_1 = 11$; thus, the control group was labeled as Group 1.

$$SS_1 = s_1^2 \times df_1$$

$$1815 = 11^2 \times df_1$$

$$df_1 = \frac{SS_1}{s_1^2} = \frac{1815}{121} = 15$$

The experimental group was labeled as Group 2:

$$784 = 7^2 \times df_2$$

$$df_2 = \frac{SS_2}{s_2^2} = \frac{784}{49} = 16$$

The critical value of F is $F_{(15,16)} = 2.35$. The rejection zone is $F > 2.35$.

Substep 3: Calculate the test statistic.

$$F = \frac{s_1^2}{s_2^2} = \frac{11^2}{7^2} = \frac{121}{49} = 2.47$$

Substep 4: Make the correct conclusion.

The calculated $F = 2.47$ was larger than the critical value of $F_{(15,16)} = 2.35$. Or the p value associated with $F = 2.47$ was .0413. Therefore, we rejected H_0. The evidence was strong enough to claim that the two variances were not equal. Therefore, the correct independent-samples *t*-test approach is to assume unequal variances.

Step 2: State the pair of hypotheses regarding group means.

The problem asked to conduct a test on whether meditation training reduces anxiety level. *Reduce* is a key word for direction; therefore, we need to conduct a one-tailed test. The control group (Group 1) did not get the mediation training so the anxiety would be higher; thus, $\mu_1 > \mu_2$. Therefore, a right-tailed test is appropriate. Be aware that the designation of μ_1 and μ_2 needs to be consistent with s_1 and s_2. In a one-tailed test, it is important to identify the correct direction in the alternative hypothesis, H_1.

H_0: $\mu_1 - \mu_2 = 0$

H_1: $\mu_1 - \mu_2 > 0$

Step 3: Identify the rejection zones for the *t* test.

We need to calculate the *df* for the *t* test to identify the rejection zone.

$$df = \frac{(w_1 + w_2)^2}{\frac{w_1^2}{n_1 - 1} + \frac{w_2^2}{n_2 - 1}}$$

where

$$w_1 = \frac{s_1^2}{n_1} = \frac{11^2}{16} = 7.5625$$

and

$$w_2 = \frac{s_2^2}{n_2} = \frac{7^2}{17} = 2.8824$$

$$df = \frac{(w_1 + w_2)^2}{\frac{w_1^2}{n_1 - 1} + \frac{w_2^2}{n_2 - 1}} = \frac{(7.5625 + 2.8824)^2}{\frac{7.5625^2}{15} + \frac{2.8824^2}{16}} = 25.18$$

The approximated $df = 25.18$ is rounded down to 25.

According to the *t* Table, when $df = 25$, $\alpha = .05$; it is a right-tailed test, thus the critical *t* value that sets the boundary of the rejection zone is 1.708. The rejection zone for a right-tailed test is $t > 1.708$. If you choose to use the smaller one of df_1 and df_2, then the *df* for the *t* test is 15 and its rejection zone is $t > 1.753$.

Step 4: Calculate the *t* statistic.

When $\sigma_1^2 \neq \sigma_2^2$, the variances need to be treated separately.

$$s_{(\bar{X}_1 - \bar{X}_2)} = \sqrt{\frac{s_1^2}{n_1} + \frac{s_2^2}{n_2}} = \sqrt{\frac{11^2}{16} + \frac{7^2}{17}} = 3.23$$

$$t = \frac{\bar{X}_1 - \bar{X}_2}{s_{(\bar{X}_1 - \bar{X}_2)}} = \frac{78 - 70}{3.23} = \frac{8}{3.23} = 2.477$$

Step 5: Make the correct conclusion.

The calculated $t = 2.477$ is within the rejection zone. Or the probability value associated with $t = 2.477$ is $p < .0001$. Therefore, we reject H_0. The evidence was strong enough to support the claim that meditation could help reduce anxiety levels.

2. b. To construct a 90% CI for $(\mu_1 - \mu_2)$, the point estimate is $(\bar{X}_1 - \bar{X}_2) = (78 - 70) = 8$ and the error margin depends on the *df* you chose for the *t* test. Since there were two methods to calculate the *df*, we would include the 90% CI under both.

If you used Satterthwaite's approximated $df = 25$, the margin of error is the multiplication of the critical value of t and the estimated standard error of the mean difference, $1.708(3.23) = 5.52$. Therefore, the 90% CI for $(\mu_1 - \mu_2)$ is $8 \pm 5.52 = [2.48, 13.52]$.

If you used the smaller of df_1 and df_2, $df = 15$, the margin of error is the multiplication of the critical value of t and the estimated standard error of the mean difference, $1.753(3.23) = 5.66$. Therefore, the 90% CI for $(\mu_1 - \mu_2)$ is $8 \pm 5.66 = [2.34, 13.66]$.

In either of the two methods, the 90% CI for $(\mu_1 - \mu_2)$ did not include 0; therefore, there was significant difference in the anxiety scores between the experimental group and the control group.

What You Learned

Chapter 9 demonstrates a practical statistical procedure to compare the means from two groups consisting of different individuals without requiring knowledge of any population parameters, such as μ or σ. When comparing two independent samples, one of two situations occurs: (1) two groups have equal variances or (2) two groups have unequal variances. Therefore, the test of the equality of variances needs to be performed first to determine whether the two groups have equal variances or not. It becomes the Step 1 of the five-step hypothesis testing process for the independent-samples *t* test.

Step 1: Test the equality of variances.

There are four substeps for conducting a *folded F test* to test the equality of variances:

Substep 1: State the pair of hypotheses.

$$H_0: \sigma_1^2 = \sigma_2^2$$

$$H_1: \sigma_1^2 \neq \sigma_2^2$$

Substep 2: Identify the rejection zone for the F test.

The larger standard deviation is s_1 and its $df_1 = n_1 - 1$; the smaller standard deviation is s_2, and its $df_2 = n_2 - 1$. The critical value of the F is $F_{(df_1, df_2)}$.

Substep 3: Calculate the statistic.

$$F = \frac{s_1^2}{s_2^2} = \frac{\text{Larger variance}}{\text{Smaller variance}}$$

Substep 4: Make the correct conclusion.

When the calculated F is larger than the critical value, $F_{(df_1, df_2)}$ or the p value associated with the F is less than the α level, we reject H_0. The evidence is strong enough to support the claim that the two variances are unequal. If the calculated F is smaller than or equal to the critical value, $F_{(df_1, df_2)}$ or the p value associated with the F is greater or equal to the α level, we fail to reject H_0. In this case, the evidence is not strong enough to support the claim that the two variances are unequal. Thus, we have to assume equal variances. Different sets of formulas for the independent-samples t test are required under different situations. The decision about which set to use is based on the results of the test for the equality of variances.

When $\sigma_1^2 = \sigma_2^2$, the following hypothesis testing procedure is appropriate.

Step 2: State the pair of hypotheses regarding group means.

Two-tailed test: H_0: $\mu_1 - \mu_2 = 0$

H_1: $\mu_1 - \mu_2 \neq 0$

Right-tailed test: H_0: $\mu_1 - \mu_2 = 0$

H_1: $\mu_1 - \mu_2 > 0$

Left-tailed test: H_0: $\mu_1 - \mu_2 = 0$

H_1: $\mu_1 - \mu_2 < 0$

Step 3: Identify the rejection zone for the t test.

Use three pieces of information: $df = df_1 + df_2 = n_1 + n_2 - 2$, α level, and two-tailed test versus one-tailed test to determine the critical t value that sets the boundary of the rejection zone.

Step 4: Calculate the t statistic.

When $\sigma_1^2 = \sigma_2^2$, the variances need to be pooled together.

$$SS_1 = s_1^2 df_1$$

$$SS_2 = s_2^2 df_2$$

$$s_p^2 = \frac{SS_1 + SS_2}{df_1 + df_2}$$

$$s_{(\bar{X}_1-\bar{X}_2)} = \sqrt{\frac{s_p^2}{n_1} + \frac{s_p^2}{n_2}}$$

$$t = \frac{\bar{X}_1 - \bar{X}_2}{s_{(\bar{X}_1-\bar{X}_2)}}$$

Step 5: Make the correct conclusion.

If the calculated t value is within the rejection zone or the p value associated with the t is less than the α level, we reject H_0. If the calculated t value is not within the rejection zone or the p value associated with the t is greater or equal to the α level, we fail to reject H_0.

When $\sigma_1^2 \neq \sigma_2^2$, the following hypothesis testing procedure is appropriate.

Step 2: State the pair of hypotheses regarding group means.

Two-tailed test: H_0: $\mu_1 - \mu_2 = 0$

H_1: $\mu_1 - \mu_2 \neq 0$

Right-tailed test: H_0: $\mu_1 - \mu_2 = 0$

H_1: $\mu_1 - \mu_2 > 0$

Left-tailed test: H_0: $\mu_1 - \mu_2 = 0$

H_1: $\mu_1 - \mu_2 < 0$

Step 3: Identify the rejection zone for the t test.

$$\text{Satterthwaite s approximated } df = \frac{(w_1 + w_2)^2}{\frac{w_1^2}{n_1 - 1} + \frac{w_2^2}{n_2 - 1}}$$

where

$$w_1 = \frac{s_1^2}{n_1}$$

and

$$w_2 = \frac{s_2^2}{n_2}$$

The calculated Satterthwaite's approximated *df* is not likely to be an integer, and it needs to be rounded down. Use the approximated *df*, α level, and the knowledge of two-tailed versus one-tailed test to determine the critical *t* value that sets the boundary of the rejection zone. Alternatively, you may choose the smaller of df_1 and df_2 as the *df* for the independent-samples *t* test.

Step 4: Calculate the *t* statistic.

When $\sigma_1^2 \neq \sigma_2^2$, it is inappropriate to pool the variances, so they are treated separately.

$$s_{(\bar{X}_1-\bar{X}_2)} = \sqrt{\frac{s_1^2}{n_1} + \frac{s_2^2}{n_2}}$$

$$t = \frac{\bar{X}_1 - \bar{X}_2}{s_{(\bar{X}_1-\bar{X}_2)}}$$

Step 5: Make the correct conclusion.

If the calculated *t* value is within the rejection zone or the *p* value associated with the *t* is less than the α level, we reject H_0. If the calculated *t* value is not within the rejection zone or the *p* value associated with the *t* is greater or equal to the α level, we fail to reject H_0.

Different sets of formulas also apply to CI and effect size calculations depending on whether equal variances are assumed or not.

When $\sigma_1^2 = \sigma_2^2$, the CI is calculated as $(1-\alpha)100\%\text{ CI for }(\mu_1-\mu_2) = (\bar{X}_1 - \bar{X}_2) \pm t_{\alpha/2}\left(\sqrt{\frac{s_p^2}{n_1} + \frac{s_p^2}{n_2}}\right)$ and the effect size is calculated as Cohen's *d*.

$$\text{Cohen's } d = \frac{\bar{X}_1 - \bar{X}_2}{s_{\text{pooled}}}$$

where

$$s_{\text{pooled}} = \sqrt{\frac{SS_1 + SS_2}{df_1 + df_2}}$$

When $\sigma_1^2 \neq \sigma_2^2$,

$$(1-\alpha)100\%\text{ CI for }(\mu_1-\mu_2) = (\bar{X}_1 - \bar{X}_2) \pm t_{\alpha/2}\left(\sqrt{\frac{s_1^2}{n_1} + \frac{s_2^2}{n_2}}\right)$$

and the effect size is calculated as Glass's Δ.

$$\text{Glass's } \Delta = \frac{\bar{X}_1 - \bar{X}_2}{s_{\text{control}}} \quad \text{or} \quad \Delta = \frac{\bar{X}_1 - \bar{X}_2}{s_{\text{larger}}}$$

Key Words

Cohen's *d*: Cohen's $d = \frac{\bar{X}_1 - \bar{X}_2}{s_{\text{pooled}}}$ is the formula to calculate effect sizes for independent-samples *t* tests when equal variances are assumed. 278

Folded *F* test: The folded *F* test means that in calculating the *F* test, the larger variance is designated as the numerator and the smaller variance as the denominator; the calculated *F* value will always be larger than 1 to avoid the left-tailed *F* values. 258

Glass's delta (Δ): Glass's $\Delta = \frac{\bar{X}_1 - \bar{X}_2}{s_{\text{control}}}$ or $\Delta = \frac{\bar{X}_1 - \bar{X}_2}{s_{\text{larger}}}$ is the formula to calculate effect sizes for independent-samples *t* tests when equal variances are not assumed. 279

Independent samples: Independent samples refer to samples selected from different populations where the values selected from one population are not related to those from the other population. Independent samples contain different individuals in each sample. 254

***p* value associated with a test statistic:** The *p* value associated with a test statistic is defined as the probability of obtaining the magnitude of the calculated test statistic assuming that H_0 is true. 259

Learning Assessment

Multiple Choice: Circle the Best Answer to Every Question

1. The folded *F* test, $F = s_1^2 / s_2^2 = \text{Larger variance} / \text{Smaller variance}$, is a test for
   a. equality of means
   b. equality of variances
   c. equality of correlations
   d. equality of probabilities
2. When $\sigma_1^2 \neq \sigma_2^2$, the correct procedure to compare means is to
   a. pool the variances together
   b. pool the degrees of freedom together

c. leave the variances the way they are when calculating the standard error of the mean difference

d. pool the means together

3. When $\sigma_1^2 = \sigma_2^2$, the correct formula for calculating the effect size for the independent-samples *t* test is
   a. Cohen's *d*
   b. Satterthwaite's *df*
   c. the folded *F* test
   d. Glass's Δ

4. When $\sigma_1^2 = \sigma_2^2$, the correct *df* for the independent-samples *t* test is
   a. $df = df_1 + df_2$
   b. Satterthwaite's *df*
   c. $df = n_1 + n_2 - 1$
   d. $df = n_1 + n_2$

5. When $\sigma_1^2 \neq \sigma_2^2$, the correct formula for calculating the effect size for the independent-samples *t* test is
   a. Cohen's *d*
   b. Satterthwaite's *df*
   c. the folded *F* test
   d. Glass's Δ

6. With $\alpha = .05$ and $df = 8$, the rejection zone for a right-tailed *t* test is $t > 1.860$. Assuming that all other factors are held constant, if the *df* increases to 20, the critical value of *t* that sets the boundary for the rejection zone would
   a. increase
   b. decrease
   c. remain the same
   d. not be possible to determine based solely on the information provided here

Free Response Questions

7. A statistics professor was interested in the effect of frequent quizzes on the retention of statistics knowledge. She taught two large sections of statistics. In one of the sections, she assigned quizzes twice a week. In the other section, she did not assign quizzes. Everything else remained the same for both sections. Students could only sign up for one or the other section but not both. Assume that statistics exam scores were normally distributed. At the end of the semester, both sections received the same final exam, and the scores from a randomly selected sample from each section were reported in Table 9.4.

TABLE 9.4 ⬢ Students' Final Exam Scores

Section With Quizzes	Section Without Quizzes
$n = 15$	$n = 13$
$\bar{X} = 80$	$\bar{X} = 75$
$s = 8.2$	$s = 12.5$

a. Conduct a proper statistical procedure to test whether giving quizzes increases the retention of statistics knowledge, as demonstrated by the average final exam scores in these two sections, using $\alpha = .05$.

b. What is the 95% CI for the difference of final exam scores between these two sections?

c. Calculate and interpret the effect size of the difference in the average final exam scores between these two sections.

8. Educators evaluated the effect of remote learning on students' grades. A group of 21 students was randomly selected among students who chose to learn in person. At the end of the semester, their average GPA was 3.4, and *SS* for the GPA was 12.8. A group of 25 students was randomly selected among students who chose to learn online. At the end of the semester, their average GPA was 3.0, and *SS* for the GPA was 19.44. Use $\alpha = .05$ to answer the following questions:

   a. Conduct a proper statistical procedure to test whether the types of learning have a significant effect on students' GPA.

   b. What is the 90% CI for the differences of GPA between these two types of learning?

   c. Calculate and interpret the effect size of the difference in students' average GPA between these two types of learning.

9. An individual's social skills are learned through face-to-face interactions over time. Researchers examined the effect of receiving a cell phone in early childhood on teenagers' social skills. Social skills scores were normally distributed. Higher scores meant better social skills. The researchers measured and compared two groups of teenagers. One group received cell phones before age 5, and the other group received cell phones after age 5. The results were reported in Table 9.5.

TABLE 9.5 ⬢ Social Skills of two Groups of Teenagers

Receiving Cell Phone Before Age 5	Receiving Cell Phone After Age 5
$n = 10$	$n = 13$
$\bar{X} = 60.5$	$\bar{X} = 73.8$
$SS = 345.96$	$SS = 1559.52$

a. Conduct a proper statistical procedure to test whether these two groups of teenagers have different social skills, using $\alpha = .05$.

b. Calculate and interpret the effect size of the difference in the teenagers' social skills between these two groups.

c. Construct a 90% CI for the difference in social skills between these two groups.

10. The BMI is the ratio of a person's weight measured in kilograms to the square of the person's height measured in meters. BMI measures are normally distributed. A group of 15 runners and a group of 16 nonrunners were randomly selected and their BMIs were reported in the following table. Use $\alpha = .05$ to answer the following questions.

    a. Use Excel to test whether the average runner's BMI differs from that of nonrunners?

    b. Construct a 95% CI for the difference of BMI between runners and nonrunners.

    c. Calculate and interpret the effect size of the difference in the BMI between runners and nonrunners.

Runners	Nonrunners
21.71	30.41
19.67	25.93
21.93	29.36
28.56	28.63
18.94	32.45
22.38	22.19
23.44	23.27
20.82	24.98
20.83	19.46
18.75	31.82
21.08	31.03
19.11	24.16
18.06	39.86
25.49	25.76
26.63	26.97
	29.08

Answers to Pop Quiz Questions

1. a
2. c
3. a
4. d
5. b
6. b

10

CORRELATION

Learning Objectives

After reading and studying this chapter, you should be able to do the following:

- Define and explain the purpose of Pearson's product moment correlation, r
- Conduct a hypothesis test on Pearson's r
- Explain the coefficient of determination
- Describe how to purify the relationship between two primary variables by keeping the third variable constant using partial correlation

WHAT YOU KNOW AND WHAT IS NEW

It makes sense to introduce the concept of *correlation* first because it is different from the various t tests that we have covered so far. The t tests are used to compare two group means, the test variable is an interval or ratio variable, and the grouping variable is a nominal variable to identify the groups (i.e., pet owners vs. non–pet owners, control group vs. experimental group, or before-measures vs. after-measures). However, the correlation is a statistical tool to measure the strength and direction of the relationship between two interval or ratio variables. Correlations allow us to see if these two variables are moving in the same direction (i.e., when one goes up, the other also goes up) or moving in opposite directions (i.e., when one goes up, the other goes down). The "up" or "down" is relative to the sample mean of that particular variable. For example, we want to study the relationship between hours spent on a treadmill and calories burned. It is logical to expect that the more hours spent on a treadmill, the more calories burned. If we spent more than average hours on a treadmill, we were likely to burn higher than average calories. If we spent less than average hours on a treadmill, we were likely to burn lower than average calories. Therefore, before we calculate a correlation, we need to calculate the means for the two variables first. Once you understand correlation conceptually, the calculation is easy.

In Chapter 3, you learned to calculate means and sum of squares (*SS*), which sets the foundation for Chapter 10. Calculating the relationship between two interval or ratio variables involves figuring out means and *SS* of both variables. The means and *SS* are the very basic statistics you already learned. Now, you simply need to know how to put means and *SS* together to show if two variables are moving in the same direction or in opposite directions. Detailed formulas for the correlation are described in the next section.

PEARSON'S CORRELATION

The *Pearson product moment correlation coefficient*, or simply called **Pearson's correlation, *r*,** is a statistical procedure that calculates a linear relationship between two interval or ratio variables. Sometimes, it is simply referred to as the correlation. The Pearson's correlation formulas provide objective mathematical operations to measure the strength and direction of the linear relationship between two interval or ratio variables. Because any two interval or ratio variables can generate a correlation coefficient, it is important to report a correlation coefficient with the pair of variables involved in the calculation clearly identified. We will introduce two different formulas to calculate Pearson's *r*, describe and interpret Pearson's *r*, and use examples to demonstrate the entire process.

Pearson's Correlation Formulas

Pearson's correlation is logically simple and mathematically designed to measure the strength and direction of the relationship between two variables, *X* and *Y*. I will explain the formula and everything that goes into it.

$$\text{Pearson's correlation } r = \frac{SP}{\sqrt{SS_X SS_Y}}$$

where

SP is the sum of cross products $= \Sigma(X - \bar{X})(Y - \bar{Y})$

$SS_X = \Sigma(X - \bar{X})^2$ is the sum of squares of X

$SS_Y = \Sigma(Y - \bar{Y})^2$ is the sum of squares of Y

The SS_X and SS_Y are familiar terms from earlier chapters, and both are always positive. Therefore, the denominator in the *r* formula is always positive. Only $SP = \Sigma(X - \bar{X})(Y - \bar{Y})$ is a new term. SP stands for the sum of cross products, also called covariance. The multiplication of $(X - \bar{X})$ and $(Y - \bar{Y})$ provides the indication of whether the variables *X* and *Y* are moving in the same direction or in opposite directions. When *X* and *Y* are moving in the same direction, they are likely to be above $\bar{X}$ and $\bar{Y}$ at the same time or below $\bar{X}$ and $\bar{Y}$ at the same time; therefore, the cross products $(X - \bar{X})(Y - \bar{Y})$ are likely to be positive. When *X* and *Y* are moving in opposite directions, one is likely to be above its mean (i.e., positive difference) and the other is likely to be below its mean (i.e., negative difference); therefore, the cross products $(X - \bar{X})(Y - \bar{Y})$ are likely to be negative. The summation (Σ) sums up the cross products for every data point and provides an overall trend of the relationship between *X* and *Y*. When SP is positive, it means that *X* and *Y* are mostly moving in the same direction; therefore, the correlation *r* is positive. When SP is negative, it means that *X* and *Y* are mostly moving in the opposite directions; therefore, the correlation *r* is negative.

Let's use two graphs as shown in Figures 10.1a and b to illustrate the SP. When we want to investigate the relationship between *X* and *Y*, we can plot the data points (X_i, Y_i) on a two-dimensional graph with each dot showing the values of the variables on both *X* and *Y* axes. Now calculate and ; then draw a line on the graph to mark the mean value on each axis. The graph is divided into four quadrants by the $\bar{X}$ and $\bar{Y}$ Let's label them as ++, +–, ––, and –+ based on the outcome of $(X - \bar{X})$ and $(Y - \bar{Y})$. When most data points are in the ++ or –– quadrants, the SP is positive and Pearson's *r* is positive as shown in Figure 10.1a. When most data points are in the +– or –+ quadrants, the SP is negative and Pearson's *r* is negative as shown in Figure 10.1b.

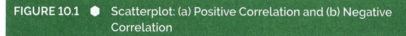

FIGURE 10.1 ● Scatterplot: (a) Positive Correlation and (b) Negative Correlation

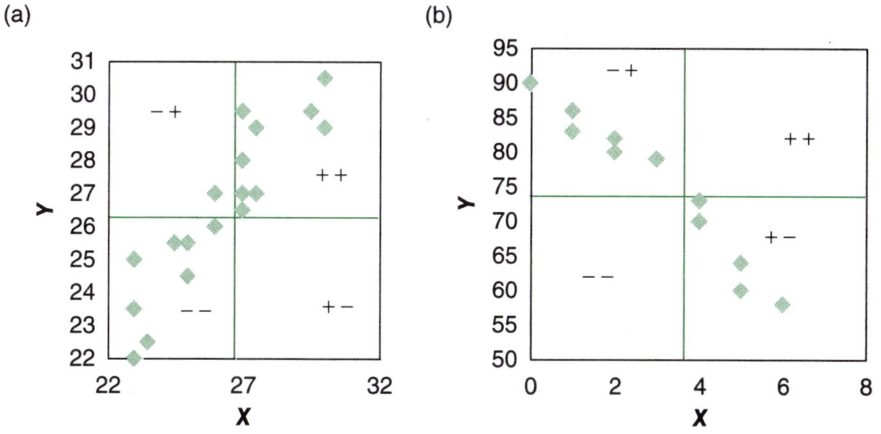

You may also calculate Pearson's correlation from Z scores (which you learned in Chapter 4). This formula is mathematically the same as the first Pearson's r formula just presented in a slightly different format. First, transform the raw scores X and Y into Z scores: Z_X and Z_Y.

Given Z scores, the Pearson correlation formula is

$$r = \frac{\sum Z_X Z_Y}{(n-1)}$$

where n = the number of paired variables, X and Y.

As discussed in Chapter 4, the Z score provides the exact location of each data point in the distribution. When the Z scores are positive, the raw values are above the mean. When the Z scores are negative, the raw values are below the mean. When calculating the correlation between variables X and Y, simply convert X and Y into Z scores, Z_X and Z_Y, multiply them to create $Z_X Z_Y$, and then sum them up and divide it by $(n - 1)$.

Pearson's r is used to investigate the linear relationship between two interval or ratio variables. Even strong correlations do not prove causation. Strong correlation shows that the two variables have a tendency of showing up together and they have shared variances. One of the most important applications of Pearson's r in psychological research is to calculate *reliability* and *validity* of a newly designed instrument (test or assessment) to measure a specific psychological construct. Reliability and validity are essential concepts in research methods. However, the focus of this book is not research methods. I will only provide a very brief introduction about these two important concepts and show how correlations can be used in this context.

Reliability measures the stability, consistency, and replicability of an instrument. When a psychological instrument is reliable, it is supposed to produce the same result repeatedly when used to measure the same person. One of the most commonly used methods to measure reliability is *test–retest reliability*. Test–retest reliability is measured by giving the instrument to a group of people once and giving the same instrument to the same group of people again, and then calculating the correlation between scores from the first test and the scores from the second test. *Validity* refers to the extent that the instrument actually measures what it is supposed to measure. One of the most commonly used validity measures that actually requires empirical evidence is criterion-related validity.

Criterion-related validity is used to demonstrate the accuracy of a new measure by correlating its results with a well-established and validated measure. Assume that a newly designed 10-minute screening instrument for depression is given to a group of volunteers with different levels of depression symptoms and the same group is also given a comprehensive depression diagnostic test that is well-established with high validity. The correlation between the results from the 10-minute screening tool and the comprehensive depression diagnostic test can be used to demonstrate the validity of the 10-minute screening tool. We will have more discussions on the interpretation of correlation in the next section.

Let's use an example to demonstrate how to conduct a Pearson's correlation. Let's investigate about the relationship between adults' height and weight. Height and weight are both ratio variables. We don't expect that there is a cause-and-effect relationship between height and weight, but we suspect that height and weight are related. Adults' height and weight are both normally distributed. A group of 20 adults was randomly selected and their height and weight measures were reported in Table 10.1.

TABLE 10.1 ● Height and Weight Measures From 20 Adults

Adult	Height (inches)	Weight (pounds)
1	63.5	170
2	67	140
3	67	178
4	67	160
5	75	225
6	66.5	135
7	59	97
8	63	108
9	66	127
10	60	120
11	74.5	180
12	74	200
13	69	170
14	64.5	118
15	68	170
16	66.5	130
17	63	170
18	64.5	130
19	53	110
20	55	134

It is not possible to figure out the relationship between height and weight by staring at this table. Although Pearson's r is one of the most common statistical tools to measure the relationship between two interval or ratio variables, it is not applicable in all situations. When the relationship between two variables does not form a straight line, Pearson's r is not the appropriate option to measure the relationship. Therefore, we are putting off the urge to jump into calculating Pearson's r until we discuss how to evaluate the appropriateness of applying Pearson's r to measure the relationship between two variables. It is also crucial to learn how to interpret the results of Pearson's r.

Describing and Interpreting Pearson's *r*

A **scatterplot** is commonly used as a preliminary tool to verify whether the relationship between two interval or ratio variables is linear or not. A scatterplot between two variables is defined as a graph of dots plotted in a two-dimensional space. Each dot contains a pair of values on the *X* and *Y* axes. In this case, simply use height as the *X*-axis and weight as the *Y*-axis. Plot each person's height and weight reported in Table 10.1 in a scatterplot as shown in Figure 10.2a. The step-by-step instructions on how to create a scatterplot in Excel are provided later in this chapter.

After creating a scatterplot for two variables, you need to learn to extract information out of it. In other words, you need to interpret the scatterplot.

FIGURE 10.2a ● Scatterplot of Height and Weight of 20 Adults

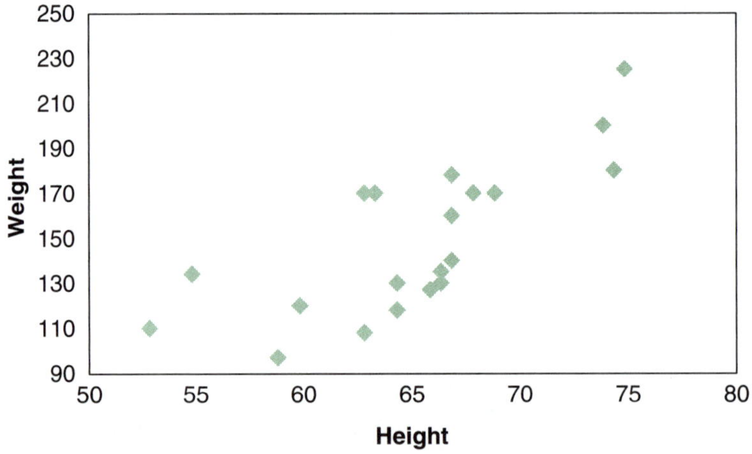

The Shape of the Relationship: Linear or Curvilinear

The way the data points spread across the plot tells us whether a straight line or a curve fits the data better. Pearson's correlation assumes a linear relationship between two variables. It is not the appropriate statistics to use when the relationship is curvilinear. How can we tell when the relationship is linear? When we draw a straight line to fit the dots in the graph, the line fits all dots fairly well. Let's examine Figure 10.2a on the shape of the relationship. Draw a straight line to fit the dots as shown in Figure 10.2b.

FIGURE 10.2b ● Draw a Straight Line to Fit the Scatterplot of Height and Weight of 20 Adults

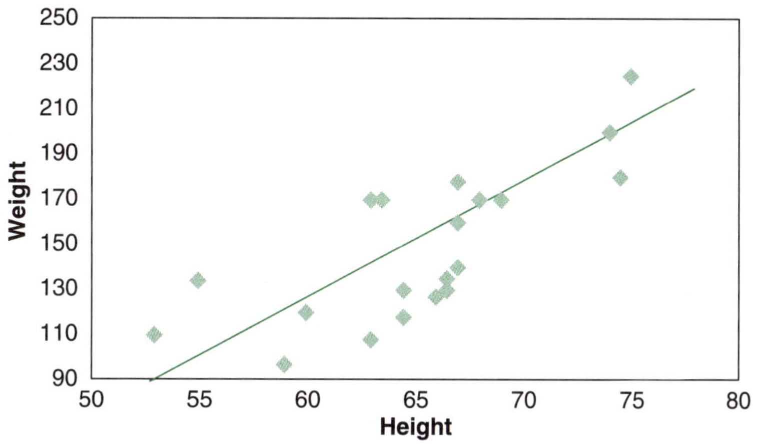

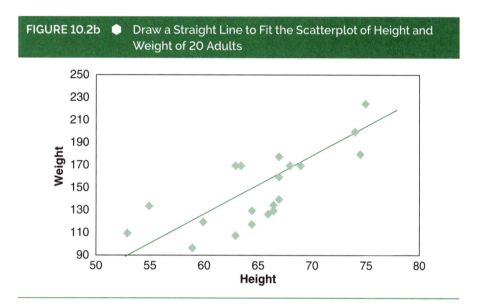

Intuitively we can see that the straight line fits the dots in the entire distribution well. This indicates that a linear relationship exists between height and weight. Scatterplots can convey a variety of information about the relationship between two variables, but you have to know what to look for. Let's look at what information can be conveyed by a number of scatterplots in the next example.

EXAMPLE 10.1

Researchers gathered measures of left arm and right arm from a randomly selected sample. Most people's left-arm measures were closely matched with their right-arm measures. The scatterplot of these two measures were closely clustered around a straight line as shown in Figure 10.3a. Another

(Continued)

(Continued)

group of researchers gathered numbers on the hours of sleep students got the night before an exam and their exam grades. Exam grades could be influenced by many factors, such as students' intelligence, study skills, test-taking skills, and a good night's sleep. Therefore, it is unlikely that there is a strong relationship between the hours of sleep and exam grades. The scatterplot of these two variables is shown in Figure 10.3b. Which one of the scatterplots shown in Figures 10.3a and b indicates that correlation is an appropriate statistic to investigate the relationship between the two variables?

FIGURE 10.3a–d ● Linear or Curvilinear Relationship

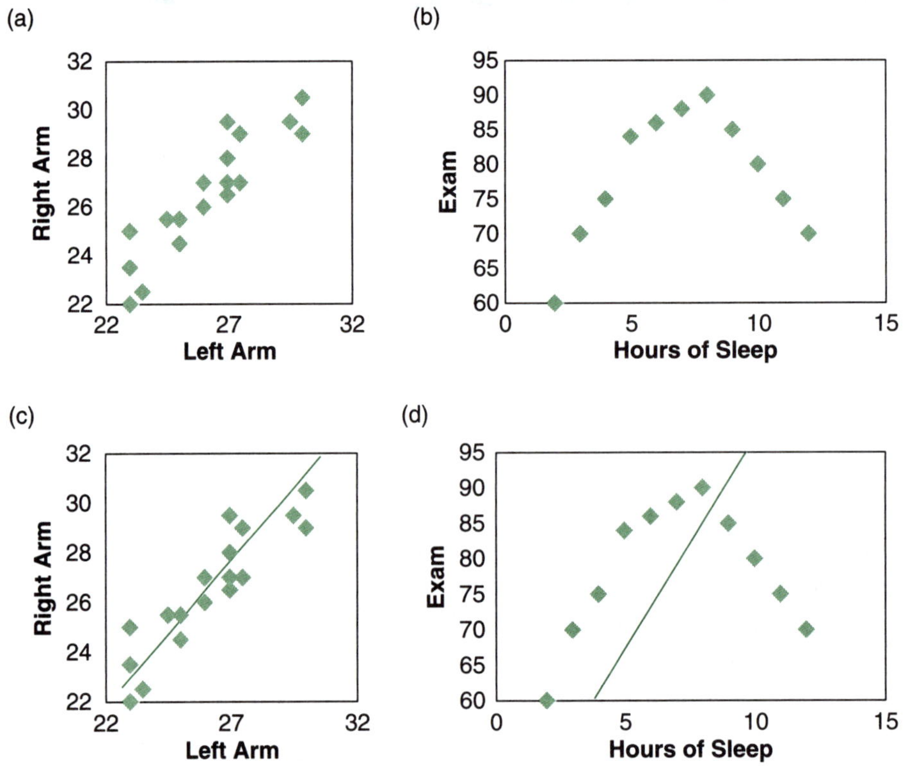

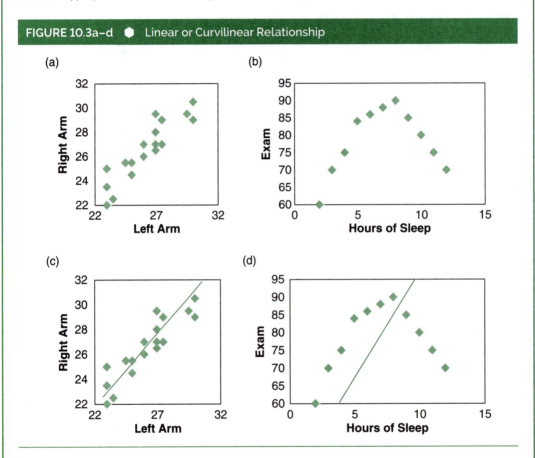

Judging by the scatterplot Figure 10.3a, the dots all lined up pretty well. People who have a long left arm also have a long right arm, and people who have a short left arm also have a short right arm. When we draw a straight line to fit the dots, the line fit the dots in the entire distribution well, as shown in Figure 10.3c. Therefore, correlation is an appropriate statistic to use to characterize the relationship between left-arm and right-arm measures.

Figure 10.3b showed a curvilinear pattern because the straight line fit the dots well only when hours of sleep were less than 8. The line was far away from the dots when hours of sleep are higher than

8 as shown in Figure 10.3d. The exam scores went up with the hours of sleep until it reached about 8 hours; then the exam scores dropped with more sleep. This pattern showed that the two variables moved in the same direction when the hours of sleep was less than 8, then they moved in the opposite directions when the hours of sleep was more than 8. Such a change of direction indicates a curvilinear relationship between the variables. Therefore, Figure 10.3b shows that correlation is not an appropriate statistic to characterize the relationship between hours of sleep and the exam scores.

If you wonder how the straight line that best fits the data points is determined, then you have to wait until Chapter 11 (Simple Regression) to satisfy your curiosity, where it is discussed in detail.

The Direction of the Relationship: Positive (+) or Negative (–)

Once a linear pattern has been established, we can judge the direction of the relationship by the way the dots spread across the graph. The direction of the relationship is either positive or negative. A positive correlation depicts an uptrend, with the data points spreading from the lower left corner to the upper right corner of the graph, as shown in Figure 10.3a. When a person has a long left arm, the person also has a long right arm. The positive correlation can be seen in that as the X values increase, the Y values tend to increase as well. In other words, the X and Y values are moving in the same direction. A negative correlation depicts a downtrend in which the data points spread from the upper left corner to the lower right corner of the graph, as shown in Figure 10.3e. The negative correlation can be seen in that as the X values increase, the Y values tend to decrease. In other words, the X and Y values are moving in opposite directions. The X-axis in Figure 10.3e represents the number of missed classes, and the Y-axis is the exam score. Each dot in the scatterplot represents a student. The negative correlation shows that students who missed higher numbers of classes got lower exam scores.

FIGURE 10.3e, f ● Strong Negative r (e) and Weak r (f)

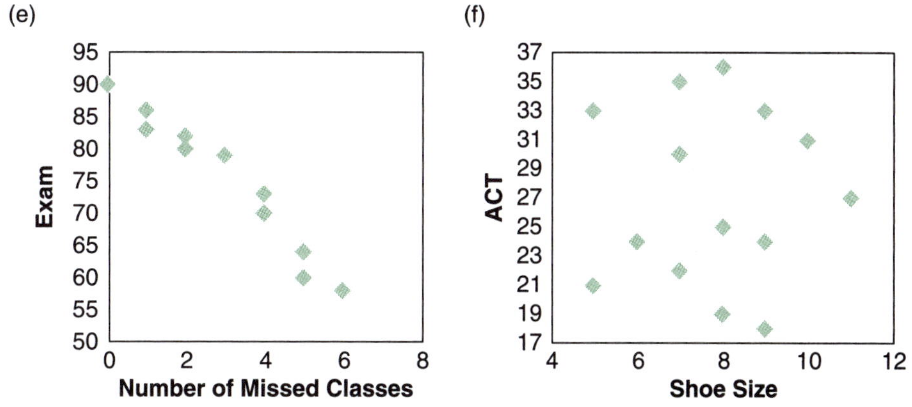

Author's Aside

This negative correlation between number of classes missed and exam grades has been observed in statistics courses repeatedly, due to the accumulative nature of the course material. Once a student missed a class or two, it created a gap in their knowledge. They fell behind in the following class sessions. The impact of missed classes showed in the exam scores. Don't blame Murphy's law that the day you miss a class is the day when your professor presents key concepts that end up in the next exam.

The Strength of the Relationship

Without actually calculating the correlation coefficient, the strength of the relationship can be roughly seen in the scatterplots by considering the aggregated distance from the data points to the straight line: The smaller the aggregated distance between the dots to the line, the stronger the correlation. In other words, when dots are close to the line, the correlation is strong. When dots are far away from the line, the correlation is weak. In Figure 10.3e, the dots fall either right on the line or very close to it. This is a strong correlation. In Figure 10.3f, the dots spread out in every direction and almost form a circle. When we draw a line to fit the dots, the aggregated distance between the dots to the line is large. The correlation in Figure10.3f is weak.

The range of correlation is $-1 \le r \le 1$. The strength of the correlation can be defined as the absolute value of the correlation. Compare Figures 10.3e and f. Figure 10.3e has a stronger correlation than Figure 10.3f. The strongest correlation occurs when $r = 1$ or $r = -1$, and the weakest correlation occurs when $r = 0$.

The Presence of Outliers

Correlation can be heavily influenced by outliers. For example, Figures 10.3g and h are identical data except for one outlier shown by a large star(✧) in the graphs. However, the location of the outlier has a huge impact on the correlation coefficients. Without the outlier, the correlation is close to 0. Figure 10.3g with the outlier on the lower left-hand side creates a positive correlation, but Figure 10.3h with the outlier on the lower right-hand side creates a negative correlation.

It is prudent to visually inspect the scatterplot to identify potential outliers. If outliers are due to data entry errors, remove those outliers before calculating correlations. The

FIGURE 10.3g, h ● Undue Influence of an Outlier on the Correlation

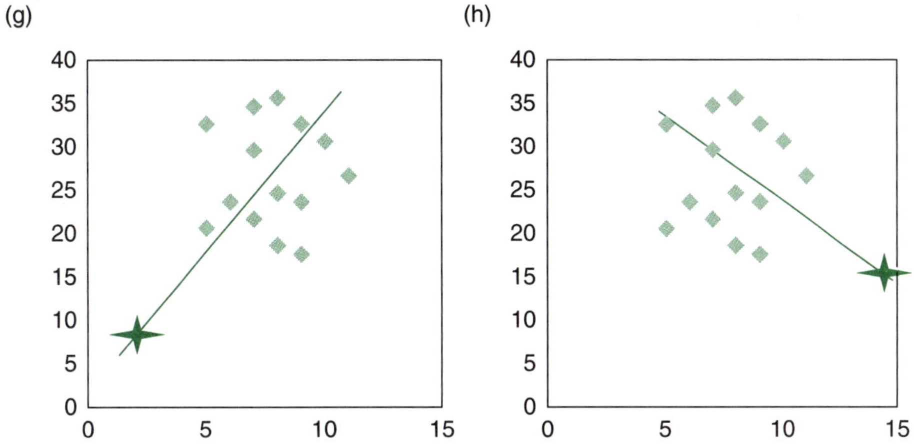

effect of other outliers can be identified by calculating the correlation with outliers first and calculating the correlation again without outliers. The difference between these two calculations is due to the effect of outliers.

Showing Pearson's Correlation in Action

We have covered the Pearson's r formula and learned how to evaluate whether it was appropriate to apply the formula in each situation. Let's use an example to go through the entire process.

EXAMPLE 10.2

A statistics course instructor is curious about the relationship between the number of students' missed classes and their final exam scores. A group of 11 students was randomly selected. Students' missed classes and final exam scores were reported in Table 10.2.

Use Table 10.2 to answer the following questions.

a. Is correlation appropriate to measure the relationship between number of missed classes and final exam scores?

(Continued)

(Continued)

b. What is the Pearson's correlation between number of missed classes and final exam scores?

a. We need to construct a scatterplot to answer Part a. The scatterplot of missed classes and final exam scores is presented in Figure 10.4. According to Figure 10.4, a straight line fit the data points in the entire distribution very well, and there were no outliers. Outliers can be visually spotted by a data point that is far away from the rest of the data points. Correlation is thus appropriate to measure the relationship between number of missed classes and final exam scores.

TABLE 10.2 ⬢ The Number of Classes Missed and Final Exam Scores

Student	Missed Classes, *X*	Final Exam Score, *Y*
1	0	93
2	1	75
3	2	86
4	2	82
5	3	80
6	3	79
7	3	75
8	4	80
9	4	75
10	5	64
11	6	58

FIGURE 10.4 ⬢ Scatterplot of Number of Missed Classes and Final Exam Scores

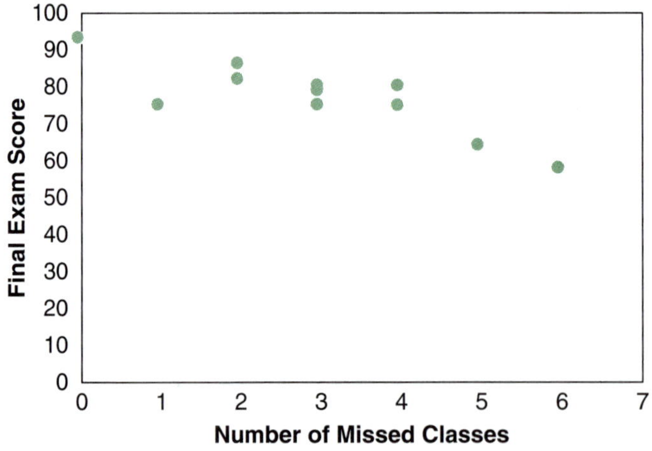

b.

$$\text{Pearson's correlation } r = \frac{SP}{\sqrt{SS_X SS_Y}}$$

where

$$SP = \Sigma(X - \bar{X})(Y - \bar{Y})$$
$$SS_X = \Sigma(X - \bar{X})^2$$
$$SS_Y = \Sigma(Y - \bar{Y})^2$$

According to Pearson's *r* formula, we need to add the following five columns to Table 10.2, $(X - \bar{X})$, $(Y - \bar{Y})$, $(X - \bar{X})$ $(Y - \bar{Y})$, $(X - \bar{X})^2$, and $(Y - \bar{Y})^2$, to demonstrate the step-by-step process. The results are shown in Table 10.2a. There were only 11 cases in this example to allow calculation by hand. In case you want to learn how to calculate Pearson's *r* with large data set, the step-by-step instruction to calculate Pearson's *r* using Excel will be provided in a later section of this chapter. You should refresh your memory and skill on how to follow the formula to construct additional columns in Excel as stated in Chapter 3.

TABLE 10.2a ● Step-by-Step Calculation of Pearson's *r* Between the Number of Classes Missed and Final Exam Scores

	Missed Classes, *X*	Final Exam Score, *Y*	$(X - \bar{X})$	$(Y - \bar{Y})$	$(X - \bar{X})(Y - \bar{Y})$	$(X - \bar{X})^2$	$(Y - \bar{Y})^2$
	0	93	−3	16	−48	9	256
	1	75	−2	−2	4	4	4
	2	86	−1	9	−9	1	81
	2	82	−1	5	−5	1	25
	3	80	0	3	0	0	9
	3	79	0	2	0	0	4
	3	75	0	−2	0	0	4
	4	80	1	3	3	1	9
	4	75	1	−2	−2	1	4
	5	64	2	−13	−26	4	169
	6	58	3	−19	−57	9	361
Total	33	847			−140	30	926

(Continued)

(Continued)

There are 11 students in the table, $n = 11$. The first step is to calculate $\bar{X}$ and $\bar{Y}$ to construct the additional five columns, and the last row of every column is the total of the column.

$$\bar{X} = \Sigma X/n = 33/11 = 3$$

$$\bar{Y} = \Sigma Y/n = 847/11 = 77$$

$$SP = \Sigma(X - \bar{X})(Y - \bar{Y}) = -140$$

$$SS_X = \Sigma(X - \bar{X})^2 = 30$$

$$SS_Y = \Sigma(Y - \bar{Y})^2 = 926$$

$$r = \frac{SP}{\sqrt{SS_X SS_Y}} = \frac{-140}{\sqrt{(30)(926)}} = -.840$$

The Pearson's r between the number of missed classes and final exam scores was –.840. As the number of missed classes increased, the final exam scores decreased. This result confirms that good grades go hand in hand with good attendance. If you want to earn good grades, trying not to miss classes would be a good start.

As you see, the differences in the ranges of variables X and Y can be large, as shown in Example 10.2. This is attributable in part to differences in the units in which their values are observed and reported. One missed class is not the same thing—not the same unit—as one missed point on an exam. However, when both variables are converted into standardized Z scores, differences in the units or range of values simply do not matter. We can demonstrate this by applying another form of the Pearson's correlation formula.

$$r = \frac{\Sigma Z_X Z_Y}{n - 1}$$

Based on the calculation from Table 10.2a, we already knew the following sample statistics.

$$\bar{X} = \Sigma X/n = 33/11 = 3$$
$$\bar{Y} = \Sigma Y/n = 847/11 = 77$$
$$SS_X = \Sigma(X - \bar{X})^2 = 30$$
$$SS_Y = \Sigma(Y - \bar{Y})^2 = 926$$

We need to calculate s_X and s_Y to convert the raw variables X and Y into standardized Z scores.

$$s = \sqrt{\frac{SS}{n-1}}$$

$$s_X = \sqrt{\frac{SS_X}{n-1}} = \sqrt{\frac{30}{10}} = 1.73$$

$$s_Y = \sqrt{\frac{SS_Y}{n-1}} = \sqrt{\frac{926}{10}} = 9.62$$

Let's use $\bar{X}$, $\bar{Y}$, S_X, and S_Y to calculate the standardized Z_X and Z_Y and create $Z_X Z_Y$ in Table 10.2b.

TABLE 10.2b ● Calculation of Pearson's *r* From *Z* Scores of Number of Classes Missed and Final Exam Scores

Missed Classes, *X*	Final Exam Score, *Y*	Z_X	Z_Y	$Z_X Z_Y$
0	93	-1.734	1.663	-2.884
1	75	-1.156	-0.208	0.240
2	86	-0.578	0.936	-0.541
2	82	-0.578	0.520	-0.300
3	80	0.000	0.312	0.000
3	79	0.000	0.208	0.000
3	75	0.000	-0.208	0.000
4	80	0.578	0.312	0.180
4	75	0.578	-0.208	-0.120
5	64	1.156	-1.351	-1.562
6	58	1.734	-1.975	-3.425
33	847			-8.412

Let me do a slow-motion walk-through on the process of calculating Z_X and Z_Y. For the first student who missed $X = 0$ classes and got $Y = 93$ on the final exam, the Z_X and Z_Y are calculated as follows:

$$Z_X = \frac{X - \bar{X}}{s_X} = \frac{0 - 3}{1.73} = -1.734$$

$$Z_Y = \frac{Y - \bar{Y}}{s_Y} = \frac{93 - 77}{9.62} = 1.663$$

Continue the process and calculate the Z_X and Z_Y for the rest of the sample. Their Z_X and Z_Y are reported in Table 10.2b. Multiply Z_X and Z_Y to create $Z_X Z_Y$; then add them up to get the summation $\Sigma Z_X Z_Y$. Now we are ready to calculate Pearson's r from the standard Z scores.

$$r = \frac{\Sigma Z_X Z_Y}{n - 1} = \frac{-8.412}{10} = -.841$$

The slight difference in the two Pearson's r values at the third place after the decimal point is due to rounding.

POP QUIZ

1. Both test-retest reliability and criterion-related validity can be calculated by
   a. *t* tests.
   b. *Z* tests.
   c. *F* tests.
   d. Pearson's *r*.
2. The direction of Pearson's *r* is determined by
   a. $SS_X = \Sigma(X - \bar{X})^2$
   b. $SS_Y = \Sigma(Y - \bar{Y})^2$
   c. $SP = \Sigma(X - \bar{X})(Y - \bar{Y})$
   d. the coefficient of determination, r^2

HYPOTHESIS TESTING FOR PEARSON'S CORRELATION

The four-step procedure for conducting a hypothesis test as discussed in Chapter 7 is still applicable in this chapter. We will apply the same basic process to conduct hypothesis tests on Pearson's correlations.

Step 1: Explicitly state the pair of hypotheses.

Step 2: Identify the rejection zone.

Step 3: Calculate the test statistic.

Step 4: Make the correct conclusion.

Let's explain this four-step hypothesis testing for Pearson's r step by step.

Step 1: Explicitly state the pair of hypotheses.

Hypothesis tests are conducted using sample statistics to make inferences about the population parameters. In hypothesis tests of Pearson's correlation, the main focus is on whether the relationship between two variables exists in the population. The population correlation is expressed as ρ (rho, the 17th letter of the Greek alphabet), and the sample correlation is expressed as r. Here is a list of hypotheses for all possible tests for Pearson's correlation.

Two-tailed test: $H_0: \rho = 0$
$H_1: \rho \neq 0$

Right-tailed test: $H_0: \rho = 0$
$H_1: \rho > 0$

Left-tailed test: $H_0: \rho = 0$
$H_1: \rho < 0$

Step 2: Identify the rejection zone.

When conducting a hypothesis test for a Pearson's correlation, the calculated r needs to be compared with the critical value of r that sets the boundary of the rejection zone using the same principle stated in Chapter 7. The critical values for Pearson's r can be found in the Critical Values of Pearson's Correlation Table, or r Table, as shown in Appendix D. The numbers in the top row of the r Table stipulate the α level for one-tailed tests, and the numbers in the second row stipulate the α level for two-tailed tests. The numbers in the first column identify the degrees of freedom ($df = n - 2$) for Pearson's r, and all the numbers inside the r Table are critical values that set the boundaries of the rejection zones as shown in Table 10.3. We first discussed the degrees of freedom in the context of identifying the rejection zone back in Chapter 8, one-sample t test when σ is unknown. The df for one-sample t tests is $(n - 1)$ because the one sample statistic is used to estimate one population parameter, and so 1 df is lost in the process, and we are left with $df = (n - 1)$ in a one-sample t test. Pearson's correlation formula requires calculation of both $\bar{X}$ and $\bar{Y}$ to estimate μ_X and μ_Y, so 2 df are lost in the process, and we are left with $df = (n - 2)$ for Pearson's correlation.

TABLE 10.3 ⬢ A Portion of the Critical Values of the Pearson's Correlation

	Level of Significance for One-Tailed Test			
	.05	.025	.01	.005
	Level of Significance for Two-Tailed Test			
$df = n - 2$	.10	.05	.02	.01
1	.988	.997	.9995	.9999
2	.900	.950	.980	.990
3	.805	.878	.934	.959
4	.729	.811	.882	.917
5	.669	.754	.833	.874
6	.622	.707	.789	.834
7	.582	.666	.750	.798
8	.549	.632	.716	.765
9	.521	.602	.685	.735
10	.497	.576	.658	.708
11	.476	.553	.634	.684
12	.458	.532	.612	.661
13	.441	.514	.592	.641
14	.426	.497	.574	.628
15	.412	.482	.558	.606
16	.400	.468	.542	.590
17	.389	.456	.528	.575
18	.378	.444	.516	.561

Notice that there are no negative numbers in the Pearson's r Table. Therefore, we need to assign the negative sign to the left-tailed tests as we did in t Table. You need to know three pieces of information to identify the critical value of r to conduct a hypothesis test: (1) a one-tailed or two-tailed test, (2) the df of the test, and (3) the α level.

Rejection zone for a two-tailed test: $|r| > r_{\alpha/2}$, with $df = n - 2$

Rejection zone for a right-tailed test: $r > r_{\alpha}$, with $df = n - 2$

Rejection zone for a left-tailed test: $r < -r_{\alpha}$, with $df = n - 2$

Step 3: Calculate the test statistic.

$$\text{Pearson's correlation } r = \frac{\text{SP}}{\sqrt{SS_X SS_Y}}$$

where

$$\text{SP} = \Sigma(X - \bar{X})(Y - \bar{Y})$$

$$SS_X = \Sigma(X - \bar{X})^2$$

$$SS_Y = \Sigma(Y - \bar{Y})^2$$

The rounding rule for calculating Pearson's correlation is to stop at the third place after the decimal point so as to conform to the way correlation coefficients are listed in the *r* Table.

Step 4: Make the correct conclusion.

The decision rules are if the calculated r is within the rejection zone, we reject H_0. Therefore, we conclude that the evidence is strong enough to support the claim that a linear relationship exists between the paired variables. If the calculated r is not within the rejection zone, we fail to reject H_0. Thus, we conclude that the evidence is not enough to support the claim that a linear relationship exists between the paired variables.

When the p value associated with the calculated r is obtainable, there is no need to use the r Table to determine the conclusion. The decision rules can be based on the p value. If the p value associated with the calculated r is less than the predetermined α level, we reject H_0 and conclude that the evidence is strong enough to support the claim that a linear relationship exists between the paired variables. If the p value associated with the calculated r is not less than the predetermined α level, we fail to reject H_0 and conclude that the evidence is not enough to support the claim of a linear relationship between the paired variables.

When the p value associated with a test statistic is used as a criterion for a hypothesis test, the decision rules are as follows:

When $p < \alpha$, we reject H_0.

When $p \geq \alpha$, we fail to reject H_0.

Now you have learned the entire hypothesis test for Pearson's correlation. Let's go through a few examples to practice the procedure. First, we will demonstrate how to conduct the process when we have to calculate the numbers by hand, so we specifically select an example with a small sample size to make the calculation manageable. Later on, we will demonstrate how to conduct the process by using Excel, so we can calculate Pearson's correlation in a large sample without being overwhelmed by the tedious calculation procedure.

EXAMPLE 10.3

TABLE 10.4 ● Height and Weight Measures of 10 Adults

Height	Weight
67	178
67	160
59	97
63	108
66	127
60	120
74.5	180
69	170
68	170
66.5	130

Adults' height and weight measures are normally distributed. A sample of 10 adults was randomly selected and their height and weight measures were reported in Table 10.4. Was there a positive linear relationship between height and weight, assuming $\alpha = .05$?

We should visually inspect the scatterplot of height and weight as shown in Figure 10.5 to make sure that a linear relationship exists between height and weight measures.

According to Figure 10.5, there were no obvious outliers and a straight line fits all data points well. Therefore, it was appropriate to apply the four-step hypothesis test for a Pearson's *r* between height and weight measures.

Step 1: Explicitly state the pair of hypotheses.

The problem statement asked to conduct a hypothesis test for a positive linear relationship between the paired variables. Therefore, a right-tailed hypothesis test was the appropriate choice.

Right-tailed test: $H_0: \rho = 0$

$H_1: \rho > 0$

FIGURE 10.5 ● Scatterplot of Height and Weight of 10 Adults

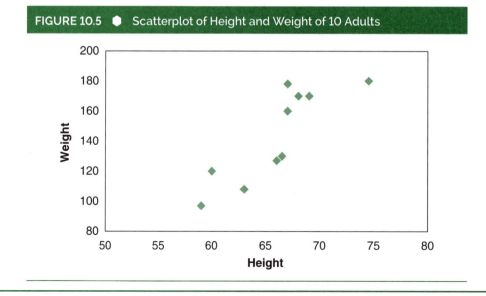

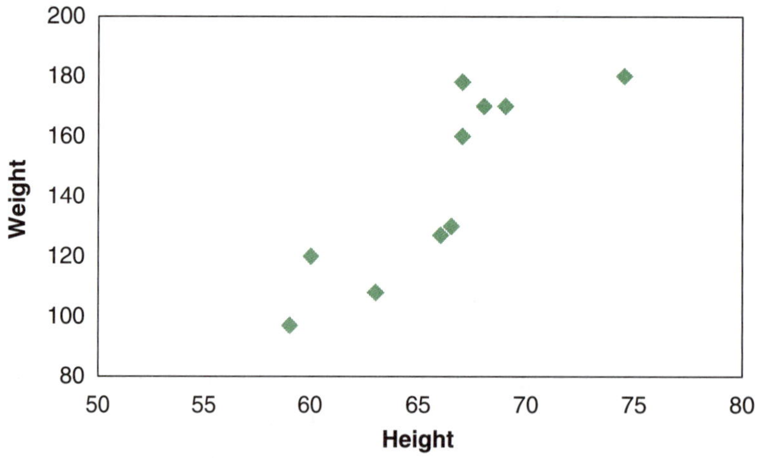

Step 2: Identify the rejection zone for the hypothesis test.

According to the r Table with a one-tailed test, $df = n - 2 = 10 - 2 = 8$, $\alpha = .05$, and the critical value of $r = .549$, the rejection zone for a right-tailed test is $r > .549$.

Step 3: Calculate the test statistic.

According to the Pearson's r formula, we need to add the following five columns to what was included earlier as Table 10.4, $(X-\bar{X})$, $(Y-\bar{Y})$, $(X-\bar{X})(Y-\bar{Y})$, $(X-\bar{X})^2$, and $(Y-\bar{Y})^2$, to demonstrate the step-by-step process. The results are shown in Table 10.4a.

TABLE 10.4a ● Step-by-Step Process to Conduct a Hypothesis Test for a Positive Linear Relationship Between Height and Weight Measures

Height, X (inches)	Weight, Y (pounds)	$(X-\bar{X})$	$(Y-\bar{Y})$	$(X-\bar{X})(Y-\bar{Y})$	$(X-\bar{X})^2$	$(Y-\bar{Y})^2$
67	178	1	34	34	1	1,156
67	160	1	16	16	1	256
59	97	-7	-47	329	49	2,209
63	108	-3	-36	108	9	1,296
66	127	0	-17	0	0	289
60	120	-6	-24	144	36	576
74.5	180	8.5	36	306	72.25	1,296
69	170	3	26	78	9	676
68	170	2	26	52	4	676
66.5	130	0.5	-14	-7	0.25	196
660	1,440			1,060	181.5	8,626

There are 10 adults in the table, $n = 10$. The first steps are to figure out $\bar{X}$ and $\bar{Y}$ to construct the additional five columns and to calculate the total of the column in the last row when necessary.

$$\bar{X} = \Sigma X/n = 660/10 = 66$$

$$\bar{Y} = \Sigma X/n = 1440/10 = 144$$

$$\text{SP} = \Sigma(X-\bar{X})(Y-\bar{Y}) = 1060$$

(Continued)

(Continued)

$$SS_X = \Sigma(X - \bar{X})^2 = 181.5$$

$$SS_Y = \Sigma(Y - \bar{Y})^2 = 8626$$

$$r = \frac{SP}{\sqrt{SS_X SS_Y}} = \frac{1060}{\sqrt{(181.5)(8626)}} = .847$$

Step 4: Make a conclusion.

The calculated $r = .847$ is within the rejection zone. Therefore, we rejected H_0. The evidence was strong enough to support the claim that there was a positive correlation between height and weight measures.

It is important to emphasize that a strong correlation such as $r = .847$ does not suggest a causal relationship. It only suggests that, as a rule, height and weight usually go together. Taller people usually weigh more than shorter people though there are exceptions to the rule. However, assuming a cause-and-effect relationship between two variables can be illogical or simply wrong. The next example shows the danger or fallacy of assuming causation between two variables due to strong positive correlations.

EXAMPLE 10.4

Some sociologists were interested in the relationship between the number of Boy Scout troops and number of violent crimes in cities. The number of Boy Scout troops in a city is a ratio variable because it has an absolute zero. Annual number of violent crimes is also a ratio variable. Both numbers can be obtained from official records of each city. A group of 12 cities was randomly selected. The pairs of numbers from each municipality were organized and reported in Table 10.5.

a. Visually inspect the data to see if there was a linear relationship between number of Boy Scout troops and annual violent crimes.

b. Conduct a hypothesis test to verify whether there was a significant relationship between number of Boy Scout troops and annual violent crimes reported in cities, using $\alpha = .05$.

a. Construct a scatterplot with number of Boy Scout troops as the *X*-axis and annual crimes as the *Y*-axis. The scatterplot is shown in Figure 10.6. The relationship was not very strong; however, there was a trend from lower left side to upper right with no outliers in this graph. We assumed that a linear relationship existed between the number of Boy Scout troops and annual violent crimes.

b. The four-step hypothesis test for a Pearson's *r* between number of Boy Scout troops and violent crimes.

TABLE 10.5 ● Number of Boy Scout Troops and Annual Number of Violent Crimes in Cities

City	Boy Scout Troops	Violent Crimes
A	15	2,601
B	22	1,150
C	17	5,871
D	70	1,799
E	59	8,380
F	49	3,813
G	83	6,958
H	87	8,535
I	63	2,970
J	50	2,229
K	25	1,020
L	12	454

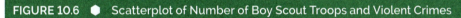

FIGURE 10.6 ● Scatterplot of Number of Boy Scout Troops and Violent Crimes

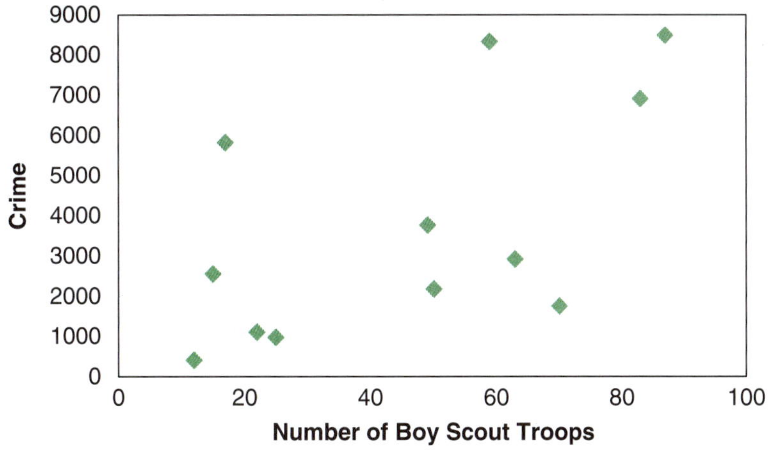

(Continued)

Step 1: Explicitly state the pair of hypotheses.

The problem statement asked to conduct a hypothesis test to verify whether there was a significant relationship between the paired variables. There was no indication of any direction. Therefore, a two-tailed hypothesis was appropriate.

Two-tailed test: $H_0: \rho = 0$

$H_1: \rho \neq 0$

Step 2: Identify the rejection zone for the hypothesis test.

According to the r Table, with a two-tailed test, $df = n - 2 = 12 - 2 = 10$, $\alpha = .05$, and the critical value of $r = .576$, the rejection zone for a two-tailed test is $|r| > .576$.

Step 3: Calculate the test statistic.

According to the Pearson's r formula, we need to add the following five columns to Table 10.5, $(X-\bar{X}),(Y-\bar{Y}),(X-\bar{X})(Y-\bar{Y}),(X-\bar{X})^2$, and $(Y-\bar{Y})^2$, to demonstrate the step-by-step process. The results are shown in Table 10.5a.

TABLE 10.5a ● Step-by-Step Process to Calculate Pearson's *r* Between Number of Boy Scout Troops and Violent Crimes

Boy Scout Troops	Crimes	$(X-\bar{X})$	$(Y-\bar{Y})$	$(X-\bar{X})(Y-\bar{Y})$	$(X-\bar{X})^2$	$(Y-\bar{Y})^2$
15	2,601	-31	-1,214	37,634	961	1,473,796
22	1,150	-24	-2,665	63,960	576	7,102,225
17	5,871	-29	2,056	-59,624	841	4,227,136
70	1,799	24	-2,016	-48,384	576	4,064,256
59	8,380	13	4,565	59,345	169	20,839,225
49	3,813	3	-2	-6	9	4
83	6,958	37	3,143	116,291	1,369	9,878,449
87	8,535	41	4,720	193,520	1,681	22,278,400
63	2,970	17	-845	-14,365	289	714,025
50	2,229	4	-1,586	-6,344	16	2,515,396
25	1,020	-21	-2,795	58,695	441	7,812,025
12	454	-34	-3,361	114,274	1,156	11,296,321
552	45,780			514,996	8,084	92,201,258

There were 12 cities in the table, n = 12. The first step is to figure out $\bar{X}$ and $\bar{Y}$, and then to construct the additional five columns. The last row of every column was the total of that column when appropriate.

$$\bar{X} = \Sigma X/n = 552/12 = 46$$

$$\bar{Y} = \Sigma Y/n = 45780/12 = 3815$$

$$\text{SP} = \Sigma(X - \bar{X})(Y - \bar{Y}) = 514{,}996$$

$$SS_X = \Sigma(X - \bar{X})^2 = 8{,}084$$

$$SS_Y = \Sigma(Y - \bar{Y})^2 = 92{,}201{,}258$$

$$r = \frac{\text{SP}}{\sqrt{SS_X SS_Y}} = \frac{514996}{\sqrt{(8084)(92201258)}} = .597$$

The Pearson correlation between number of Boy Scout troops and annual number of violent crimes was .597.

Step 4: Make a conclusion.

The calculated $r = .597$ was within the rejection zone. Therefore, we rejected H_0. The evidence was strong enough to support the claim that there was a significant correlation between the number of Boy Scout troops and number of violent crimes.

The correlation between the number of Boy Scout troops and the number of violent crimes only suggested that in a city, these two variables seemed to move together. When the number of Boy Scout troops went up, the number of violent crimes also went up. It would be a very unfortunate mistake to say that participating in Boy Scout activities causes violent crimes. The apparent correlation between two variables may be caused by a third variable, which may or may not be included in the study. We will address the interpretations and assumptions of Pearson's correlation in the next section.

POP QUIZ

3. The degrees of freedom for a Pearson's r is
   a. n.
   b. $n - 1$.
   c. $n - 2$.
   d. $n - 3$.

INTERPRETATIONS AND ASSUMPTIONS OF PEARSON'S CORRELATION

Strong correlations are commonly and mistakenly interpreted as cause-and-effect relationships. They are not. Establishing causality requires strict control over all other variables. It also requires systematic manipulation and random assignment of different levels of the independent variable, so as to isolate and observe its effect on the dependent variable. A correlation can be calculated on any two interval or ratio variables with linear relationship. For example, a correlation can be calculated between number of sun spots and the Dow Jones Industrial Average in a given day over several months. Sun spots are natural astrophysical phenomena and the Dow Jones Industrial Index is a human-made economic index based on the changes in trading prices of selected stocks. These two numbers can go in the same direction or in opposite directions, and it may even be statistically significant. It is inaccurate for anyone to state that the number of sun spots causes Dow Jones Index to go up or down because of a correlation. Another example is a correlation can be calculated between income and monthly fee paid on a cell phone plan, where both variables can be obtained by participants' self-reports. There are no controls and/or manipulation on these two variables. There are many other variables that may have an impact on the income or monthly cell phone bill but may not have been measured or reported in the study. Therefore, it is incorrect to make an assertion of causality solely based on a correlation.

Interpretations of Pearson's Correlation

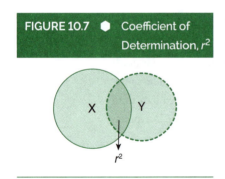

FIGURE 10.7 ● Coefficient of Determination, r^2

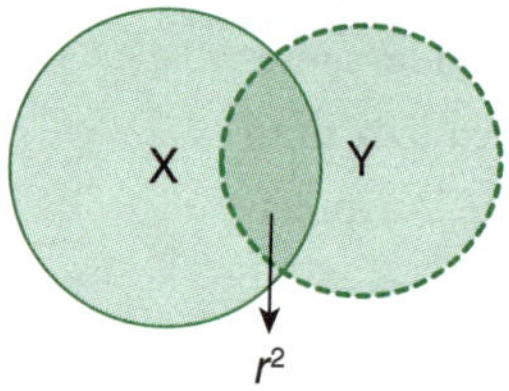

Correlation can also be used to measure the percentage of variance that overlaps between X and Y. The **coefficient of determination**, r^2, which is defined as the percentage of the variance in Y that overlap with X, is shown in Figure 10.7.

Strong correlations do not indicate causality. As in Example 10.4, it simply does not make sense to interpret that the number of Boy Scout troops in a city causes the number of violent crimes. No reasonable person would think that participating in Boy Scout activities causes people to commit violent crimes. In this case, researchers simply recorded the values of both variables as reported in each city without any attempt to manipulate these variables. Causality can only be achieved by strict experimental design to allow randomly assigning research participants to different levels of the independent variable, manipulating the independent variable, and controlling all other variables. Such a research design makes it possible to isolate the effect of the independent variable on the dependent variable to reflect the proper cause-and-effect relationship. However, a calculated positive correlation simply reflects the numerical fact that two variables move in the same direction. As the number of Boy Scout troops increased, so did the number of violent crimes. What was a potential explanation for this positive correlation? Was there a third variable that might explain why these two variables move in the same direction?

Scientific studies are usually done with careful planning in research designs. The more variables that get considered and measured, the fewer surprises researchers encounter at the end. Oftentimes, studies are conducted to test particular models with complicated and complex relationships among many variables. If a third variable can be found to have a significant relationship with both variables involved in the correlation, this third variable is a **confounding variable**. A confounding variable is an extraneous variable in a study that correlates with both the independent variable and the dependent variable. Or if there is no pertinent distinction between independent and dependent variables in a given situation, then the confounding variable is one that correlated with the two primary variables of interest. The population of a city might be a confounding variable for the strong correlation between the number of Boy Scout troops and the number of violent crimes. All else held equal, it seemed reasonable to expect to find more Boy Scout troops *and* more violent crimes in cities with larger populations.

Assumptions of Pearson's Correlation

When using sample correlation to make inferences about the population correlation, the following assumptions need to be considered.

1. Variables X and Y form a bivariate normal distribution, which means that X values are normally distributed, Y values are normally distributed, and for every value of X, Y values are normally distributed.
2. The scatterplot confirms the existence of a linear relationship between two variables.
3. The scatterplot does not show obvious outliers. If outliers are attributed to data entry errors, remove those outliers before calculating correlations. The effect of other outliers can be identified by first calculating correlations with the outliers and calculating correlation again without outliers. The difference between these two calculations is attributable to outliers.
4. Besides the regular normal distribution for the paired variables X and Y, there is an additional requirement called *homoscedasticity*. Try saying that 10 times in a row fast! **Homoscedasticity** is defined as the condition in which the variance of Y at any given value of X is the same across the entire range of X. There is equal spread of Y values for every value of X.

If any of these four assumptions is violated, Pearson's r is not an appropriate measure of association for the paired variables.

POP QUIZ

4. The coefficient of determination measures
   a. the percentage of overlap between the variables *X* and *Y*.
   b. the cause-and-effect relationship between *X* and *Y*.
   c. the mean difference between *X* and *Y*.
   d. the pooled variance of *X* and *Y*.

PARTIAL CORRELATION

We have learned that Pearson's r correlation can be calculated between two interval or ratio variables. It is simple to examine the relationship between two variables. However, in reality, the relationship between two variables does not happen in isolation. Oftentimes, there are other variables overlapping with the two variables of interest as shown in Figure 10.8. Wouldn't it be great if we could purify the relationship between two variables by eliminating the impact of a third variable? A partial correlation is the appropriate statistical tool to accomplish that goal.

FIGURE 10.8 ● Relationships Among Three Variables

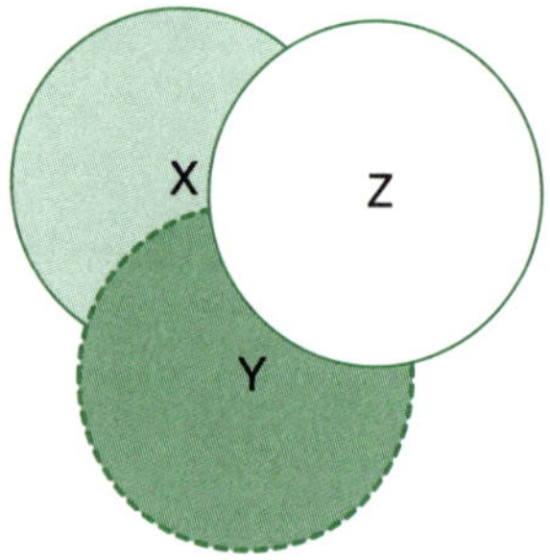

A **partial correlation** is defined as the purified correlation between two variables while controlling a third variable by holding it constant. Figure 10.8 shows the relationships among three variables: X, Y, and Z. Due to the fact that two variables can generate a correlation, there are three different correlations that can be generated by three variables: r_{XY}, r_{XZ}, and r_{YZ}.

The partial correlation between X and Y, while keeping Z constant, is expressed as $r_{XY.Z}$.

$$r_{XY.Z} = \frac{r_{XY} - (r_{XZ}\, r_{YZ})}{\sqrt{(1-r_{XZ}^2)(1-r_{YZ}^2)}}$$

All variables involved in the partial correlations still need to meet the assumptions of Pearson's r. Let's go back to Example 10.4, where the number of Boy Scout troops and number of violent crimes showed significant positive correlation. It seemed odd that these two variables move in the same direction. Whenever two variables display a surprising positive correlation, it is highly likely that a third variable is making these two variables move in the same direction. Let's explore that in the next example.

EXAMPLE 10.5

We calculated the correlation between number of Boy Scout troops (X) and number of violent crimes (Y), $r_{XY} = .597$ in Example 10.4. It seemed highly unlikely that participating in Boy Scout activities had any direct relationship with violent crimes. However, there must be a logical explanation for such a positive correlation between these two variables. Therefore, more variables were explored in this study. One such variable was the population in each city. The population in each city, Z, was obtained, and the correlation between the number of Boy Scout troops and the city population was calculated as $r_{XZ} = .751$. It makes sense that when the population goes up in the city, the number of Boy Scout troops also goes up. The correlation between the number of violent crimes and the city population was also calculated, and it was $r_{YZ} = .768$. It also makes sense that when the population goes up in the city, the number of violent crimes goes up. What was the partial correlation between the number of Boy Scout troops and the number of violent crimes, while holding city population constant?

The city population was highly correlated with both the number of Boy Scout troops $r_{XZ} = .751$ and the number of violent crimes $r_{YZ} = .768$.

To purify the relationship between the number of Boy Scout troops and violent crimes while keeping city population constant, the partial correlation is as follows:

$$r_{XY.Z} = \frac{r_{XY} - (r_{XZ} r_{YZ})}{\sqrt{(1 - r_{XZ}^2)(1 - r_{YZ}^2)}} = \frac{.597 - (.751)(.768)}{\sqrt{(1 - .751^2)(1 - .768^2)}} = \frac{.597 - .577}{\sqrt{(.436)(.410)}} = \frac{.02}{.423} = .047$$

Once the city population was held constant, the correlation between the number of Boy Scout troops and the number of violent crimes became .047, and it was a very weak relationship. Therefore, the city population was a confounding variable that caused both the number of Boy Scout troops and the number of violent crimes to move in the same direction. Once the city population was kept constant, the relationship between the number of Boy Scout troops and the number of violent crimes was no longer significant. Partial correlation is a great tool that might shed some light into particularly puzzling correlations you might encounter in the future.

POP QUIZ

5. Which one of the following correlations is used to purify the relationship between two variables while keeping the third variable constant?
    a. Pearson's r
    b. Spearman's rank correlation
    c. Partial correlation
    d. Point biserial correlation

EXCEL STEP-BY-STEP INSTRUCTION FOR CONSTRUCTING A SCATTERPLOT

This Excel step-by-step instruction is designed to provide extra help for you to construct a scatterplot. Excel is very useful when dealing with a large sample size. For illustration purposes, we use Excel to analyze the height and weight measures from 20 adults as shown in Table 10.1. First you have to enter the data in Excel as shown in Figure 10.9.

FIGURE 10.9 ● Excel Screenshot for Height and Weight of 20 Adults

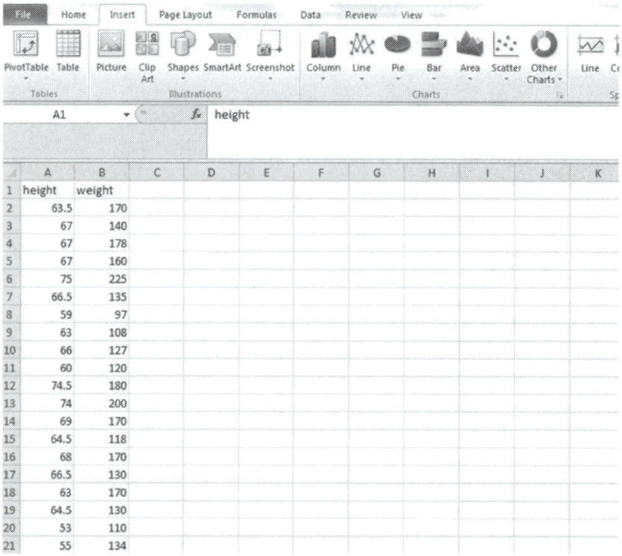

	A	B
1	height	weight
2	63.5	170
3	67	140
4	67	178
5	67	160
6	75	225
7	66.5	135
8	59	97
9	63	108
10	66	127
11	60	120
12	74.5	180
13	74	200
14	69	170
15	64.5	118
16	68	170
17	66.5	130
18	63	170
19	64.5	130
20	53	110
21	55	134

To create a scatterplot in Excel, highlight the columns where the two variables are located. To highlight both the variables, move the cursor to the top of column A and do left click to highlight the entire column, then hold down **shift** key and the right arrow to highlight column B. Once both columns are highlighted, click on the **Insert** tab, and click **Scatter**; then select the first option in the drop-down menu of **Scatter**. A scatterplot appears inside the Excel worksheet as shown in Figure 10.10. Once the scatterplot shows up, you may fine-tune the scatterplot by left clicking on the + sign on the upper right corner of the chart, then the **Chart Elements** menu appears to allow you to modify the chart title and axes titles.

FIGURE 10.10 ● Scatterplot of Height and Weight Measurements of 20 Adults

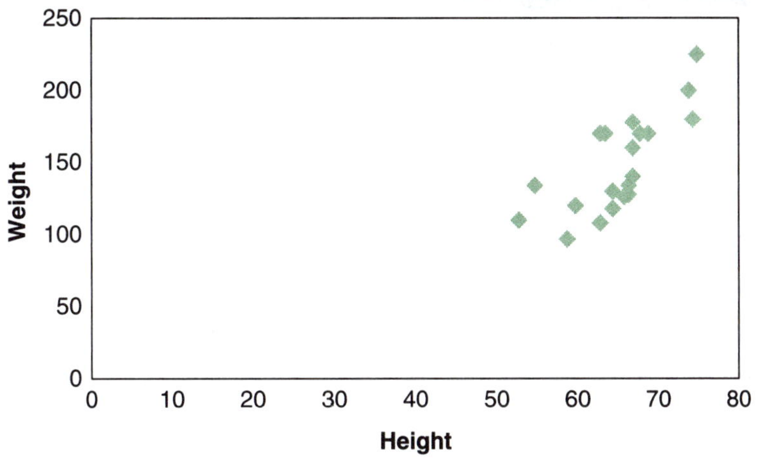

EXCEL STEP-BY-STEP INSTRUCTION FOR CALCULATING PEARSON'S *r*

First, it is important to show how to get all the numbers needed in a formula in Excel. Once you learn the skill to program math operations in Excel, you can program any statistical formula in Excel. Therefore, let's go through the step-by-step process to calculate Pearson's $r = \text{SP} / \sqrt{SS_X SS_Y}$. According to the formula, you need to add the $(X - \bar{X})$, $(Y - \bar{Y})$, $(X - \bar{X})(Y - \bar{Y})$, $(X - \bar{X})^2$, and $(Y - \bar{Y})^2$ columns to demonstrate the step-by-step process.

Type the number of classes missed and final exam scores from Table 10.2 in Excel as shown in Figure 10.11.

The following instructions refer to the Excel screenshot shown in Figure 10.11.

1. Move the cursor to **A13**, and type "**=SUM(A2:A12)**". Then hit **Enter**. The answer for the total missed classes grades is $\sum X = 33$. In **A14**, type in "**=A13/11**" to calculate Mean $\bar{X} = \frac{\sum X}{n} = \frac{33}{11} = 3$.
2. Move the cursor to **B13**, and type "**=SUM(B2:B12)**". Then hit **Enter**. The answer for the total final exam grades is $\sum Y = 847$. In **B14**, type in "**=B13/11**" to calculate Mean $\bar{Y} = \frac{\sum Y}{n} = \frac{847}{11} = 77$.

FIGURE 10.11 ● Screenshot for the First Two Added Columns: $(X - \bar{X})$ and $(Y - \bar{Y})$

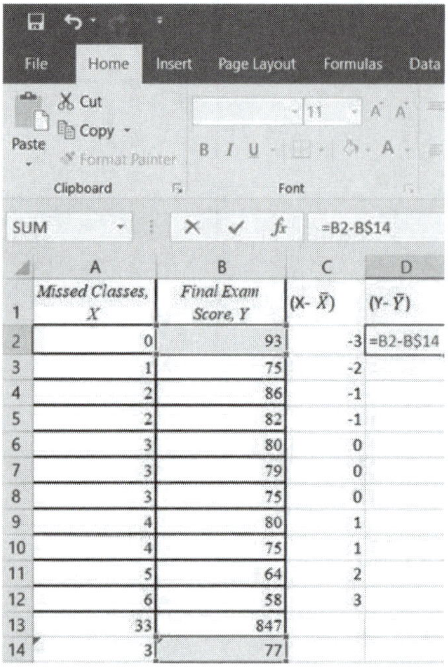

	A	B	C	D
1	Missed Classes, X	Final Exam Score, Y	(X- X̄)	(Y- Ȳ)
2	0	93	-3	=B2-B$14
3	1	75	-2	
4	2	86	-1	
5	2	82	-1	
6	3	80	0	
7	3	79	0	
8	3	75	0	
9	4	80	1	
10	4	75	1	
11	5	64	2	
12	6	58	3	
13	33	847		
14	3	77		

3. In **C1**, label the column as $(X - \bar{X})$. In **C2**, type in "**=A2-A$14**", which shows the operation of first $(X - \bar{X})$. The **$** in the Excel program allows the operation to fix the value of $\bar{X}$ in **A14** for the subsequent operations because every value of X needs to minus the sample mean, $\bar{X}$. Then, once the value of mean is fixed in the first math operation, it is possible to copy and paste the operation to produce the rest of $(X - \bar{X})$. Move the cursor to the lower right corner of **C2** until + shows up. Hold the left click and move the mouse to drag it to **C12**; then release the mouse. You will see the rest of $(X - \bar{X})$ automatically show up in **C3** to **C12**.

4. In **D1**, label the column as $(Y - \bar{Y})$. In **D2**, type in "**=B2-B$14**", which shows the operation of first $(Y - \bar{Y})$ as shown in Figure 10.11. The **$** in the Excel program allows the operation to fix the value of $\bar{Y}$ in **B14** as you move from row to row down the column for the subsequent operations. Then, it is possible to copy the operation to produce the rest of $(Y - \bar{Y})$. Move the cursor to the lower right corner of **D2** until + shows up. Hold the left click and move the mouse to drag it to **D12**; then release the mouse. You will see the rest of $(Y - \bar{Y})$ automatically show up in **D3** to **D12**.

5. The following instructions refer to the Excel screenshot shown in Figure 10.12. In **E1**, label the column as $(X - \bar{X})(Y - \bar{Y})$. In **E2**, type in "**=C2*D2**", which shows the operation of first $(X - \bar{X})(Y - \bar{Y})$. Then move the cursor to the lower right corner of **E2** until + shows up. Hold the left click and move the mouse to drag it to **E12**; then release the mouse. You will see the rest of $(X - \bar{X})(Y - \bar{Y})$ automatically show up in **E3** to **E12**.

6. In **F1**, label the column as $(X - \bar{X})^2$. In **F2**, type in "**=C2^2**", which shows the operation of first $(X - \bar{X})^2$. The "^" symbol can be created by holding down the **Shift** key and **6** at the same time. Then move the cursor to the lower right corner of **F2** until + shows up. Hold the left click and move the mouse to drag it to **F12**; then release the mouse. You will see the rest of $(X - \bar{X})^2$ automatically show up in **F3** to **F12**.

7. In **G1**, label the column as $(Y-\bar{Y})^2$. In **G2**, type in "**=D2^2**", which shows the operation of first $(Y-\bar{Y})^2$. Then move the cursor to the lower right corner of **G2** until + shows up. Hold the left click and move the mouse to drag it to **G12**; then release the mouse. You will see the rest of $(Y-\bar{Y})^2$ automatically show up in **G3** to **G12**.

8. Perform summation at **E14** by typing in "**=SUM(E2:E12)**" to get SP = -140. Follow the same principle to perform summation at **F14** and **G14** or move the cursor to the lower right corner of **E14** until + shows up. Hold down the left click key and drag it across **F14** and **G14**; then let go to get **SS**x = 30 and $\mathbf{SS}_Y$ = 926 as shown in Figure 10.12.

FIGURE 10.12 ● Screenshot for the Last Three Added Columns: $(X-\bar{X})(Y-\bar{Y})$, $(X-\bar{X})^2$, and $(Y-\bar{Y})^2$, as Well as the Results for SP, SS_X, and SS_Y

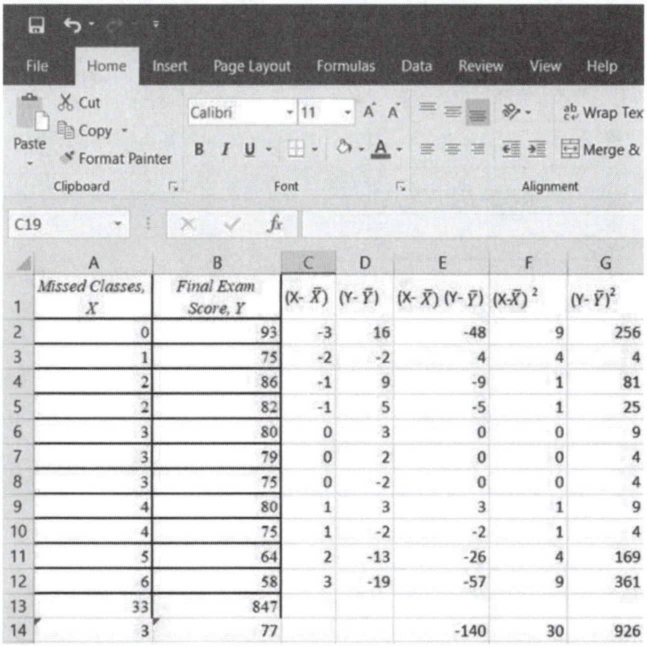

	A	B	C	D	E	F	G
1	Missed Classes, X	Final Exam Score, Y	(X- X̄)	(Y- Ȳ)	(X- X̄) (Y- Ȳ)	(X-X̄)²	(Y- Ȳ)²
2	0	93	-3	16	-48	9	256
3	1	75	-2	-2	4	4	4
4	2	86	-1	9	-9	1	81
5	2	82	-1	5	-5	1	25
6	3	80	0	3	0	0	9
7	3	79	0	2	0	0	4
8	3	75	0	-2	0	0	4
9	4	80	1	3	3	1	9
10	4	75	1	-2	-2	1	4
11	5	64	2	-13	-26	4	169
12	6	58	3	-19	-57	9	361
13	33	847					
14	3	77			-140	30	926

Now, you have all the numbers you need to calculate Pearson's *r*.

$$r = \frac{\text{SP}}{\sqrt{SS_X SS_Y}} = \frac{-140}{\sqrt{(30)(926)}} = .-840$$

9. You may verify your answer with the Excel built-in formula by labeling B16 as "Pearson's r" then moving the cursor to **C16** as shown in Figure 10.13 and typing in "**=PEARSON(A2:A12,B2:B12)**"; then hit **Enter**. This command directs Excel to calculate the Pearson correlation between two arrays of values in columns A and B and the answer is –.840. This shortcut is easy and convenient, but it does not show the step-by-step process as demonstrated from Steps 1 to 8.

FIGURE 10.13 ● Screenshot for Using Excel Built-In Formula to Obtain Pearson's *r*

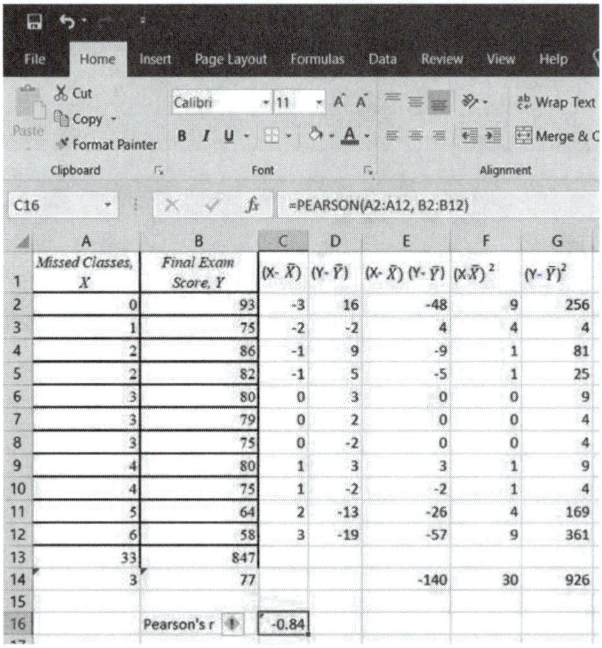

	A	B	C	D	E	F	G
1	Missed Classes, X	Final Exam Score, Y	$(X-\bar{X})$	$(Y-\bar{Y})$	$(X-\bar{X})(Y-\bar{Y})$	$(X-\bar{X})^2$	$(Y-\bar{Y})^2$
2	0	93	-3	16	-48	9	256
3	1	75	-2	-2	4	4	4
4	2	86	-1	9	-9	1	81
5	2	82	-1	5	-5	1	25
6	3	80	0	3	0	0	9
7	3	79	0	2	0	0	4
8	3	75	0	-2	0	0	4
9	4	80	1	3	3	1	9
10	4	75	1	-2	-2	1	4
11	5	64	2	-13	-26	4	169
12	6	58	3	-19	-57	9	361
13	33	847					
14	3	77			-140	30	926
15							
16		Pearson's r	-0.84				

EXERCISE PROBLEMS

1. A group of 14 people was randomly selected and their cell phone bills and monthly income were reported in Table 10.6. Assume that income and cell phone bills are normally distributed.
   a. Construct a scatterplot to see if there was a linear relationship between cell phone bill and monthly income.
   b. Conduct a hypothesis test for a positive relationship between cell phone bill and monthly income, using $\alpha = .05$.

TABLE 10.6 ● Cell Phone Bill and Monthly Income

Cell Phone Bill	Monthly Income
101	2,450
65	4,167
90	1,007
105	1,750
33	202
95	1,506
55	3,000
44	2,101
49	1,943
87	4,784
79	3,556
84	3,025
106	2,877
85	2,450

TABLE 10.7 ● Results From the 5-Minute Rapid Test and the PCR Test

5-Minute Screening Test	PCR Test
20	47
21	46
19	54
22	45
22	50
23	45
23	43
24	49
24	63
25	60
26	69
27	65

2. Beach resorts had annual reports on numbers of drowning accidents, ice cream sales, and numbers of visitors. Assume that the numbers of drowning accidents, ice cream sales, and visitors were all normally distributed. The Pearson's correlation between drowning accidents and ice cream sales was .678, the Pearson correlation between drowning accidents and number of visitors was .895, and the Pearson correlation between ice cream sales and number of visitors was .743. What was the correlation between drowning accidents and ice cream sales while keeping number of visitors constant?

3. A newly designed 5-minute rapid COVID-19 test was given to a group of 12 volunteers who were also given a PCR test that was well-established with high accuracy. The results of these tests for the 12 volunteers on both the rapid test and the PCR test were reported in Table 10.7. Assume the scores of the 5-minute rapid test and the PCR test are normally distributed. Conduct a hypothesis test to see if there is a positive relationship between these two measures.

Solutions

1.a. Construct a scatterplot between cell phone bill and monthly income as shown in Figure 10.14. According to the scatterplot, it was reasonable to assume a weak linear relationship between cell phone bill and monthly income.

FIGURE 10.14 ⬢ A Scatterplot of Cell Phone Bills and Monthly Income

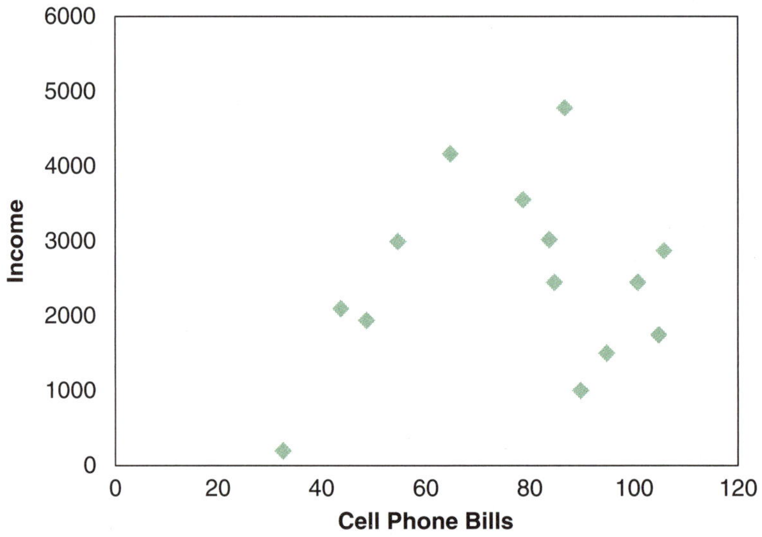

1. b. Assume that cell phone bills and monthly income are normally distributed. The four-step hypothesis test for Pearson's r is appropriate to answer this question.

Step 1: State the pair of hypotheses.

The problem statement asked if there was a positive relationship between cell phone bill and monthly income. A right-tailed test is the appropriate test.

Right-tailed test: H_0: $\rho = 0$

H_1: $\rho > 0$

Step 2: Identify the rejection zone.

The critical value of Pearson's r for a right-tailed test with $df = n - 2 = 14 - 2 = 12$ and $\alpha = .05$, $r = .458$, the rejection zone is $r > .458$.

Step 3: Calculate the statistic.

$$r = \frac{\text{SP}}{\sqrt{SS_X SS_Y}}$$

The step-by-step calculation is shown in Table 10.6a with added columns $(X-\bar{X}), (Y-\bar{Y}), (X-\bar{X})(Y-\bar{Y}), (X-\bar{X})^2$, and $(Y-\bar{Y})^2$.

$$\bar{X} = \Sigma X / n = 1{,}078/14 = 77$$

$$\bar{Y} = \Sigma Y / n = 34818/14 = 2487$$

$$SP = \Sigma(X-\bar{X})(Y-\bar{Y}) = 78{,}530$$

$$SS_x = \Sigma(X-\bar{X})^2 = 7{,}348$$

$$SS_Y = \Sigma(Y-\bar{Y})^2 = 19310908$$

TABLE 10.6a ● Calculation of Pearson's *r* Between Cell Phone Bill and Monthly Income

Cell Phone Charge	Monthly Income	$(X-\bar{X})$	$(Y-\bar{Y})$	$(X-\bar{X})(Y-\bar{Y})$	$(X-\bar{X})^2$	$(Y-\bar{Y})^2$
101	2,450	24	-37	-888	576	1,369
65	4,167	-12	1,680	-20,160	144	2,822,400
90	1,007	13	-1,480	-19,240	169	2,190,400
105	1,750	28	-737	-20,636	784	543,169
33	202	-44	-2,285	100,540	1,936	5,221,225
95	1,506	18	-981	-17,658	324	962,361
55	3,000	-22	513	-11,286	484	263,169
44	2,101	-33	-386	12,738	1,089	148,996
49	1,943	-28	-544	15,232	784	295,936
87	4,784	10	2,297	22,970	100	5,276,209
79	3,556	2	1,069	2,138	4	1,142,761
84	3,025	7	538	3,766	49	289,444
106	2,877	29	390	11,310	841	152,100
85	2,450	8	-37	-296	64	1,369
1,078	34,818			78,530	7,348	19,310,908

$$r = \frac{SP}{\sqrt{SS_X SS_Y}} = \frac{78530}{\sqrt{(7348)(19310908)}} = .208$$

Step 4: Make the correct conclusion.

The calculated Pearson $r = .208$, which was not within the rejection zone. Therefore, we failed to reject the null hypothesis. The evidence was not strong enough to support the claim that a positive relationship exists between cell phone bill and monthly income.

2. The partial correlation calculates the relationship between X and Y, while keeping Z constant.

$$r_{XY.Z} = \frac{r_{XY} - (r_{XZ} r_{YZ})}{\sqrt{(1 - r_{XZ}^2)(1 - r_{YZ}^2)}}$$

where

$r_{XY} = .678$

$r_{XZ} = .895$

$r_{YZ} = .743$

$$r_{XY.Z} = \frac{r_{XY} - (r_{XZ} r_{YZ})}{\sqrt{(1 - r_{XZ}^2)(1 - r_{YZ}^2)}} = \frac{.678 - (.895)(.743)}{\sqrt{(1 - .895^2)(1 - .743^2)}} = \frac{.678 - .665}{\sqrt{(.199)(.488)}} = \frac{.013}{.299} = .043$$

The partial correlation between drowning accidents and ice cream sales, while keeping number of visitors constant, was .043. The relationship was very weak.

3. The four-step hypothesis test for Pearson's r is appropriate to answer this question.

Step 1: State the pair of hypotheses.

The problem statement asks you to test a positive relationship between the 5-minute rapid test and the PCR test. This process is likely to be used to establish the validity of the 5-minute rapid test. A right-tailed test is appropriate.

Right-tailed test: $H_0: \rho = 0$

$H_1: \rho > 0$

Step 2: Identify the rejection zone.

The critical value of Pearson's r for a right-tailed test with $df = n - 2 = 12 - 2 = 10$ and $\alpha = .05$ is .497. The rejection zone is $r > .497$.

Step 3: Calculate the statistic.

$$r = \frac{SP}{\sqrt{SS_X SS_Y}}$$

The step-by-step calculation is shown in Table 10.7a with added columns $(X-\bar{X})$, $(Y-\bar{Y})$, $(X-\bar{X})(Y-\bar{Y})$, $(X-\bar{X})^2$, and $(Y-\bar{Y})^2$.

$$\bar{X}=\sum X/n=276/12=23$$

$$\bar{Y}=\sum Y/n=636/12=53$$

$$\text{SP}=\Sigma(X-\bar{X})(Y-\bar{Y})=155$$

$$SS_X=\Sigma(X-\bar{X})^2=62$$

$$SS_Y=\Sigma(Y-\bar{Y})^2=888$$

TABLE 10.7a ● Calculation of Pearson's *r* Between the 5-Minute Rapid Test and the PCR Test

5-Minute Rapid Test	PCR Test	$(X-\bar{X})$	$(Y-\bar{Y})$	$(X-\bar{X})(Y-\bar{Y})$	$(X-\bar{X})^2$	$(Y-\bar{Y})^2$
20	47	-3	-6	18	9	36
21	46	-2	-7	14	4	49
19	54	-4	1	-4	16	1
22	45	-1	-8	8	1	64
22	50	-1	-3	3	1	9
23	45	0	-8	0	0	64
23	43	0	-10	0	0	100
24	49	1	-4	-4	1	16
24	63	1	10	10	1	100
25	60	2	7	14	4	49
26	69	3	16	48	9	256
27	65	4	12	48	16	144
276	**636**			**155**	**62**	**888**

$$r=\frac{\text{SP}}{\sqrt{SS_X SS_Y}}=\frac{155}{\sqrt{(62)(888)}}=.661$$

Step 4: Make the correct conclusion.

The calculated Pearson $r = .661$, which is within the rejection zone. Therefore, we rejected the null hypothesis. The evidence was strong enough to support the claim that a positive relationship existed between the 5-minute rapid test and the PCR test. As we discussed before, the correlation between the 5-minute rapid test and the well-established and validated PCR test was the criterion-related validity for the 5-minute rapid test.

What You Learned

Pearson's correlation is a statistical procedure that quantifies the extent that two variables move in the same direction or opposite directions. Pearson's *r* provides both the direction and strength of the relationship. The range of Pearson's *r* is $-1 \leq r \leq 1$.

$$\text{Pearson's } r = \frac{SP}{\sqrt{SS_X SS_Y}}$$

or

$$\text{Pearson's } r = \frac{\sum Z_X Z_Y}{(n-1)}$$

A partial correlation is defined as the purified correlation between two variables while controlling a third variable by holding it constant.

$$r_{XY.Z} = \frac{r_{XY} - (r_{XZ} r_{YZ})}{\sqrt{(1 - r_{XZ}^2)(1 - r_{YZ}^2)}}$$

Key Words

Coefficient of determination: The coefficient of determination is defined as r^2, which is the percentage of variance overlap between *X* and *Y* or the percentage of variance of *Y* explained by *X*. 322

Confounding variable: A confounding variable is an extraneous variable in a study that correlates with both the independent variable and the dependent variable. 323

Homoscedasticity: Homoscedasticity is the condition under which the variance of *Y* is the same across all possible values of *X*. In other words, there is equal spread of *Y* for every value of *X*. 323

Partial correlation: A partial correlation is the purified correlation between two variables while holding a third variable constant. 324

Pearson's correlation, *r*: Pearson's *r* is a statistical procedure that quantifies the extent that two variables move in the same direction or opposite directions. It provides direction and strength of the relationship. 298

Scatterplot: A scatterplot between two variables is a graph of dots plotted on a two-dimensional chart. Each dot is a data point that contains a pair of values on the *X* and *Y* axes. 302

Learning Assessment

Multiple Choice: Circle the Best Answer to Every Question

1. Which of the following pairs of variables is most likely to be positively correlated?
   a. Years after retirement and bank account amount
   b. Age and running speed
   c. Number of hours in the sun and severity of sunburn
   d. Number of missed classes and exam grades
2. Which of the following pairs of variables is most likely to be negatively correlated?
   a. Years on the job and salary
   b. Educational attainment and income
   c. Number of hours in the sun and severity of sunburn
   d. Number of missed classes and exam grades
3. What is the most likely direction of the relationship between time running on a treadmill and calories burned?
   a. Positive correlation
   b. Negative correlation
   c. No relationship
   d. None of the above
4. How is the strength of Pearson's product moment correlation coefficient, *r*, expressed?
   a. The calculated value of Pearson's *r*
   b. The absolute value of Pearson's *r*
   c. The square of Pearson's *r*
   d. The square root of Pearson's *r*

5. What is the range of Pearson's product moment correlation coefficient, r?
   a. $-1 < r < 1$
   b. $0 < r < 1$
   c. $-1 \leq r \leq 1$
   d. $0 \leq r \leq 1$

6. The Pearson's r between age and reaction time is .784. What is the percentage of variance in reaction time that is overlapped with age?
   a. 78.4%
   b. 21.6%
   c. 4.6%
   d. 61.5%

7. The partial correlation is calculated
   a. to show association between an interval or ratio variable and a dichotomous variable
   b. to show association between two dichotomous variables
   c. to show association between two primary variables while keeping a third variable constant
   d. to show association between two interval or ratio variables

8. The coefficient of determination is defined as
   a. the direction of Pearson's r
   b. the absolute value of Pearson's r
   c. the absolute value of Spearman's rank correlation
   d. the square of Pearson's r

Free Response Questions

9. A group of 10 incoming freshmen was randomly selected. Their ACT composite scores and SAT scores were reported in Table 10.8. Assume that ACT and SAT scores are normally distributed.
   a. Construct a scatterplot to visually inspect the data to see if there was a linear relationship between ACT and SAT scores.
   b. Conduct a hypothesis test to see whether there was a significant relationship between ACT and SAT scores, using $\alpha = .05$.

TABLE 10.8 ● ACT and SAT Scores of 10 Incoming Freshmen	
ACT	**SAT**
27	1,890
26	1,450
31	2,050
34	2,150
30	2,250
21	1,380
20	1,450
19	1,600
18	1,350
24	1,680

10. A group of eight adult males was randomly selected. Their exercise logs recorded hours of exercise over a 2-month period. The total hours of exercise were calculated from their exercise logs, and their weight changes were measured at the end of 2 months as reported in Table 10.9. Assume that weight changes and hours of exercise were normally distributed.
    a. Construct a scatterplot to visually inspect the data to see if a linear relationship existed between exercise and weight change.
    b. Conduct a hypothesis test to see whether there was a significant relationship between hours of exercise and weight change, using $\alpha = .10$.

TABLE 10.9 ● Hours of Exercise and Weight Change for Eight Adult Males	
Hours of Exercise	**Weight Change**
27	−7
52	−18
39	18
58	−25
21	41
54	−29
20	3
49	−39

11. Assume that the amount of saturated fat in a person's diet, cholesterol level, and weight were all normally distributed. The Pearson's correlation between the amount of saturated fat in a person's diet and cholesterol level was .768, the Pearson's correlation between the amount of saturated fat in a person's diet and weight was .803, and the Pearson correlation's between cholesterol level and weight was .643. What was the correlation between the amount of saturated fat in a person's diet and cholesterol level while keeping weight constant?

Answers to Pop Quiz Questions

1. d
2. c
3. c
4. a
5. c

11

SIMPLE REGRESSION

Learning Objectives

After reading and studying this chapter, you should be able to do the following:

- Describe how the regression line is mathematically determined
- Explain how to calculate the regression slope and *Y*-intercept
- Identify the four-step hypothesis test for a simple regression
- Describe the statistical assumptions for regression
- Distinguish and explain the differences between Pearson's correlation and the simple regression

WHAT YOU KNOW AND WHAT IS NEW

We learned how to calculate Pearson's correlation in Chapter 10, which is designed to estimate a linear relationship between two interval or ratio variables. It does not specify which one is the independent variable (i.e., predictor) or the dependent variable (i.e., criterion). *X* and *Y* are completely on an equal footing with each other. Based on the Pearson's *r* formula, you know if *X* and *Y* switch positions the correlation stays the same, $r_{XY} = r_{YX}$.

$$\text{Pearson's } r = \frac{\text{SP}}{\sqrt{SS_X SS_Y}}$$

Although the linear relationship in Pearson's r is indicated by the direction and strength of the correlation coefficient, it does not explicitly identify a line to best fit the scatterplot of X and Y. Locating the line that best fits the X and Y values is the main focus of this chapter.

In a simple regression, there is only one independent variable (or predictor), X, and one dependent variable (or criterion), Y. We explicitly identify the line that fits the scatterplot of X and Y with the smallest sum of the squared vertical distances between the actual Y values and the line. While it is possible to draw many lines to fit the data points, but there is only one best fitting line. The best fitting line is depicted by a straight line that goes through $(\bar{X}, \bar{Y})$ to minimize the vertical distances between the actual Y value and the predicted value of Y, which is denoted as $\hat{Y}$. The value of $\hat{Y}$, is calculated based on the relationship between X and Y to figure out how $\hat{Y}$, changes along with X values. Such predictions of Y based on X values are the essential differences between regression and correlation.

Y-INTERCEPT AND SLOPE

Simple regression is the most commonly used linear predictive analysis when making quantitative predictions of the dependent variable, Y, based on the values of the independent variable, X. Many straight lines could be drawn to fit the data points. The best fitting straight line is the regression line that produces the minimal sum of squared distances. The solid line in Figure 11.1 is the regression line, while the dotted line is not. The vertical black line represents the distance between the actual Y value and the predicted $\hat{Y}$, and it is expressed as **error** $= Y - \hat{Y}$ as shown in Figure 11.1. An error can be calculated for every data point.

FIGURE 11.1 ● Straight Lines to Fit the Scatterplot of *X* and *Y*

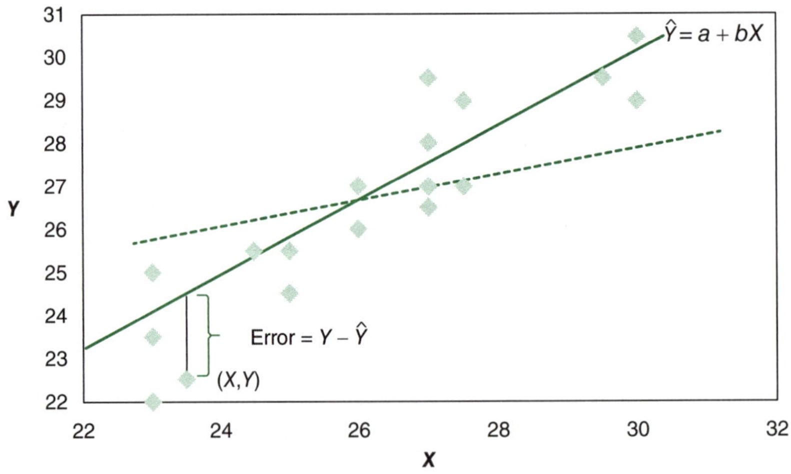

The "best fitting" is mathematically defined by the smallest sum of squared errors, $\Sigma(Y-\hat{Y})^2$, across all data points. The **regression equation** is the equation for the regression line, which is expressed as $\check{Y} = a + bX$, where a is the **Y-intercept** and b is the **slope**. The Y-intercept is the value of Y when $X = 0$. The slope is the value change in Y when X increases by one unit. They are mathematically determined by the following formulas.

$$b = \frac{SP}{SS_X}$$

$$a = \bar{Y} - b\bar{X}$$

This method of estimating a regression line is referred to as the **ordinary least squares (OLS)**. OLS depicts the best fitting line by a straight line that goes through $(\bar{X}, \bar{Y})$ and minimizes the vertical distance between the actual Y values and the predicted $\hat{Y}$.

Let's use an example to illustrate how to calculate the Y-intercept and slope from the two variables X and Y. Both SP and SS_X are familiar terms from Chapter 10. You should feel confident with these calculations.

EXAMPLE 11.1

Physical activities burn calories. Both hours of exercise and calories burned are normally distributed. The relationship between hours of exercise and calories burned can be analyzed. A group of 10 adults was randomly selected. Their hours of exercise and calories burned are reported in Table 11.1. What is the regression line to predict calories burned by using hours spent on exercising?

Similar to calculating Pearson's r, we need to calculate $\bar{X}$ and $\bar{Y}$ first, then create the following four columns in Table 11.1a: $(X-\bar{X})$, $(Y-\bar{Y})$, $(X-\bar{X})(Y-\bar{Y})$, and $(X-\bar{X})^2$ to demonstrate the step-by-step calculation process.

$$\bar{X} = \Sigma X / n = 37/10 = 3.7$$

$$\bar{Y} = \Sigma Y / n = 7100/10 = 710$$

TABLE 11.1 ● Hours of Exercise and Calories Burned for 10 Adults

Hours of Exercise, X	Calories Burned, Y
1	300
3	460
6	990
2.5	850
4.5	940
5	1,100
4	750
2	610
3.5	450
5.5	650

(Continued)

(Continued)

TABLE 11.1a ● Step-by-Step Calculation for Regression Line Between Hours of Exercise and Calories Burned for 10 Adults

Hours of Exercise, X	Calories Burned, Y	$(X-\bar{X})$	$(Y-\bar{Y})$	$(X-\bar{X})(Y-\bar{Y})$	$(X-\bar{X})^2$
1	300	-2.7	−410	1,107	7.29
3	460	−0.7	−250	175	0.49
6	990	2.3	280	644	5.29
2.5	850	−1.2	140	−168	1.44
4.5	940	0.8	230	184	0.64
5	1,100	1.3	390	507	1.69
4	750	0.3	40	12	0.09
2	610	−1.7	−100	170	2.89
3.5	450	−0.2	−260	52	0.04
5.5	650	1.8	−60	−108	3.24
37	**7,100**			**2,575**	**23.1**

$$\text{SP} = \Sigma(X-\bar{X})(Y-\bar{Y}) = 2575$$

$$SS_X = \Sigma(X-\bar{X})^2 = 23.1$$

$$b = \frac{\text{SP}}{SS_X} = \frac{2575}{23.1} = 111.47$$

$$a = \bar{Y} - b\bar{X}$$

$$a = 710 - 111.47(3.7) = 297.55$$

The regression line between the hours of exercise and calories burned is $\hat{Y} = 297.55 + 111.47X$.

$\hat{Y}$ is the estimated value of Y. For every value of X, the corresponding $\hat{Y}$ can be calculated by plugging in the value of X into the regression equation. The sum of squared distance between every $\hat{Y}$ and the actual Y value, $\Sigma(Y-\hat{Y})^2$, is minimized by the regression line.

When we draw the regression line to fit the scatterplot of hours of exercise and calories burned, the result is shown in Figure 11.2.

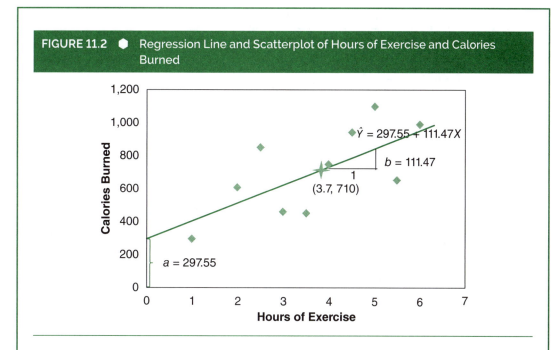

FIGURE 11.2 ● Regression Line and Scatterplot of Hours of Exercise and Calories Burned

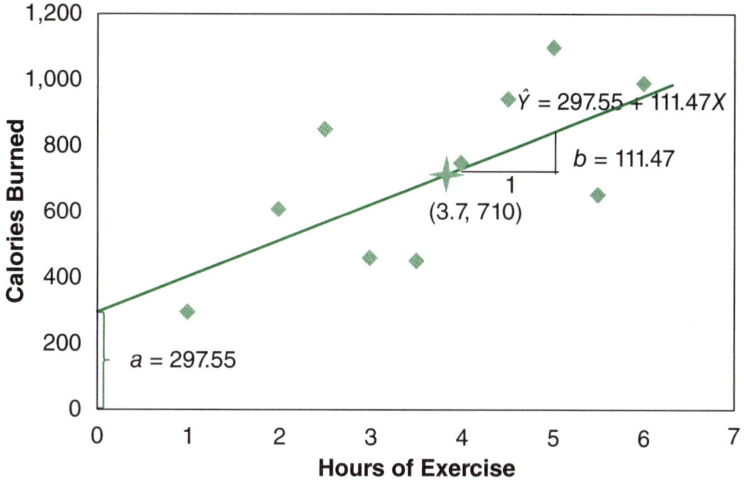

In Figure 11.2, the Y-intercept 297.55 is the value of Y when $X = 0$. The slope is the value change in Y when X increases by one unit. In Example 11.1, for every additional hour exercised, 111.47 more calories were burned. The blue star in the middle of Figure 11.2 is the data point representing $(\bar{X}, \bar{Y}) = (3.7, 710)$. The regression line always goes through $(\bar{X}, \bar{Y})$. This is true for all simple regressions.

POP QUIZ

1. Which one of the following data points is always on the regression line?
   a. (0, 0)
   b. (a, b)
   c. (10, 10)
   d. $(\bar{X}, \bar{Y})$

HYPOTHESIS TESTING FOR SIMPLE REGRESSION

Predictions are used in everyday activities. Educators want to predict students' academic performance, corporations want to predict the market trends, and employers want to predict new hires' job performance, just to name a few examples. Not all predictions are created equal. We need to know how to judge whether the independent variable really helps predicting the values of the dependent variable; therefore, we need to learn how to conduct hypothesis tests on regressions. Simple regression is to predict the values of Y by using the values of X. The focus of regression is the dependent variable Y. The independent variable X is used in a supporting role to provide information, so we can do a better job of predicting Y. Let's break down the process on how to use the information of X to predict Y. Without any information on X, the best prediction for Y is $\bar{Y}$. As you recall from Chapter 3, the mean is the most frequently used central tendency measure for representing the center of a distribution, assuming there are no outliers. By using the extra information from X, we can calculate a regression line $\hat{Y} = a + bX$ to predict the values of Y based on X. With the information of X, the best prediction for Y is $\hat{Y}$. Thus, the difference between $\hat{Y}$ and $\bar{Y}$, $(\hat{Y} - \bar{Y})$, could be viewed as the contribution from regression. Let's contrast the differences among Y, $\bar{Y}$, and $\hat{Y}$.

$$\text{Deviation of } Y = (Y - \bar{Y})$$

$$\text{Explained deviation by regression} = (\hat{Y} - \bar{Y})$$

$$\text{Unexplained deviation by regression} = (Y - \hat{Y})$$

Putting all three deviations together, we get

Deviation of Y = Explained deviation by regression + Unexplained deviation by regression

$$(Y - \bar{Y}) = (\hat{Y} - \bar{Y}) + (Y - \hat{Y})$$

The unexplained deviation is what is missed by the regression. It is also referred to as error or residual. Some books use these two terms interchangeably. I prefer to use error throughout this book. Errors represent the difference or distance between the actual Y and the predicted $\hat{Y}$. Errors do not mean mistakes are made. It means the predicted values, $\hat{Y}$, are not exactly the same as the actual Y values. The smaller the errors, the better the predictions.

$$\text{Error} = (Y - \hat{Y})$$

Sometimes the actual Y is higher than $\hat{Y}$ (positive error), and sometimes the actual Y is lower than $\hat{Y}$ (negative error). To avoid positive errors completely canceling out the negative errors, we square the errors, $\text{Squared errors} = (Y - \hat{Y})^2$. This is the same as we discussed the SS in Chapter 3 where we squared the deviations, then added them up to create the useful sum of squares.

To figure out the total squared errors across all values, simply add them up. When the slope and intercept are calculated the way we just learned, the sum of the squared errors is known as the least squares. This is because of all the possible lines that could be used to predict Y given X, the slope and intercept produce the smallest (or least) amount of errors.

$$\text{Sum of squared errors} = \text{SSE} = \Sigma(Y - \hat{Y})^2$$

Apply the same principle to $(\hat{Y} - \bar{Y})$. We square the $(\hat{Y} - \bar{Y})$ and then add them up. We get a new term, which is labeled the *sum of squares due to regression*, $\text{SSR} = \Sigma(\hat{Y} - \bar{Y})^2$.

Again, apply the same principle to $(Y - \bar{Y})$. We square $(Y - \bar{Y})^2$ and then add them up. We get a familiar sum of squared deviations of Y, $SS_Y = \Sigma(Y - \bar{Y})^2$. Since Y is the main focus of a regression analysis, the SS_Y is also labeled as SST or the sum of squares total. The relationship among all three sum of squared terms in regression is $\text{SST} = \text{SSR} + \text{SSE}$.

Your head might be spinning at this point. A visual presentation of what I just explained might help. In Figure 11.3, SST is the Y circle, SSR is the overlap between X circle and Y circle, and SSE is the part of Y that does not overlap with X. The overlap between X and Y can be calculated by one simple formula, the coefficient of determination r^2, as first introduced in Chapter 10. The total variation of Y is SS_Y and the overlap between X and Y is r^2; therefore, the explained variation of Y is $\text{SSR} = r^2 SS_Y$. The unexplained variation of Y is $\text{SSE} = (1 - r^2)SS_Y$. When we add the explained variation with the unexplained variation of Y, we get the total variation of Y.

$$\text{SSR} = r^2 SS_Y$$

$$\text{SSE} = (1 - r^2)SS_Y$$

Total variation of Y = Explained variation in Y + Unexplained variation in Y

$$SS_Y = r^2 SS_Y + (1 - r^2)SS_Y$$

$$\text{SST} = \text{SSR} + \text{SSE}$$

To evaluate whether the X values help predict the Y values, we partition the total variances of Y into two separate parts: (1) variances due to regression and (2) variances due to errors.

FIGURE 11.3 ● Explained and Unexplained Variation of Y

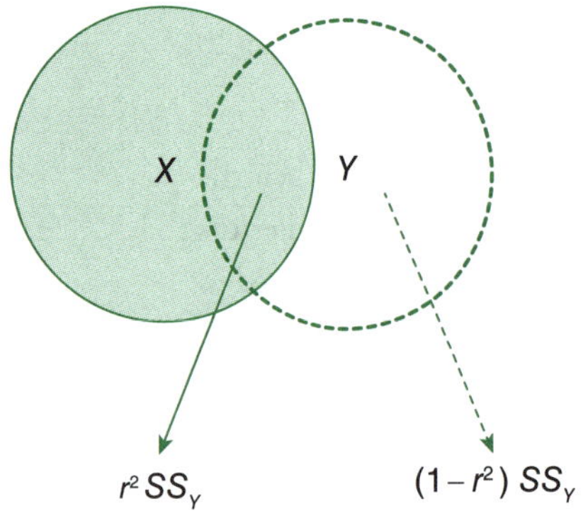

Variance was first introduced in Chapter 3, and it is calculated as $s^2 = SS/df$. This formula is still relevant in this chapter. Partitioning variances due to different sources is called the **analysis of variances (ANOVA)**, which is discussed in depth in Chapter 12. In simple regression, we construct an ANOVA summary table to test the significance of the regression model. The idea is to compare and contrast the variances due to regression and the variances due to error. If the variances due to regression are much bigger than the variances due to error, the model is effective. If the variances due to regression are not much bigger than the variances due to error, the model is not effective. We will go over all the formulas to construct the ANOVA summary table for regression.

We know variance is calculated as $s^2 = SS/df$. We already learned *SS* terms in regression: SST, SSR, and SSE. Now, we need to figure out degrees of freedom for these terms. The **degrees of freedom due to regression** are defined as the number of predictors in the regression equation. There is only one predictor in a simple regression. Therefore, the degree of freedom for a simple regression is 1. In the language of ANOVA, variances are also labeled as mean squares (MS). Therefore, the variances due to regression are also labeled as mean squares due to regression, MSR.

$$\text{MSR} = \frac{\text{SSR}}{df_{\text{R}}} = \frac{\text{SSR}}{1}$$

The **degrees of freedom due to error** are defined as $(n - 2)$. This is because in the process of calculating a simple regression, $\bar{X}$ and $\bar{Y}$ are used to estimate two population parameters, μ_X and μ_Y. Every time we estimate a parameter, we lose 1 *df*. Estimating two parameters means losing 2 *df*. The variances due to error are also labeled as mean squares error, or *MSE*.

$$MSE = \frac{\text{SSE}}{df_{\text{E}}} = \frac{\text{SSE}}{(n-2)}$$

The significance of the regression is calculated by an *F* test. The calculated *F* value demonstrates the ratio of variance explained by the regression to the variance left unexplained by the errors.

$$F = \frac{\text{MSR}}{MSE}$$

The critical values of F that set the boundaries of the rejection zones can be found using $F_{(df_1, df_2)}$ in the same F Table we used to determine whether equal variances are assumed in Chapter 9. This table is provided in Appendix C. Each critical F value is determined by two different degrees of freedom, where df_1 is the degrees of freedom for the numerator and df_2 is the degrees of freedom for the denominator. In the case of a simple regression, df_1 is the degrees of freedom due to regression, $df_1 = df_R = 1$ and df_2 is the degrees of freedom due to errors, $df_2 = df_E = (n - 2)$. Therefore, the critical values of F can be identified by $F_{(1,n-2)}$ in the F Table for simple regression.

Now, we are ready to put everything together and conduct the four-step hypothesis test for a simple regression.

Step 1: State the pair of hypotheses.

The purpose of a simple regression is to use the values of X to predict Y. Therefore, one of the most important parts of the regression equation is the slope, which describes how the value of Y changes when X increases by one unit. The slope is also called the regression coefficient. It is denoted as b when the data are from a sample and as β when describing a population. The hypotheses always describe the population parameters. If knowing the values of X does not help predict the values of Y, then $\beta = 0$. On the other hand, if knowing the values of X better predicts the values of Y, then $\beta \neq 0$. Accordingly, the pair of hypotheses for a simple regression is expressed as follows:

H_0: $\beta = 0$

H_1: $\beta \neq 0$

You probably noticed that there is only one pair of two-tailed hypotheses listed here. Let's explain the reason. The test statistic for a simple regression is an F test, calculated as $F = \text{MSR}/MSE$, where MSR is the variance explained by the regression, and MSE is the variance due to error. Variances are always positive. Therefore, the calculated F values are always positive. Unlike the Z tests or t tests in previous chapters, F tests only produce positive values. This makes it impossible to use F for one-tailed tests. One-tailed or directional tests are *possible* only when the test statistics can be either positive or negative. Therefore, hypothesis tests for simple regression using F tests can only be done in two-tailed tests.

Step 2: Identify the rejection zone.

The critical values of F that set off the boundaries of the rejection zones are identified by $F_{(1,n-2)}$ in the F Table. The rejection zone is identified as calculated $F > F_{(1,n-2)}$.

Step 3: Calculate the test statistic.

The process of calculating F for a simple regression as discussed earlier can be summarized in Table 11.2.

TABLE 11.2 ● The ANOVA Summary Table for a Simple Regression

Source	SS	df	MS	F
Regression	$SSR = r^2 SS_Y$	1	$MSR = \frac{SSR}{1}$	$\frac{MSR}{MSE}$
Error	$SSE = (1 - r^2) SS_Y$	$n - 2$	$MSE = \frac{SSE}{(n-2)}$	
Total	SS_Y or SST	$n - 1$		

A large F value indicates that the variances explained by the regression are relatively large compared to the variances left unexplained (i.e., error variances). A small F value indicates that the variances explained by the regression are relatively small compared to the variances left unexplained (i.e., error variances). However, the standard to judge what is considered to be "large" is in the critical values from the F Table as specified in Step 2.

Step 4: Make the correct conclusion.

If the calculated F from Step 3 is within the rejection zone, we reject H_0. If the calculated F is not within the rejection zone, we fail to reject H_0.

Finally, the process of conducting a hypothesis test on a simple regression is complete. It is time to use an example to go through the four-step hypothesis test for a simple regression. In Example 11.1, we calculated the regression line, and the results are shown in Table 11.1a. Now we can complete the process by conducting a hypothesis test on the overall significance of the simple regression, using $\alpha = .05$.

$$\bar{X} = \Sigma X / n = 37 / 10 = 3.7$$

$$\bar{Y} = \Sigma Y / n = 7100 / 10 = 710$$

Step 1: State the pair of hypotheses.

H_0: $\beta = 0$

H_1: $\beta \neq 0$

Step 2: Identify the rejection zone.

The critical value of F under $\alpha = .05$ is $F_{(1,8)} = 5.32$. The rejection zone is $F > 5.32$.

Step 3: Calculate the statistic.

Conducting a hypothesis test on a simple regression requires r^2. Therefore, one additional column, $(Y-\bar{Y})^2$, needs to be added to Table 11.1a to complete this process. This is shown in Table 11.1b.

$$r = \frac{\text{SP}}{\sqrt{SS_X SS_Y}} = \frac{2575}{\sqrt{(23.1)(616400)}} = .682$$

$$r^2 = .682^2 = .466$$

TABLE 11.1b ● Step-by-Step Calculation for Conducting a Hypothesis Test on a Simple Regression Between Hours of Exercise and Calories Burned for 10 Adults

Hours of Exercise, X	Calories Burned, Y	$(X-\bar{X})$	$(Y-\bar{Y})$	$(X-\bar{X})(Y-\bar{Y})$	$(X-\bar{X})^2$	$(Y-\bar{Y})^2$
1	300	−2.7	−410	1107	7.29	168,100
3	460	−0.7	−250	175	0.49	62,500
6	990	2.3	280	644	5.29	78,400
2.5	850	−1.2	140	−168	1.44	19,600
4.5	940	0.8	230	184	0.64	52,900
5	1,100	1.3	390	507	1.69	152,100
4	750	0.3	40	12	0.09	1,600
2	610	−1.7	−100	170	2.89	10,000
3.5	450	−0.2	−260	52	0.04	67,600
5.5	650	1.8	−60	−108	3.24	3,600
37	7,100			2,575	23.1	616,400

$$SS_Y = 616{,}400$$

$$\text{SSR} = r^2 SS_Y = .466(616{,}400) = 287242.4$$

$$\text{SSE} = (1 - r^2)SS_Y = (1 - .466)(616{,}400) = 329157.6$$

Put all the numbers into the ANOVA summary table for a simple regression, as shown in Table 11.1c.

TABLE 11.1c ● The ANOVA Summary Table for the Simple Regression Between Hours of Exercise and Calories Burned

Source	SS	df	MS	F
Regression	287242.4	1	287242.4	6.98
Error	329157.6	8	41144.7	
Total	616,400	9		

Step 4: Make the correct conclusion.

The calculated $F = 6.98$ was within the rejection zone; therefore, we rejected H_0. The evidence was strong enough to support the claim that hours of exercise predicted calories burned by this simple regression line $\hat{Y} = 297.55 + 111.47X$.

Once we reject H_0, we know that using X values helps predict Y values. We may use the regression equation to calculate $\hat{Y}$ for any value of X within the value range in the sample. For example, we can calculate the predicted calories burned, $\hat{Y}$, when the hours of exercise is 4 by plugging $X = 4$ into the regression equation.

$$\hat{Y} = 297.55 + 111.47(4) = 743.43$$

The predicted calories burned with 4 hours of exercise are 743.43 according to the regression equation. Such predictions work well for X values within the range of the original X values in the sample but might not work for X values outside the range of X in the sample.

There is a quick and easy way to double-check our calculations in this summary table by using the basic principle SST = SSR + SSE. In this example, SSR = 287242.4 and SSE = 329157.6, if and only if these two add up to SST = 616,400, we can be certain that our calculations are correct. The process to conduct a hypothesis test for the simple regression is quite complicated. You can certainly benefit from doing more practice.

As college tuitions are rising every year, there are heated discussions and debates about whether education is a worthy investment for the future. Let's use an example to predict annual salary based on education.

EXAMPLE 11.2

To answer the question "whether education is a worthy investment", a group of 24-year-olds was randomly selected. Their years of schooling and annual salaries (in thousands of dollars) were reported in Table 11.3. Assume that years of schooling and annual salaries for 24-year-olds are normally distributed. Use the numbers in Table 11.3 to answer the following issues.

a. Construct a scatterplot to verify if a linear relationship exists between years of schooling and annual salaries.
b. Identify the best fitting line for years of schooling and annual salaries.
c. Conduct a hypothesis test to verify if years of schooling help predict annual salaries.

a. The scatterplot of years of schooling and annual salaries is shown in Figure 11.4.

TABLE 11.3 ● Years of Schooling and Annual Salaries for 24-Year-Olds

Years of Schooling	Annual Salary ($1,000)
12	25
13.5	19
14	61
15	27
16.5	35
17	34
18.5	36
19	33
20.5	75
21	55

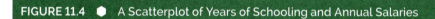

FIGURE 11.4 ● A Scatterplot of Years of Schooling and Annual Salaries

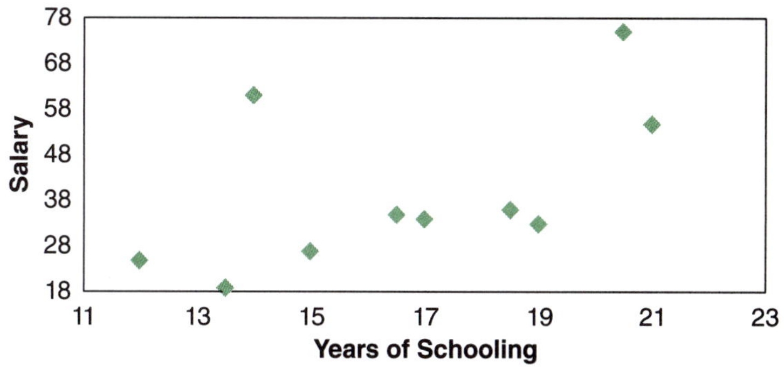

(Continued)

(Continued)

According to the scatterplot, a linear relationship existed between years of schooling and annual salaries. There were no obvious outliers.

b. According to the formulas for the slope and Y-intercept for the regression line,

$$b = \frac{SP}{SS_X}$$

$$a = \overline{Y} - b\overline{X}$$

We need to figure out the $\overline{X}$ and $\overline{Y}$.

$$\overline{X} = \Sigma X / n = 167 / 10 = 16.7$$

$$\overline{Y} = \Sigma Y / n = 400 / 10 = 40$$

Then add the following five columns to Table 11.3, $(X - \overline{X}), (Y - \overline{Y}), (X - \overline{X})(Y - \overline{Y}), (X - \overline{X})^2$, and $(Y - \overline{Y})^2$, to demonstrate the step-by-step calculation process as shown in Table 11.3a.

TABLE 11.3a ● Step-by-Step Calculation to Predict Annual Salaries Using Years of Schooling for 24-Year-Olds

Years of Schooling	Annual Salary ($1,000)	$(X - \overline{X})$	$(Y - \overline{Y})$	$(X - \overline{X})(Y - \overline{Y})$	$(X - \overline{X})^2$	$(Y - \overline{Y})^2$
12	25	−4.7	−15	70.5	22.09	225
13.5	19	−3.2	−21	67.2	10.24	441
14	61	−2.7	21	−56.7	7.29	441
15	27	−1.7	−13	22.1	2.89	169
16.5	35	−0.2	−5	1	0.04	25
17	34	0.3	−6	−1.8	0.09	36
18.5	36	1.8	−4	−7.2	3.24	16
19	33	2.3	−7	−16.1	5.29	49
20.5	75	3.8	35	133	14.44	1,225
21	55	4.3	15	64.5	18.49	225
167	400			276.5	84.1	2,852

$$SP = \Sigma(X - \bar{X})(Y - \bar{Y}) = 276.5$$

$$SS_X = \Sigma(X - \bar{X})^2 = 84.1$$

$$b = \frac{SP}{SS_X} = \frac{276.5}{84.1} = 3.288$$

$$a = \bar{Y} - b\bar{X} = 40 - 3.288(16.7) = -14.910$$

Therefore, the regression line to predict annual salary using years of schooling is

$$\hat{Y} = -14.910 + 3.288X$$

On average, for every additional year of schooling, annual salary increases by 3.288 thousand dollars.

c. Conduct a hypothesis test of the regression.

Step 1: State the pair of hypotheses.

H_0: $\beta = 0$

H_1: $\beta \neq 0$

Step 2: Identify the rejection zone.

The critical value of F in the F Table under $\alpha = .05$ is $F_{(1,8)} = 5.32$. The rejection zone is $F > 5.32$.

Step 3: Calculate the statistic.

$$SS_Y = \Sigma(Y - \bar{Y})^2 = 2852$$

$$r = \frac{SP}{\sqrt{SS_X SS_Y}} = \frac{276.5}{\sqrt{(84.1)(2852)}} = .565$$

$$r^2 = .565^2 = .319$$

$$SSR = r^2 SS_Y = .319(2{,}852) = 909.788$$

$$SSE = (1 - r^2)SS_Y = (1 - .319)(2{,}852) = 1942.212$$

Put all the numbers into the ANOVA summary table for a simple regression to predict annual salary using years of schooling as shown in Table 11.3b.

(Continued)

(Continued)

TABLE 11.3b ● The ANOVA Summary Table for the Simple Regression Using Years of Schooling to Predict Annual Salary

Source	*SS*	*df*	*MS*	*F*
Regression	909.788	1	909.788	3.747
Error	1942.212	8	242.777	
Total	2,852	9		

Step 4: Make the correct conclusion.

The calculated $F = 3.747$ was not within the rejection zone; therefore, we failed to reject H_0. The evidence was not strong enough to support the claim that the years of schooling helped predict the annual salary.

Standard Error of the Estimate

Another way to demonstrate the accuracy of the regression line is to calculate the **standard error of the estimate**. The standard error of the estimate is defined as the standard deviation of the prediction errors $(Y - \hat{Y})$. In other words, the standard error of the estimate measures the standard distance between the regression line and the actual *Y* values. If the standard error of the estimate is large, it means that $\hat{Y}$ is far away from the actual value of *Y*. If the standard error of the estimate is small, it means that $\hat{Y}$ is close to the actual value of *Y*. To obtain the standard error of the estimate, follow the same operation as we do to create standard deviation (*s*) from sum of squared (*SS*) deviations for a sample,

$$s^2 = \frac{SS}{df} \text{ and } s = \sqrt{\frac{SS}{df}}$$

The sum of squared deviations of estimate is $\Sigma(Y - \hat{Y})^2$; therefore,

$$\text{The standard error of estimate} = \sqrt{\frac{\Sigma(Y - \hat{Y})^2}{df}} = \sqrt{\frac{\text{SSE}}{(n-2)}} = \sqrt{MSE}$$

The degrees of freedom for standard error of estimate are $n - 2$ because in the process of estimating $\hat{Y}$, we calculate two sample means to estimate two population means. Every time we estimate a population parameter, we lose 1 *df*. Estimating two parameters means losing 2 *df*. Therefore, the degrees of freedom for the standard error of estimate are $n - 2$.

The Mathematical Relationship Between Pearson's *r* and the Regression Slope, *b*

Pearson's *r* formula is $r = \text{SP} / \sqrt{SS_X SS_Y}$. The regression slope is $b = \text{SP} / SS_X$.

You can see that they are very similar to each other. Indeed, there is a simple mathematical relationship between r and b.

$$b = r\frac{\sqrt{SS_Y}}{\sqrt{SS_X}}$$

Because $s_X = \sqrt{SS_X / df}$ and $s_Y = \sqrt{SS_Y / df}$, the mathematical relationship between r and b can be simplified to

$$b = r\frac{s_Y}{s_X}$$

This means that the regression slope b is equal to Pearson's r multiplied by the standard deviation of Y, then divided by the standard deviation of X. Pearson's r is calculated based on the standard scores (Z scores) of both X and Y; thus, both variables are on the same Z scale. Therefore, Pearson's r has a definite range of $-1 \leq r \leq 1$. In contrast, the regression slope, b, is calculated on the original scales of the variables X and Y, which may vary dramatically. Thus, there is no definite range for a regression slope. A regression slope is defined as the value change in Y when X increases by one unit. As demonstrated in previous examples, the regression slope can be the calories burned for each additional hour of exercise or the salary for each additional year of schooling. The values of the calculated regression slopes do not confine themselves within a certain range. The bridge between the Z scale and the original scale is the standard deviation of the variable, so it makes perfect sense that the regression slope b equals Pearson's r multiplied by the standard deviation of Y and divided by the standard deviation of X.

POP QUIZ

2. The F test is used to evaluate the quality of the simple regression model, $F = MSR / MSE$. What is the relationship between the calculated F value and the standard error of the estimate $\sqrt{MSE}$?
   a. The larger the F value, the smaller the standard error of the estimate.
   b. The larger the F value, the larger the standard error of the estimate.
   c. The F value is the square of the standard error of the estimate.
   d. The F value is the square root of the standard error of the estimate.

STATISTICAL ASSUMPTIONS FOR SIMPLE REGRESSION

There are a few statistical assumptions we need to pay attention to in order to ensure that results of OLS regression analysis are accurate. These assumptions are commonly referred to as Gauss–Markov assumptions, named after two mathematicians, Carl Friedrich Gauss and Andrey Markov, who were credited with developing the OLS method to come up with the *best linear unbiased estimator*, abbreviated BLUE.

1. There is a linear relationship between variables X and Y. If the relationship between X and Y is not linear, regression analysis will produce biased estimates of the actual relationship. A scatterplot of X and Y can verify whether the relationship is linear.
2. Error (or residual) is calculated by the distance between the predicted $\widehat{Y}$ and the actual Y values, $(Y-\widehat{Y})$. The variances of errors are the same across all values of X. The variances of Y are also the same across all values of X. This is denoted as $\text{Var}(\varepsilon) = \text{Var}(Y) = \sigma^2$, which is commonly referred to as the homoscedasticity in regression. Homoscedasticity is defined as homogeneity of variances of Y across all values of X. It is a good word to throw around to impress people with your knowledge in statistics. Being able to say homoscedasticity is much more impressive than saying homogeneity of variances. Many of you are visual learners, so seeing a chart to demonstrate the concept of homoscedasticity will be beneficial to your learning. A scatterplot of errors and $\widehat{Y}$ can verify whether the homoscedasticity is violated, as shown in Figure 11.5. In Figure 11.5, you can see a roughly even spread of residuals (i.e., errors) across all $\widehat{Y}$, and most errors are within −100 and +100. There is a slightly larger range of errors on the left side of the graph than on the right side of the graph. Such a slight variation of homoscedasticity has little effect on the OLS regression, but a severe violation of homoscedasticity will lead to serious distortion of the results.

FIGURE 11.5 ● Homoscedasticity as Shown by a Scatterplot of Residuals and Fitted Values

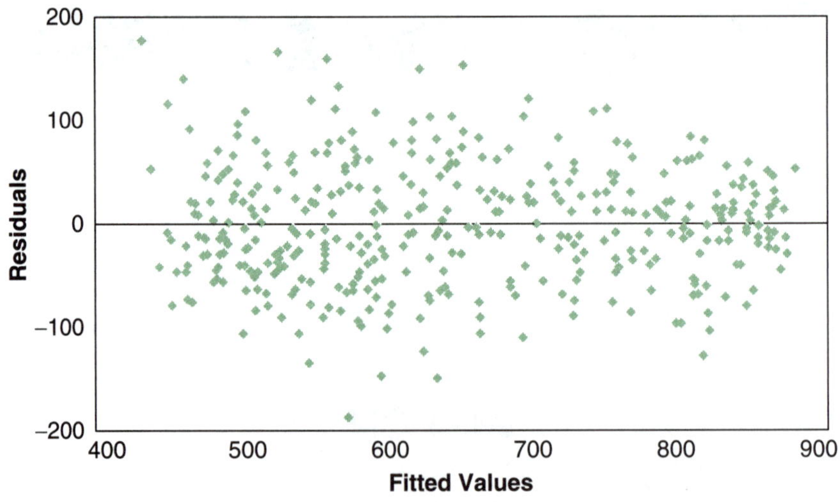

Note: Chen, X., Ender, P., Mitchell, M. and Wells, C. (2003). *Regression with Stata*. Institute for Digital Research and Education, University of California at Los Angeles. https://stats.idre.ucla.edu/stata/webbooks/reg/chapter2/stata-webbooksregressionwith-statachapter-2-regression-diagnostics/

When the scatterplot of errors and $\hat{Y}$ assumes a funnel shape with errors on one side of the graph significantly larger than errors on the other side of the graph, as shown in Figure 11.6, it is called **heteroscedasticity**. Figure 11.6 shows that the errors on the left side of the graph are spread out much further (i.e., between 10 and −10) than the errors on the right side (i.e., between 1 and −1). It is a funnel shape with the larger opening to the left. The homogeneity of error variances assumption is violated. If the errors on the left side of the graph are much smaller than the errors on the right side, the larger opening of the funnel is to the right. It is also called heteroscedasticity.

FIGURE 11.6 ● Heteroscedasticity as Shown by a Scatterplot of Residuals and Fitted Values

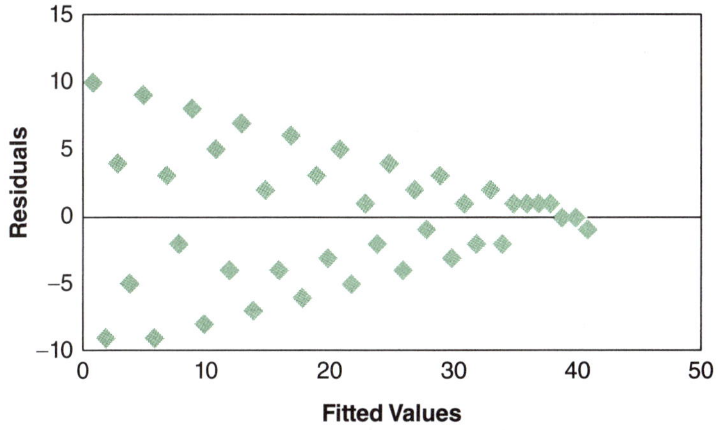

3. The mean (or expected value) of the error term is zero. This is denoted as $E(\varepsilon) = 0$. This is due to one of the mathematical constraints that the sum of all error terms adds to zero; it is denoted as $\Sigma(\varepsilon) = \Sigma(Y - \hat{Y}) = 0$. All positive errors cancel out the negative errors.
4. Error terms are independent of one another. They are not correlated.

In simple regression, there is only one predictor in the regression equation. We can extend the same principles further to consider using multiple predictors at the same time to predict the criterion when we conduct multiple regressions. For example, in evaluating student applicants for a graduate program admission, multiple pieces of information can be considered in the decision-making process, such as GPA from undergraduate degrees, GRE scores, the quality of recommendation letters, and the relevance of the personal statement to the program's curriculum. The calculations of multiple regressions are usually done by statistical software packages, which provide the significance test for every

Author's Aside

Students who are interested in getting more details on the assumptions of regression can do so by reading Berry (1993).

predictor included in the regression analysis and also evaluate the effectiveness of the entire regression model. Multiple regressions are highly useful and practical statistical procedures, however, it is not covered in this introductory statistics book. I would like to point out the proper scales of measurement for the predictors in multiple regressions. The predictors in a multiple regression can be either interval or ratio variables or dichotomous variables coded as 0 or 1. Although a dichotomous variable when coded as 0 or 1 can be considered as a predictor in a simple regression, this is rarely done. A simple regression using a dichotomous predictor generates the same result as one would get when conducting a *t* test. Most statisticians would opt to conduct a *t* test instead of conducting a simple regression with a dichotomous predictor.

Another cautionary note on regression is that outliers have a strong impact on regressions, the same way as they do on correlations. Remove outliers that are known to be caused by mistakes in measurement or data collection procedures. Examining the impact of remaining outliers can be conducted by running regression analysis with and without outliers.

To make it clear, a simple regression analysis reveals the relationship between the independent variable (*X*) and the dependent variable (*Y*) by specifying how the *Y* value changes when *X* increases by one unit. A significant simple regression does not prove causation. It is the same situation as discussed in correlations. Causation can only be established by strict experimental designs, not by any statistical procedure.

STEP-BY-STEP INSTRUCTIONS OF SIMPLE REGRESSION IN EXCEL

You may follow the exact same Excel step-by-step instruction shown in Chapter 10 to calculate SP, SS_X, and SS_Y. Then calculate the slope and *Y*-intercept for the regression line by using the OLS formulas. Or you can use Excel's **Analysis ToolPak** to conduct a simple regression. The instruction to install this Excel Add-Ins is provided in Chapter 3 of this book.

1. We use school ratings data to conduct a simple regression in Excel. Public school districts are mandated by law to provide students' proficiency test results on social studies, language, and math from selected grades in K–12 schools. Proficiency tests along with other performance criteria are included in the school report cards every year. To demonstrate a simple regression, we used the median household income as the independent variable and the number of standards

met by the school districts as the dependent variable in a school rating data set. Assume that the median household income and the number of standards met were normally distributed. A random sample of 31 school districts was selected. The median household income is reported in thousands of dollars in **Column A**, and the number of standards met by the school district is in **Column B**. To start conducting a simple regression, click on **Data** tab, then click **Data Analysis** at the top right corner, a pop-up menu options of Data Analysis appears as shown in Figure 11.7. Click on **Regression**, then click **OK**. If you did not see **Data Analysis**, you need to reinstall the **Analysis ToolPak Add-Ins.**

FIGURE 11.7 ● Using School Ratings Dawta to Conduct a Simple Regression in Excel

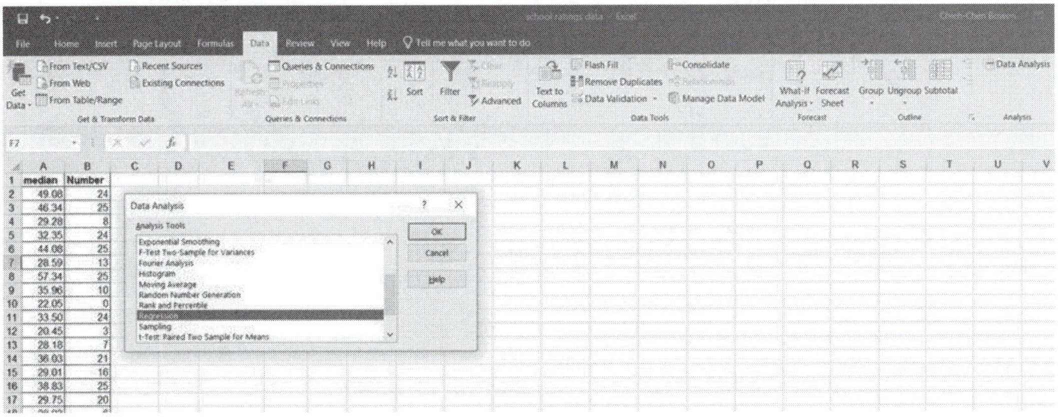

2. A pop-up window appears to ask for **Input Y Range, Input X Range**. Such a procedure is needed to identify the data range for the dependent variable and the independent variable in the simple regression. In the **Input Y Range**, click on the , the screen switch to the data. Move the cursor to **B1**, then hold the shift key and move all the way to **B32** and let go. The Input window showed the **Input Y Range** as **$B$1:$B$32**. In the **Input X Range**, click on the , the screen switch to the data. Move the cursor to **A1** then hold the shift key and move all the way to **A32** and let go. The Input window showed the **Input X Range** as **$A$1:$A$32** as shown in Figure 11.8. The data range include the variable names in **A1** and **B1** so make sure the box for **Labels** is checked, then click **OK**. Now Excel calculates the simple regression and reports the results in a different worksheet.

FIGURE 11.8 ● Excel Simple Regression With "Number of Standards Met" as the Dependent Variable and Median Household Income as the Independent Variable

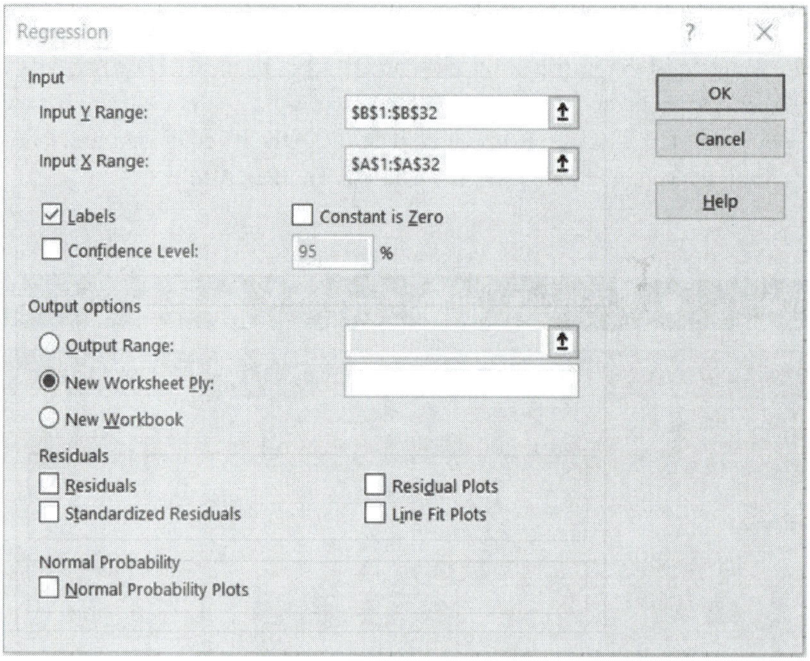

3. The simple regression results are shown in a separate worksheet as shown in Figure 11.9. The top portion of the **Summary Output** includes **Regression Statistics, ANOVA and Coefficients**. The R^2 is .528, which indicates the median household income explained 52.8% of variance in the number of standards met in the school district. The standard error of the estimate was 5.511, which was calculated as

$$\text{Standard error of estimate} = \sqrt{\frac{\Sigma(Y-\hat{Y})^2}{df}} = \sqrt{\frac{\text{SSE}}{(n-2)}} = \sqrt{MSE}.$$

FIGURE 11.9 ● Simple Regression Results as Reported by Excel

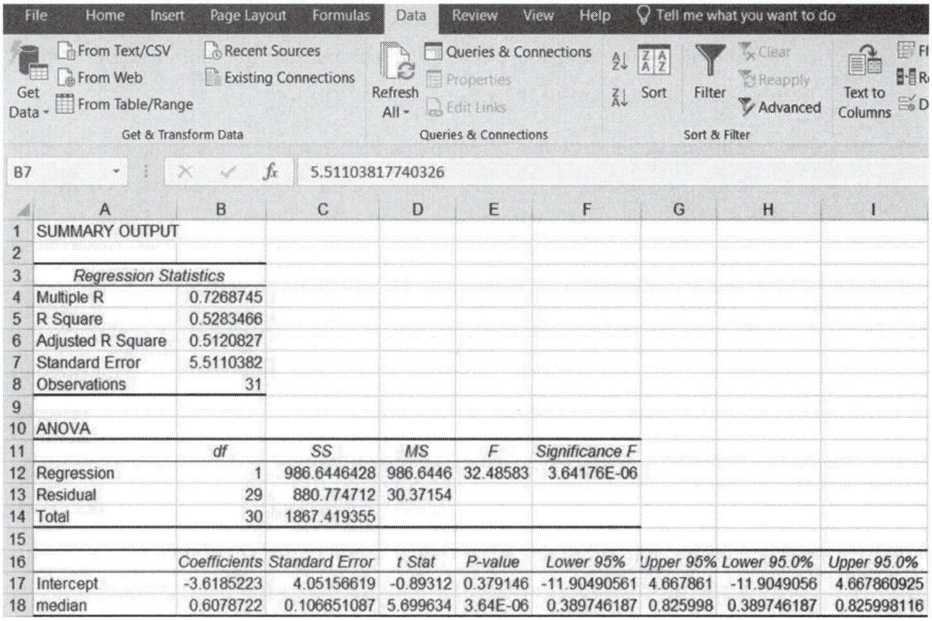

	A	B	C	D	E	F	G	H	I
1	SUMMARY OUTPUT								
2									
3	*Regression Statistics*								
4	Multiple R	0.7268745							
5	R Square	0.5283466							
6	Adjusted R Square	0.5120827							
7	Standard Error	5.5110382							
8	Observations	31							
9									
10	ANOVA								
11		*df*	*SS*	*MS*	*F*	*Significance F*			
12	Regression	1	986.6446428	986.6446	32.48583	3.64176E-06			
13	Residual	29	880.774712	30.37154					
14	Total	30	1867.419355						
15									
16		*Coefficients*	*Standard Error*	*t Stat*	*P-value*	*Lower 95%*	*Upper 95%*	*Lower 95.0%*	*Upper 95.0%*
17	Intercept	-3.6185223	4.05156619	-0.89312	0.379146	-11.90490561	4.667861	-11.9049056	4.667860925
18	median	0.6078722	0.106651087	5.699634	3.64E-06	0.389746187	0.825998	0.389746187	0.825998116

You could find the SSE, *df*, and *MSE* in the ANOVA summary table to verify the accuracy of the standard error, $\sqrt{\dfrac{SSE}{(n-2)}} = \sqrt{\dfrac{880.77}{29}} = \sqrt{30.372} = 5.511.$

4. The ANOVA summary table presents the outcome of the hypothesis test for the simple regression. The calculated F value demonstrates the ratio of variance explained by the regression to the variance left unexplained by the errors. The calculated $F = 32.486$ means that the variance explained by the regression is 32.486 times the variance left unexplained by the errors. The *significance* level of the calculated F value in the last column is the calculated p value associated with the F test. The p value gives the probability of obtaining the observed test statistic assuming that the null hypothesis is true. When the p value associated with a test statistic is used as a standard for a hypothesis test, the decision rules are as follows:

   When $p < \alpha$, we reject H_0.

   When $p \geq \alpha$, we fail to reject H_0.

This is the same set of decision rules as discussed in Chapter 9. The decision rules for a hypothesis test using p values are the same across all statistical procedures. In this example, the calculated p was 3.641E-06 which was the scientific expression of $p = 3.641 \times 10^{-6}$, or .000003641, therefore, we rejected H_0. The median household income predicted the number of standards met in the school district very well.

5. The last table in Figure 11.9 provided the regression coefficients to specify the regression equation, $\widehat{Y} = a + bX$. We used the unstandardized coefficients to specify the regression equation, since both variables were measured in their original measurement units. The standardized regression coefficients, or beta values, are used when variables are converted to standard Z scores. The first row in the table provided the constant, which referred to the intercept a, and the second row gives the unstandardized regression coefficient for median household income, which referred to the slope. Therefore, the regression line was $\widehat{Y} = -3.619 + 0.608X$ as shown in Figure 11.9.

The correct way to interpret this regression equation was that for every 1,000-dollar increase in the median household income in a school district, the number of standards met increased by 0.608. The ***t* Stat** column indicates the t test for the slope in the regression model. The pair of hypotheses for slope are:

$H_0: \beta_1 = 0$

$H_1: \beta_1 \neq 0$

$$\text{The } t \text{ is calculated by } t = \frac{\text{Unstanardized coefficient}}{\text{Standard error of the coefficient}} = \frac{.60787}{.10665} = 5.6996$$

The *p value* associated with the t test was 3.64E-06, therefore, we rejected H_0. The median household income is highly significant in predicting number of standards met. Both the ANOVA table and the regression coefficient table provide identical results. The F test in ANOVA provides significance for the entire model. The t test is used to test the significance for each predictor. Since there is only one predictor in a simple regression, the F test produces the same result as the t test. A unique mathematical relationship exists between the F test and the t test under the simple regression: $F = t^2$. Let's verify this relationship with the reported t and F values. When you square the t value $5.699^2 = 32.485$ and compare it with the reported F value 32.486, the slight discrepancy is due to rounding.

EXERCISE PROBLEMS

1. A car dealer specializing in used Corvettes had a large inventory on the lot. A group of 10 used Corvettes was randomly selected. Ages of the cars and their prices were reported in Table 11.4. Assume that the ages of the cars and their prices were normally distributed. Use the numbers in Table 11.4 to answer the following questions.
   a. Construct a scatterplot of age and price of used Corvettes to see if a linear relationship exists between these two variables.
   b. Identify the regression line using used Corvette's age to predict its price.
   c. What is the predicted price for a 6-year-old Corvette?
   d. Use $\alpha = .05$ to evaluate if using the ages of the cars helps predict the prices of used Corvettes.

TA BLE 11.4 ● Age and Price of Used Corvettes

Age, X	Price, Y ($1,000)
1	62.8
2	58.5
4	46.2
5	45
7	35.4
10	43.5
11	40.3
12	33
13	25.3
15	26

2. A group of nine students was randomly selected from statistics courses. Their number of hours of studying for an exam and grades were reported in Table 11.5. Assume that the hours of studying and the exam grades were normally distributed.
   a. Construct a scatterplot of hours of studying and grades to see if a linear relationship exists between these two variables.
   b. Identify the regression line using the hours of studying to predict grades.
   c. Use $\alpha = .05$ to evaluate if hours of studying help predict grades.

TABLE 11.5 ● Hours of Studying and Exam Grades

Hours of Studying, X	Grade, Y
2	60
3	71
3	83
4	72
4	75
4	65
5	83
5	74
6	92

3. A group of 20 employees was randomly selected from a large company. Employees' years of experience on the job and performance ratings are reported in Table 11.6. Higher ratings mean better job performance.
   a. Construct a scatterplot of years of experience and performance ratings to see if a linear relationship exists between these two variables.
   b. Identify the regression line using years of experience to predict job performance.

TABLE 11.6 ⬢ Employees' Years of Experience and Job Performance Ratings

Years, X	Job Performance, Y
1	51
3	75
4	57
5	63
7	69
9	75
10	80
12	88
13	88
15	90
16	70
17	90
18	87
19	84
21	81
23	77
24	74
26	70
27	87
29	60

Solutions

1. a. The scatterplot of age and price of used Corvettes is shown in Figure 11.10. According to the scatterplot, a straight line with downhill trend fits the data points well. Therefore, a linear relationship exists between the age and price of used Corvettes.

FIGURE 11.10 ⬢ Scatterplot of Age and Price of Used Corvettes

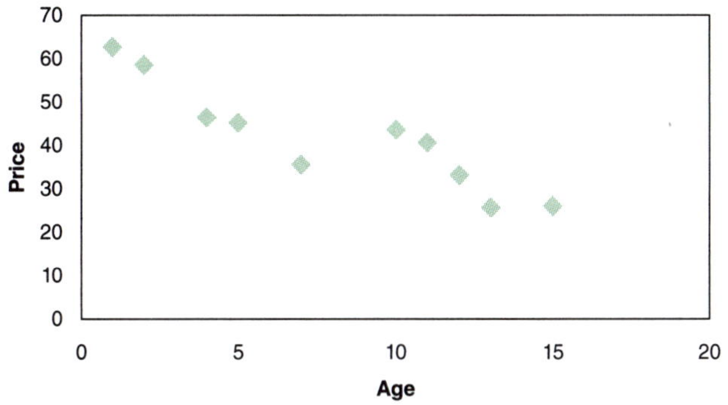

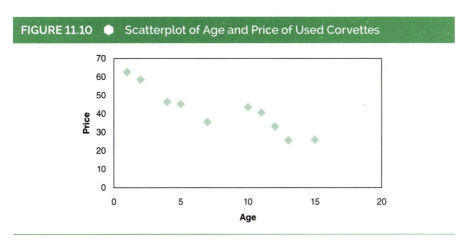

1. b. We need to figure out the $\bar{X}$ and $\bar{Y}$ then add the following five columns to Table 11.4a, $(X-\bar{X}), (Y-\bar{Y}), (X-\bar{X})(Y-\bar{Y}), (X-\bar{X})^2$, and $(Y-\bar{Y})^2$, to demonstrate the step-by-step process.

TABLE 11.4a ⬢ Step-by-Step Calculation Using Age to Predict Price of Used Corvettes

Age, X	Price, Y ($1,000)	$(X-\bar{X})$	$(Y-\bar{Y})$	$(X-\bar{X})(Y-\bar{Y})$	$(X-\bar{X})^2$	$(Y-\bar{Y})^2$
1	62.8	−7	21.2	−148.4	49	449.44
2	58.5	−6	16.9	−101.4	36	285.61
4	46.2	−4	4.6	−18.4	16	21.16
5	45	−3	3.4	−10.2	9	11.56
7	35.4	−1	−6.2	6.2	1	38.44
10	43.5	2	1.9	3.8	4	3.61
11	40.3	3	−1.3	−3.9	9	1.69
12	33	4	−8.6	−34.4	16	73.96
13	25.3	5	−16.3	−81.5	25	265.69
15	26	7	−15.6	−109.2	49	243.36
80	416			−497.4	214	1,394.52

$$\bar{X} = \Sigma X / n = 80/10 = 8$$

$$\bar{Y} = \Sigma Y / n = 416/10 = 41.6$$

$$\text{SP} = \Sigma(X - \bar{X})(Y - \bar{Y}) = -497.4$$

$$\text{SS}_X = \Sigma(X - \bar{X})^2 = 214$$

$$b = \frac{\text{SP}}{SS_X} = \frac{-497.4}{214} = -2.32$$

$$a = \bar{Y} - b\bar{X}$$

$$a = 41.6 - (-2.32)(8) = 60.16$$

The regression line for using age to predict price is $\hat{Y} = 60.16 - 2.32X$. For each additional year of the used Corvette, the price decreased by 2.32 thousand dollars.

1. c. When $X = 6$, the predicted The predicted $\hat{Y} = 60.16 - 2.32(6) = 46.24$. price for a 6-year-old Corvette was 46.24 thousand dollars.

1. d. Here is the four-step hypothesis test for a simple regression.

Step 1: State the pair of hypotheses.

H_0: $\beta = 0$

H_1: $\beta \neq 0$

Step 2: Identify the rejection zone.

The critical value of F in the F Table under $\alpha = .05$ is $F_{(1,8)} = 5.32$. The rejection zone is $F > 5.32$.

Step 3: Calculate the test statistic.

Conducting a hypothesis test on a simple regression requires r^2 and SS_Y.

$$r = \frac{\text{SP}}{\sqrt{SS_X SS_Y}} = \frac{-497.4}{\sqrt{(214)}\sqrt{(1394.52)}} = -.910$$

$$r^2 = (-.910)^2 = .828$$

$$SS_Y = 1394.52$$

$$\text{SSR} = r^2 SS_Y = .828(1394.52) = 1154.66$$

$$\text{SSE} = (1 - r^2)SS_Y = (1 - .828)(1394.52) = 239.86$$

Put all the numbers into the ANOVA summary table for age and price of used Corvettes as shown in Table 11.4b.

TABLE 11.4b ● ANOVA Summary Table for the Simple Regression Between the Age and Price of Used Corvettes

Source	SS	df	MS	F
Regression	1154.66	1	1154.66	38.514
Error	239.86	8	29.98	
Total	1394.52	9		

Step 4: Draw the correct conclusion.

The calculated $F = 38.514$ is within the rejection zone, so we reject H_0. The ages of the cars help predict the prices of used Corvettes.

2. a. The scatterplot of hours of studying and grades is shown in Figure 11.11. According to the scatterplot, a straight line with uphill trend fits the data points well. Therefore, a linear relationship exists between these two variables.

2. b. We need to figure out the $\bar{X}$ and $\bar{Y}$ then add the following five columns to Table 11.5a, $(X-\bar{X}), (Y-\bar{Y}), (X-\bar{X})(Y-\bar{Y}), (X-\bar{X})^2$, and $(Y-\bar{Y})^2$, to demonstrate the step-by-step calculation for the regression line.

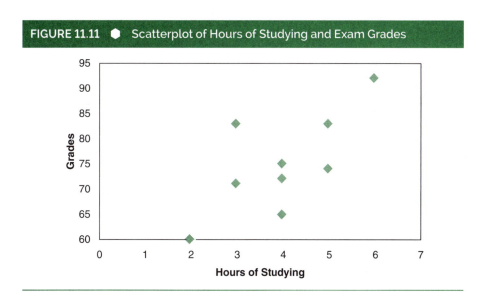

FIGURE 11.11 ● Scatterplot of Hours of Studying and Exam Grades

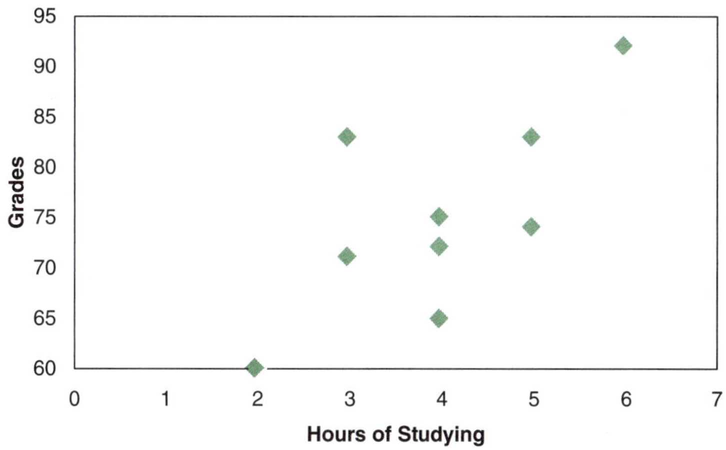

TABLE 11.5a ● Step-by-Step Calculation Using Hours of Studying to Predict Exam Grades

Hours of Studying, X	Grade, Y	$(X-\bar{X})$	$(Y-\bar{Y})$	$(X-\bar{X})(Y-\bar{Y})$	$(X-\bar{X})^2$	$(Y-\bar{Y})^2$
2	60	−2	−15	30	4	225
3	71	−1	−4	4	1	16
3	83	−1	8	−8	1	64
4	72	0	−3	0	0	9
4	75	0	0	0	0	0
4	65	0	−10	0	0	100
5	83	1	8	8	1	64
5	74	1	−1	−1	1	1
6	92	2	17	34	4	289
36	675			67	12	768

$$\bar{X} = \Sigma X / n = 36/9 = 4$$

$$\bar{Y} = \Sigma Y / n = 675/9 = 75$$

$$\text{SP} = \Sigma(X-\bar{X})(Y-\bar{Y}) = 67$$

$$\text{SS}_X = \Sigma(X-\bar{X})^2 = 12$$

$$b = \frac{\text{SP}}{SS_X} = \frac{67}{12} = 5.58$$

$$a = \bar{Y} - b\bar{X}$$

$$a = 75 - (5.58)(4) = 52.68$$

The regression line for using hours of studying to predict exam grade is $\hat{Y} = 52.68 + 5.58X$. For each additional hour of studying, the exam grade increased by 5.58. The lesson learned here is that you need to study longer to get higher grades.

2. c. Here is the four-step hypothesis test for a simple regression.

Step 1: State the pair of hypotheses.

H_0: $\beta = 0$

H_1: $\beta \neq 0$

Step 2: Identify the rejection zone.

The critical value of F in the F Table under $\alpha = .05$ is $F_{(1,7)} = 5.59$. The rejection zone is $F > 5.59$.

Step 3: Calculate the test statistic.

Conducting a hypothesis test on a simple regression requires r^2 and SS_Y.

$$r = \frac{\text{SP}}{\sqrt{SS_X SS_Y}} = \frac{67}{\sqrt{(12)}\sqrt{(768)}} = .698$$

$$r^2 = (.698)^2 = .487$$

$$SS_Y = 768$$

$$\text{SSR} = r^2 SS_Y = .487(768) = 374.016$$

$$\text{SSE} = (1 - r^2)SS_Y = (1 - .487)(768) = 393.984$$

Put all the numbers into the ANOVA summary table between the hours of studying and grades as shown in Table 11.5b.

TABLE 11.5b ● ANOVA Summary Table for the Simple Regression Between the Hours of Studying and Grades

Source	*SS*	*df*	*MS*	*F*
Regression	374.016	1	374.016	6.645
Error	393.984	7	56.283	
Total	768	8		

Step 4: Draw the correct conclusion.

The calculated $F = 6.645$ was within the rejection zone, so we rejected H_0. The hours of studying predicted the exam grades well.

3. a. The scatterplot of the years of experience and performance ratings is shown in Figure 11.12. According to the scatterplot, it is impossible for a straight line to fit all data points across the entire distribution equally well. The relationship between the years of experience and performance ratings varied at different stages. At the beginning of one's career, performance increased with years of experience. In the range 1 to 15 years, the relationship between years of experience and performance is positively correlated. Near the middle, around 17 years of experience, the performance went down with increasing years of experience. In the range 17 to 30 years, the relationship between years of experience and performance switched to a negative correlation.

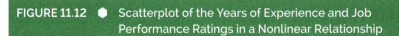

FIGURE 11.12 ● Scatterplot of the Years of Experience and Job Performance Ratings in a Nonlinear Relationship

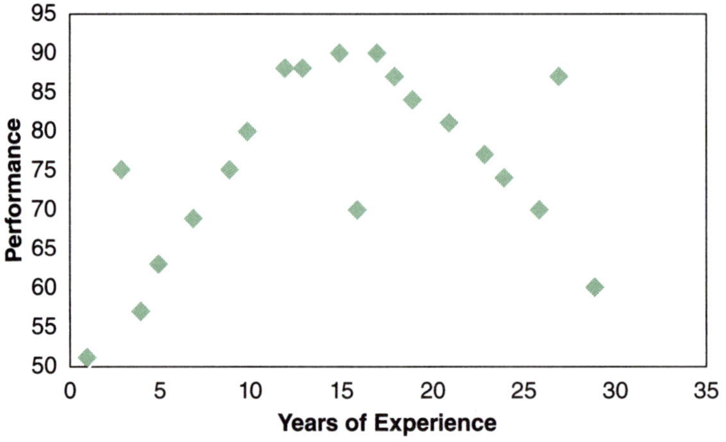

3. b. You would have a hard time trying to fit a straight line to the data points in the scatterplot. The relationship between years of experience and job performance started in a positive direction up to 17 years; then it switched to a negative direction. A linear relationship does not change direction. A relationship between two variables that changes directions is not a linear relationship. Thus, it was not appropriate to use the simple linear regression equation as discussed in this chapter to describe a nonlinear relationship.

What You Learned

There are a couple of important distinctions between correlation and regression.

1. A correlation coefficient can be generated by any two interval or ratio variables without distinguishing between an independent variable and a dependent variable; because $r_{XY} = r_{YX}$, there is no need for such a distinction. Regression, on the other hand, requires explicitly identifying both the independent variable and the dependent variable in the analysis. The purpose of regression is to predict the values of *Y* using the values of *X*. Therefore, *Y* is the main variable and *X* is the supporting variable; they are not interchangeable.

2. All correlation coefficients fall within the range of $-1 \leq r \leq 1$. There is no range for the unstandardized regression coefficients.

An OLS regression line is mathematically determined to produce the smallest squared distance between the line $\hat{Y}$ and the actual Y values.

$$\text{Regression line } \hat{Y} = a + bX$$

$$b = \frac{\text{SP}}{SS_X}$$

$$a = \overline{Y} - b\overline{X}$$

The four-step hypothesis test for a simple regression is as follows.

Step 1: State the pair of hypotheses.

H_0: $\beta = 0$

H_1: $\beta \neq 0$

Step 2: Identify the rejection zone.

The critical values of F that set off the boundaries of the rejection zones are identified by $F_{(1,n-2)}$ in the F Table. The rejection zone is identified as calculated $F > F_{(1,n-2)}$.

Step 3: Calculate the test statistic.

The process of calculating the F test for a simple regression is summarized and shown in Table 11.2, which includes all the operations that lead to the calculated F.

$$F = \frac{\text{MSR}}{MSE}$$

Step 4: Make the correct conclusion.

If the calculated F from Step 3 is within the rejection zone, we reject H_0. If the calculated F is not within the rejection zone, we fail to reject H_0.

Key Words

Analysis of variance (ANOVA): ANOVA is a statistical procedure that partitions variances due to different sources. 348

Degrees of freedom due to error: The degrees of freedom due to error in a simple regression are calculated as $(n - 2)$. 348

Degrees of freedom due to regression: The degrees of freedom due to regression are the number of predictors in the regression equation. In simple regression, the degree of freedom due to regression is 1. 348

Error: Error in regression is the difference between the actual Y and the predicted $\hat{Y}$, Error $= (Y - \hat{Y})$. 342

Heteroscedasticity: Heteroscedasticity refers to the condition when the variances of the errors are not the same across all values of X or across all $\hat{Y}$. This is commonly referred to as violations of the homogeneity of variance of Y across all values of X. 359

Ordinary least squares (OLS): OLS depicts that the regression line is the best fitting line that goes through $(\bar{X}, \bar{Y})$ and minimizes the vertical distances between the actual Y values and the predicted $\hat{Y}$. 343

Regression equation: The regression equation is the equation for the regression line, which is expressed as $\hat{Y} = a + bX$, where a is the Y-intercept and b is the slope. 343

Simple regression: The simple regression is the most commonly used linear predictive analysis when making quantitative predictions of the dependent variable, Y, based on the values of the independent variable, X. 342

Slope: The regression slope is the value change in Y when X increases by one unit. 343

Standard error of the estimate: The standard error of the estimate is the standard deviation of the prediction errors. It is to measure the standard distance between the regression line and the actual Y values. 356

Y-intercept: The Y-intercept is the value of Y when X equals 0. 343

Learning Assessment

Multiple Choice: Circle the Best Answer to Every Question

1. The simple regression equation is expressed as $\hat{Y} = a + bX$. $\hat{Y}$ is the predicted value of Y according to the regression. Errors are defined as the differences between actual Y values and the predicted $\hat{Y}$. Therefore, the sum of squared errors is expressed as
   a. $\text{SSE} = \Sigma(Y - \hat{Y})^2$
   b. $\text{SSE} = \Sigma(\hat{Y} - \bar{Y})^2$
   c. $\text{SSE} = \Sigma(Y - \bar{Y})^2$
   d. $\text{SSE} = \Sigma(X - \bar{X})^2$
2. Besides $(\bar{X}, \bar{Y})$, which other data point is also on the regression line $\hat{Y} = a + bX$?
   a. (0, 1)
   b. (a, b)

   c. (0, a)
   d. (b, 0)

3. The regression line is $\hat{Y} = a + bX$. The intercept a is defined as
   a. the value of Y when $X = 0$
   b. the value of X when $Y = 0$
   c. the value changes in Y when X increases by one unit
   d. the value changes in X when Y increases by one unit

4. Which one of the following statements is correct?
   a. The correct range for the correlation coefficient is $-1 \le r \le 1$
   b. The correct range for the regression coefficient is $-1 \le b \le 1$
   c. The correct range for probability is $-1 \le p \le 1$
   d. The correct range for the probability of committing a Type I error is $-1 \le \alpha \le 1$

5. The coefficient of determination measures the percentage of variance from the dependent variable Y that can be explained by the independent variable X, and is calculated by
   a. r
   b. r^2
   c. b
   d. b^2

6. What are the degrees of freedom when conducting an F test for a simple regression analysis?
   a. $(1, n - 1)$
   b. $(1, n - 2)$
   c. $(2, n - 1)$
   d. $(2, n - 2)$

7. Which of the following statistics will always be positive?
   a. t tests
   b. Z tests
   c. Pearson correlations
   d. F tests

8. What is the coefficient that describes the value changes in Y when X increases by one unit?
   a. Intercept
   b. Slope
   c. F value
   d. Pearson's r

Free Response Questions

9. The car curb weight measures the weight of a car without including cargo, driver, passengers, or any other item. When comparing cars with the same type of engine, the lighter the car weighs, the more fuel efficient the car is. Fuel efficiency is measured by averaging highway and city driving. The number is reported as miles per gallon, usually abbreviated as mpg. A group of 15 cars was randomly selected. Assume that the fuel efficiency and car weight were normally distributed. Their weights and fuel efficiency numbers are reported in Table 11.7.
   a. Construct a scatterplot of car weight and fuel efficiency to see if a linear relationship exists between these two variables.
   b. Identify the regression line using weight to predict fuel efficiency.
   c. What is the predicted fuel efficiency when a car weighs 4,500 pounds?
   d. Use $\alpha = .05$ to evaluate if car weights predict fuel efficiency.

TABLE 11.7 ● Car Weight and Fuel Efficiency

Weight (pounds)	Fuel Efficiency (mpg)
3,190	28.8
3,572	23
2,888	28.5
3,777	20
3,208	18.5
3,393	25.6
4,653	20
3,495	25
3,208	24
4,345	22.5
2,616	31.5
5,950	14
2,235	31.5
4,039	15.7
6,500	13.4

10. A retail company investigated the relationship between number of customers' complaints and number of sales transactions among its sales associates. A group of 10 sales associates was randomly selected from this company. Assume number of sales and customers' complaints were normally distributed. Their number of sales and number of customers' complaints are reported in Table 11.8.

    a. Construct a scatterplot of the number of sales and number of customers' complaints to see if a linear relationship exists between these two variables.

    b. Identify the regression line using the number of customers' complaints to predict the number of sales transactions.

    c. Use $\alpha = .05$ to evaluate if customers' complaints predict number of sales.

TABLE 11.8 ● Number of Customers' Complaints and Number of Sales Transactions

Customers' Complaints	Sales Transactions
12	126
13	298
19	444
20	277
16	356
17	340
26	361
30	482
25	384
22	582

Answers to Pop Quiz Questions

1. d
2. a

12

ONE-WAY ANALYSIS OF VARIANCE

Learning Objectives

After reading and studying this chapter, you should be able to do the following:

- Describe the purpose for the one-way analysis of variance (ANOVA)
- Describe the sources of variance as between-group versus within-group
- Calculate sum of squares total (SST), sum of squares between (SSB), and sum of squares within (SSW)
- Construct and interpret an ANOVA summary table
- Conduct the hypothesis test for ANOVA
- Conduct post hoc comparisons after a significant ANOVA result with three or more groups
- Calculate and interpret the effect size for ANOVA

WHAT YOU KNOW AND WHAT IS NEW

Over the course of earlier chapters of this textbook, you have learned different types of *t* tests that are suitable in different situations. Independent-samples *t* tests are used to compare means from two different groups. Dependent-sample *t* tests are applied to compare the means from pretest and posttest scores of the same group or to compare means from paired or matched samples. *ANOVA* can also be applied to independent samples as well as dependent samples.

In ANOVA terms, the distinction between comparisons among groups consisting of different people versus comparisons among the same group of individuals measured repeatedly is referred to as "*between-subject*" design versus "*within-subject*" design. The **between-subject design** refers to research in which participants are assigned to different groups. Every participant is assigned to only one group. For example, marital status was used as an independent variable to study happiness. Research participants were classified into "single," "married," or "divorced" groups. Then their reported levels of happiness were measured and compared. In this case, different groups consisted of different individuals; therefore, this research used a between-subject design.

The **within-subject design** refers to when research participants are assigned to all levels of the independent variables. Every participant experiences all levels of the independent variable and is measured repeatedly. The within-subject design is also referred to as *repeated measures*. For example, researchers studied the effect of driving while texting compared with other impairments such as driving under the influence of alcohol. Participants' driving mistakes were measured via a driving simulator under all three conditions: (1) driving without distraction, (2) driving while texting, and (3) driving under the influence of alcohol.

Mastering the ability to distinguish within-subject designs from between-subject designs is a vital first step in learning ANOVA, because different statistical formulas are required to calculate them. Due to the nature of this introductory statistics book, the focus of this chapter is on between-subject design one-way ANOVA. In the previous chapter, simple regression, we used an ANOVA summary table to evaluate the effectiveness of the regression model. In the ANOVA table, the variance due to regression and the variance due to error are partitioned separately, and the variance due to regression divided by the variance due to error is the calculated F value. Similarly, in one-way ANOVA, the sources of variance are partitioned out to between-group factors and within-group factors, and the F is calculated by the variance due to between-group factors divided by the variance due to within-group factors. As a reminder that F test is the ratio of two variances, so it is always positive. Therefore, the hypothesis tests for one-way ANOVA are limited to only two-tailed tests. The details are provided in the next section.

INTRODUCING ONE-WAY ANOVA

One-way ANOVA is a statistical method to test the equality of group means for two or more groups by partitioning variances into different sources. You might have noticed that both independent-samples t tests and one-way ANOVA are statistical methods to test the equality of group means. The only difference between the independent-samples t tests and one-way ANOVA is that t tests are only applied to comparison between two groups, while ANOVA is applied to comparisons between two or more groups. When one-way ANOVA is used in comparing means from two populations, it generates identical results to those from t tests, $F = t^2$. Therefore, to demonstrate the unique utility of ANOVA, we are going to use it to compare means in three or more populations.

One-way ANOVA is defined as an ANOVA method that uses only one factor to classify data into different groups. Not to go too wild about how many groups we can compare, the examples in this chapter are mostly three groups, as illustrated in Table 12.1. Keep in mind that comparisons among more groups are logically and practically possible. The subscript i refers to individuals in each group, and the subscript j refers to j groups. When comparing means among three groups, j = 1, 2, or 3. In Table 12.1, $\bar{X}_1$, $\bar{X}_2$, and $\bar{X}_3$ (as shown in the last row) symbolize the mean for Group 1, the mean for Group 2, and the mean for Group 3, respectively. $\bar{\bar{X}}$ at the lower-right corner refers to the mean for the entire sample. The number of participants in each group is denoted by n_1, n_2, and n_3. The entire sample size is equal to the sum of the three group sizes, $n = n_1 + n_2 + n_3$. The number of participants in each group does not need to be the same. Let's lay out all the numbers for an ANOVA with three groups in Table 12.1 to visualize the **sum of squared total deviations (SST)**, $SST = \Sigma\Sigma(X_{ij} - \bar{\bar{X}})^2$. The first subscript refers to the row (ith individual) and the second subscript refers to the column (jth group). The different styles of dash and solid blue lines all converge on the mean for the entire sample, $\bar{\bar{X}}$. The spiderweb–like lines show how the SST is calculated. First, calculate the difference between every individual value in every group, X_{ij}, and the overall sample mean, $\bar{\bar{X}}$. That is why all the lines point to $\bar{\bar{X}}$. As we discussed in Chapter 3, deviations are calculated by the difference between the individual values and the sample mean. To avoid positive deviations completely canceling out the negative deviations, the deviations need to be squared. When you add up all the squared deviations, you create $SST = \Sigma\Sigma(X_{ij} - \bar{\bar{X}})^2$.

TABLE 12.1 ● Visualization of Sum of Squares Total (SST) in Three Groups

Individual i	Group 1	Group 2	Group 3	
1	X_{11}	X_{12}	X_{13}	
2	X_{21}	X_{22}	X_{23}	
3	X_{31}	X_{32}	X_{33}	
...	...	...	...	
I	X_{i1}	X_{i2}	X_{i3}	
	$\bar{X}_1$	$\bar{X}_2$	$\bar{X}_3$	$\bar{\bar{X}}$

It is not difficult to understand how to calculate SST conceptually. In practice, however, it might take a while to conduct such a calculation by hand. I suggest that you use Excel to do such calculations.

$$SST = \Sigma\Sigma(X_{ij} - \bar{\bar{X}})^2$$

In simple regression, we construct an ANOVA summary table to evaluate the effectiveness of the regression model by partitioning the sum of squares total (SST) into the sum of squares due to regression (SSR) and the sum of squares due to error (SSE). It is denoted as SST = SSR + SSE. The focus of a regression is on predicting *Y*; therefore, SS_Y is also labeled as SST. To refresh your memory, the ANOVA summary table is shown in Table 12.2.

TABLE 12.2 ● The ANOVA Summary Table for Calculating *F* for a Simple Regression

Source	*SS*	*df*	*MS*	*F*
Regression	$SSR = r^2 SS_Y$	1	$MSR = \frac{SSR}{1}$	$\frac{MSR}{MSE}$
Error	$SSE = (1 - r^2) SS_Y$	*n* - 2	$MSE = \frac{SSE}{(n-2)}$	
Total	SSY or SST	*n* - 1		

Similarly, in an ANOVA, we partition the total sum of squares into SSB and SSW, SST = SSB + SSW. Both SSB and SSW will be discussed in detail in the next section.

BETWEEN-GROUP VARIANCE AND WITHIN-GROUP VARIANCE

You learned in Chapter 3 that $\text{Variance} = SS / df$. When you calculate the variance of a variable, you need the *SS* (sum of squares) and the degrees of freedom (*df*) of that particular variable. When we partition the total variances into variances due to different sources, we need to know the *SS* and *df* for each source as shown in Table 12.2. The ANOVA summary table in simple regression shows that the total variance is partitioned into different sources: variance due to regression and variance due to error. Then the calculated F value shows the ratio of variance due to regression to the variance due to error. Similarly, in ANOVA, the variances are partitioned into different sources: the between-group variance and the within-group variance. The discussions will involve identifying SSB, SSW, the degrees of freedom from between groups (df_B), and the degrees of freedom from within groups (df_W). The conceptual meanings and mathematical formulas of SSB, SSW, df_B, and df_W will be clearly explained one by one.

Sum of Squares Between

The **sum of squares between groups (SSB)** is the focus of this subsection. SSB shows the variability between different groups. Different people are assigned to different groups. The group assignment is based on different levels of the independent variables. In a one-way ANOVA, there is only one independent variable. Different levels of the independent variable reflect different treatment conditions in experimental designs. The differences between group means are largely attributable to systematic differences due to the study effects. Mathematically, SSB is calculated by taking the squared differences between individual group means, $\bar{X}_j$, and the sample mean, $\bar{\bar{X}}$, multiplied by the number of participants in each group, and then summing them up. Let's visualize the $\text{SSB} = \Sigma n_j(\bar{X}_j - \bar{\bar{X}})^2$ in Table 12.3. At the bottom of Table 12.3, the horizontal blue lines demonstrate the comparison between the group means and the overall sample mean. The squared difference between group mean and sample mean needs to multiply by the group size before sums across different groups.

$$\text{SSB} = \Sigma n_j(\bar{X}_j - \bar{\bar{X}})^2$$

where

$\bar{X}_j$ = the mean in Group j

n_j = the number of participants in Group j

$\bar{\bar{X}}$ = the sample mean

$$\text{SSB} = \Sigma n_j(\bar{X}_j - \bar{\bar{X}})^2$$

Let's illustrate the calculation of SSB with an example. A software company designed a new application to conduct electronic financial transactions without using a credit card. The new application can be used in smartphone app linked with a checking account, smartphone app linked with a credit card account, or an app holding its own balance

TABLE 12.3 ● Visualization of Sum of Squares Between (SSB) in Three Groups

Individual *i*	Group 1	Group 2	Group 3	
1	X_{11}	X_{12}	X_{13}	
2	X_{21}	X_{22}	X_{23}	
3	X_{31}	X_{32}	X_{33}	
...	...	...	...	
i	X_{i1}	X_{i2}	X_{i3}	
	$\bar{X}_1$	$\bar{X}_2$	$\bar{X}_3$	$\bar{\bar{X}}$
	n_1	n_2	n_3	

without linking to either a bank account or a credit card account. The company wants to explore whether there are differences in user satisfaction among the three types of applications. The company randomly selected 30 people and assigned 10 of them into each group to pilot test the application linked with a checking account, an app linked with a credit card account, or an app holding its own balance. The levels of user satisfaction were measured and reported as $\bar{X}_1 = 3.4$, $\bar{X}_2 = 4$, and $\bar{X}_3 = 4.3$, respectively. In a three-group situation, SSB can be calculated easily.

$$\text{SSB} = n_1(\bar{X}_1 - \bar{\bar{X}})^2 + n_2(\bar{X}_2 - \bar{\bar{X}})^2 + n_3(\bar{X}_3 - \bar{\bar{X}})^2$$

where $n_1 = n_2 = n_3 = 10$ and $\bar{\bar{X}}$ is the entire sample mean, which is calculated as mean of all group means.

$$\bar{X} = (3.4 + 4 + 4.3) / 3 = 3.9$$

$$\text{SSB} = 10(3.4 - 3.9)^2 + 10(4 - 3.9)^2 + 10(4.3 - 3.9)^2 = 4.2$$

According to the SSB formula, when there are large differences between group means and the sample mean, the SSB is large. When group size is large, SSB is large. However, the magnitude of SSB provides useful information only when it is compared with the variability from the other source, SSW.

Sum of Squares Within

The **sum of squares within groups (SSW)** reflects the variability within groups, which is attributable to the random individual differences or unsystematic measurement errors that are beyond the control of the researchers. Mathematically, the calculation is done within each group. The SSW is calculated by computing the squared differences between the individual scores, X_{ij}, in Group j and its group mean, $\bar{X}_j$, and then summing them up.

$$SSW = \Sigma\Sigma(X_{ij} - \bar{X}_j)^2$$

Random individual differences are the preexisting conditions research participants bring along with them. In the example of the new software applications for financial transactions introduced above, such differences would include technology savviness, socioeconomic status, family structure, occupation, personality traits, past experiences with electronic transactions, and so on. Measurement errors are unsystematic errors that happen sometimes to some people. The sources for the unsystematic measurement errors can be due to the ambiguously phrased questions on the questionnaires, the data collection procedures, or the research participants. Some of these random individual differences and unsystematic measurement errors might have unknown effects on the target variable X. Let's visualize the $SSW = \Sigma(X_{ij} - X_j)^2$ in Table 12.4. The SSW is the sum of the squared differences between individual scores in Group j and the Group j mean.

TABLE 12.4 ● Visualization of Sum of Squares Within (SSW) in Three Groups

Individual i	Group 1	Group 2	Group 3	
1	X_{11}	X_{12}	X_{13}	
2	X_{21}	X_{22}	X_{23}	
3	X_{31}	X_{32}	X_{33}	
...	...	...	...	
i	X_{i1}	X_{i2}	X_{i3}	
	$\bar{X}_1$	$\bar{X}_2$	$\bar{X}_3$	$\bar{\bar{X}}$

Judging from the number of blue lines involved in calculating SSW, it is a tedious process to get all the values calculated. You can learn to calculate SSW using Excel. However, you can easily calculate the value of SSW when SST and SSB are known.

$$SST = SSB + SSW$$

$$SSW = SST - SSB$$

Since ANOVA is a statistical procedure for testing the equality of group means by partitioning variances into different sources, we need to calculate variances from different sources. We have discussed SSB and SSW. Next, we need to figure out both df_B and df_W in order to calculate the variance between groups and variance within groups.

Degrees of Freedom for Between Groups (df_B) and Within Groups (df_W)

ANOVA is a statistical measure to partition variances due to different sources. As we mentioned earlier, $\text{Variance} = SS/df$. When the total *SS* is partitioned into SSB and SSW, we get SST = SSB + SSW. Every *SS* has a *df* particularly linked to it. The partition of the *df* follows the same pattern as the partition of the *SS*, and we get $df_T = df_B + df_W$. The *df* represents the number of values that are free to vary. The *df* linked with a particular *SS* is calculated by the number of values that go into the calculation minus 1.

For example, $SST = \Sigma\Sigma(X_{ij} - \bar{\bar{X}})^2$, so there are *j* groups in the entire sample. The sample size $n = n_1 + n_2 + n_3 + ... + n_j$, so *n* values go into the calculation of SST; hence the $df_T = n - 1$.

Similarly, we can figure out the *df* between groups by $SSB = \Sigma n_j(\bar{X}_j - \bar{\bar{X}})^2$.

The SSB is calculated by squaring the differences between each group mean and the sample mean, multiplying by the group size, and then adding them up. The number of values that goes into the calculation of SSB equals the number of groups. The df_B is defined as the number of groups minus 1. Assume that the number of groups is *k*; then df_B is $k - 1$. Apply the general formula $\text{Variance} = SS / df$ in this situation, and we get the **between-group variance** (also referred to as the mean squares between groups, MSB in ANOVA terminology).

$$MSB = \frac{SSB}{df_B} = \frac{SSB}{(k-1)}$$

Next, we can identify the df_W by $SSW = \Sigma\Sigma(X_{ij} - \bar{X}_j)^2$. The SSW is calculated within each group by squaring the differences between individual values and their group mean then adding them up. Assume that the number of groups is *k*; then the *df* within the group for Group 1 is $(n_1 - 1)$, for Group 2 is $(n_2 - 1)$, for Group 3 is $(n_3 - 1)$, and so on, and for Group k is $(n_k - 1)$. The df_W is the sum of df for each group.

$$
\begin{aligned}
df_W &= (n_1 - 1) + (n_2 - 1) + (n_3 - 1) + \ldots + (n_k - 1) \\
&= (n_1 + n_2 + n_3 + \ldots + n_k) - k \\
&= n - k
\end{aligned}
$$

The degrees of freedom within groups (df_W) are defined as $(n - k)$. Once we know the degrees of freedom within groups, we can calculate the **within-group variance** (also referred to as the mean squares within groups, MSW, in ANOVA terminology).

$$\text{MSW} = \frac{\text{SSW}}{df_W} = \frac{\text{SSW}}{n-k}$$

When SSB and SSW add to be SST, the dfs associated with these terms also add up.

$$
\begin{gathered}
df_T = df_B + df_W \\
(n - 1) = (k - 1) + (n - k)
\end{gathered}
$$

Now, we can summarize everything in this section into one figure and one table. In summary, ANOVA is a statistical method that tests the equality of group means by partitioning variances into different sources. The partitioning process is shown in Figure 12.1, where the sources of variance are partitioned into between-group variance and within group variance. Such a partitioning process applies to both the SS and the df.

Similar to the hypothesis test for regression $F = \text{MSR} / \text{MSE}$, the hypothesis test for ANOVA is calculated as $F = \text{MSB} / \text{MSW}$. We will discuss the four-step hypothesis test for ANOVA in more detail in the next section.

FIGURE 12.1 ● Partitioning the Sum of Squares and the Degrees of Freedom in One-Way ANOVA

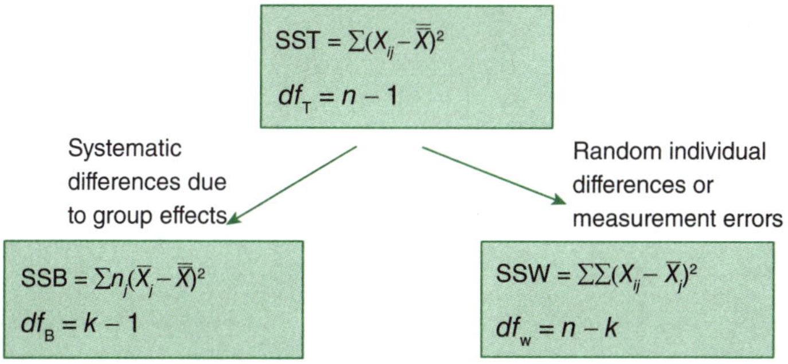

We can also organize and summarize the results of partitioning *SS*, *df*, and MS from different sources in the ANOVA summary table as shown in Table 12.5. The ANOVA summary table shows in a streamlined manner the process of calculating the *F* value.

TABLE 12.5 ● The ANOVA Summary Table

Source	*SS*	*df*	*MS*	*F*
Between groups	SSB	$df_B = k - 1$	$MSB = \frac{SSB}{(k-1)}$	$F = \frac{MSB}{MSW}$
Within groups	SSW	$df_W = n - k$	$MSW = \frac{SSW}{(n-k)}$	
Total	SST	$df_T = n - 1$		

POP QUIZ

Researchers studied the effect of distracted driving on driving behavior. Driving behavior was measured by the number of mistakes recorded on a simulator. There were 36 participants who were randomly assigned to one of the three groups: (1) driving while eating, (2) driving while texting, and (3) normal driving without distraction. There were 12 participants in each group, and the group means were 22.45, 25.39, and 16.87, respectively, SST = 2,016.45. Use this information to answer the following three questions.

1. What was the SSB based on the information provided here?
   a. 21.57
   b. 35
   c. 449.48
   d. 1566.97
2. What was the SSW based on the information provided here?
   a. 21.57
   b. 35
   c. 449.48
   d. 1566.97
3. What was the degrees of freedom within-group variance (df_W)?
   a. 2
   b. 3
   c. 33
   d. 35

HYPOTHESIS TESTING FOR ONE-WAY ANOVA

We have discussed all the elements such as *SS*, *df*, and MS involved in the ANOVA summary table. It is time to put all elements together to conduct hypothesis tests with ANOVA. The general principles of the four-step hypothesis testing procedure with ANOVA are explicitly stated, then followed by several examples to illustrate the process.

Step 1: State the pair of hypotheses for ANOVA.

H_0: $\mu_1 = \mu_2 = \mu_3 = \ldots = \mu_k$

H_1: Not all μ_k are equal.

The null hypothesis, H_0, states that all group means are the same. The alternative hypothesis, H_1, states that not all group means are the same. At least one of the equal signs in the null hypothesis is not true. If the outcome of an ANOVA is to fail to reject H_0, no further analysis needs to be done. If the outcome of an ANOVA is to reject H_0, additional **post hoc comparisons** need to be done in order to identify exactly where the significant difference comes from.

Step 2: Identify the rejection zone.

The critical values of F that set the boundaries of the rejection zones can be found using the $F_{(df_1, df_2)}$ in the same F Table we used to determine whether equal variances are assumed, in Chapter 9, and whether a regression analysis is significant, in Chapter 11. This table is provided in Appendix C. Each critical F value is determined by two different *dfs*, where df_1 is the *df* associated with the numerator and df_2 is the *df* associated with the denominator. In the case of ANOVA, df_1 is the degrees of freedom associated with between groups, $df_B = k - 1$, and df_2 is the degrees of freedom associated with within groups, $df_W = (n - k)$. Therefore, the critical values of F can be identified by $F_{(k-1,\ n-k)}$ in the F Table for ANOVA. The rejection zone is identified as the calculated $F > F_{(k-1,\ n-k)}$.

Step 3: Calculate the F test for the ANOVA.

The F test can be easily calculated by completing Table 12.5 with all the *SS* terms and *dfs* associated with them. Once the ANOVA summary table is complete,

$$F = \frac{\text{MSB}}{\text{MSW}}$$

Step 4: Make the correct conclusion.

If the calculated F is within the rejection zone, we reject H_0. If the calculated F is not within the rejection zone, we fail to reject H_0. In the case of failing to reject H_0, no additional analysis is needed. In the case of rejecting H_0, we need to conduct post hoc comparisons to identify exactly where the significant differences come from.

Post hoc comparisons are conducted by making all possible pairwise comparisons to identify exactly which equal sign in the null hypothesis is not true. In the case of $k = 3$, there are three groups: Groups 1, 2, and 3. When the calculated $F > F_{(k-1,\ n-k)}$, we reject H_0. Now, we have to compare all possible pairwise comparisons. The formula to calculate the number of all possible pairs is $C = k(k - 1)/2$. Thus, when there are three groups, $k = 3$, there are three possible pairwise comparisons because $C = k(k - 1)/2 = 3(3 - 1)/2 = 3$.

For three groups, these are all of the possible pairwise comparisons: μ_1 versus μ_2, μ_1 versus μ_3, and μ_2 versus μ_3. These comparisons need to be conducted simultaneously. Assume that the risk of committing a Type I error is α in each pairwise comparison. When conducting the same type of tests $k(k-1)/2$ times, the Type I error rate accumulates. The accumulated Type I error rate from conducting similar comparisons multiple times is called the **familywise error rate (FWER)**. The FWER is the probability of committing a Type I error when conducting $k(k-1)/2$ possible pairwise comparisons at the same time. When the Type I error for one individual pairwise comparison is α, the FWER is calculated as committing Type I errors in the multiple comparisons.

$$\text{FWER} = 1-(1-\alpha)^{\frac{k(k-1)}{2}}$$

Assume that $\alpha = .05$ and $k = 3$; then $k(k-1)/2 = 3$, so the FWER $= 1 - (1 - .05)^3 = .1426$.

The point is that when the Type I error rate for each pairwise comparison $\alpha = .05$, we conduct three pairwise comparisons simultaneously, then we have inflated the FWER. An inflated FWER may increase the risk of Type I error, which is to falsely claim an effect when it actually does not exist. **Bonferroni correction** formula is a method to control such inflated Type I error during post hoc comparisons.

$$\text{Bonferroni correction formula, } \alpha_B = \frac{\alpha}{C}$$

where

α_B = Bonferroni correction

α = risk of committing Type I error for one test

$C = k(k - 1)/2$ number of all possible pairwise comparisons

Now we have completed the hypothesis test for ANOVA, we will illustrate how to apply the entire process using multiple examples.

EXAMPLE 12.1

A software company designed new applications to conduct electronic financial transactions without using an actual credit card. The new smart phone payment application can be linked with an existing checking account, a new account with required deposited balance, or linked with a credit card. The company wants to find out whether there are differences in user satisfaction among these three options. The company randomly selected 30 people and assigned 10 of them into each group to conduct the pilot test. The user's satisfaction was measured on a Likert scale of 1 to 5, with 1 = *very dissatisfied* and 5 = *very satisfied*, and reported in Table 12.6.

$$SST = \Sigma\Sigma(X_{ij} - \bar{\bar{X}})^2 = 38.7$$

TABLE 12.6 ● User Satisfaction in Three Groups

Existing Checking Account	New Account	Credit Card
3	5	5
4	4	4
5	3	3
2	3	5
4	4	4
5	4	5
5	5	3
1	3	4
1	4	5
4	5	5
$\bar{X}_1 = 3.4$	$\bar{X}_2 = 4$	$\bar{X}_3 = 4.3$

Conduct a hypothesis test to compare the user satisfaction levels across three options. Are there significant differences in user satisfaction across three options?

As discussed before, calculating the SST is a tedious process to square the differences between every individual value and the overall sample mean and then add them up. The overall sample mean is easily calculated by the average of three groups mean because the equal group size, overall sample mean = 3.9.

(Continued)

(Continued)

You may calculate the SST on your own by following the formula $SST = \Sigma\Sigma(X_{ij} - \bar{\bar{X}})^2$ or use the number SST = 38.7 provided in the problem statement. The step-by-step instruction of conducting a one-way ANOVA in Excel will be provided at the end of this chapter.

Step 1: State the pair of hypotheses.

H_0: $\mu_1 = \mu_2 = \mu_3 = \ldots = \mu_k$

H_1: Not all μ_k are equal.

Step 2: Identify the rejection zone.

There were three groups, $k = 3$, 10 people in each group, $n_1 = n_2 = n_3 = 10$, and the sample size $n = 30$. The critical value of F is identified by $F_{(2,27)} = 3.35$ in the F Table for ANOVA. The rejection zone is identified as the calculated $F > 3.35$.

Step 3: Calculate the F test.

Under equal group size in all three groups, the sample mean is the mean of three group means.

$$\bar{\bar{X}} = (3.4 + 4 + 4.3)/3 = 3.9$$

$$SSB = \Sigma n_j(\bar{X}_j - \bar{\bar{X}})^2 = 10(3.4 - 3.9)^2 + 10(4 - 3.9)^2 + 10(4.3 - 3.9)^2 = 4.2$$

$$SSW = SST - SSB = 38.7 - 4.2 = 34.5$$

Now we have all the numbers to complete the ANOVA summary table as shown in Table 12.6a.

TABLE 12.6a ● The ANOVA Summary Table for User Satisfaction Across Three Groups

Source	SS	df	MS	F
Between groups	4.2	$df_B = 2$	MSB = 2.1	$F = 1.64$
Within groups	34.5	$df_W = 27$	MSW = 1.28	
Total	38.7	$df_T = 29$		

Step 4: Make the correct conclusion.

The calculated $F = 1.64$ was not within the rejection zone, so we failed to reject H_0. The evidence was not strong enough to claim that there were significant differences in user satisfaction for the electronic financial transaction application across three groups. No further analysis needs to be conducted.

When the calculated F value is not significant, the evidence is not strong enough to claim inequality among group means. Since no difference in group means can be inferred from the data, no further analyses are needed. In the following example, we are going to demonstrate, after the calculated F value in ANOVA turns out to be significant, what we should do next.

EXAMPLE 12.2

A food manufacturing company conducted three focus groups to test different advertising strategies on customers' perceptions of food quality. The food quality ratings were normally distributed. Three advertising strategies were tested in this study: (1) weight control, (2) all natural ingredients, and (3) environmental friendliness. The participants were randomly selected from a large panel and compensated for their time. All participants were provided exactly the same food product. The differences in food quality ratings were assumed to be influenced by the different advertising strategies. There were eight participants in each focus group. The company wants to find out whether significant differences exist in the evaluation of the food quality among the three groups. The ratings of the food quality were measured and reported in Table 12.7, with higher values representing better quality.

$$SST = 47.833$$

TABLE 12.7 ● Food Quality Ratings in Three Focus Groups

Weight Control	All Natural	Environmental Friendliness
6	9	9
5	8	8
3	6	8
7	7	7
8	7	9
6	8	6
6	6	7
7	5	8
$\bar{X}_1 = 6$	$\bar{X}_2 = 7$	$\bar{X}_3 = 7.75$

Conduct a hypothesis test to verify the equality of the food quality ratings among three focus groups. Are there significant differences in food quality ratings based on the three advertising strategies?

Step 1: State the pair of hypotheses.

H_0: $\mu_1 = \mu_2 = \mu_3 = \ldots = \mu_k$
H_1: Not all μ_k are equal.

Step 2: Identify the rejection zone.

There are three groups, $k = 3$, with eight people in each group, $n_1 = n_2 = n_3 = 8$, and the sample size $n = 24$. The critical value of F is identified by $F_{(2,21)} = 3.47$ in the F Table for ANOVA. The rejection zone is identified as the calculated $F > 3.47$.

(Continued)

(Continued)

Step 3: Calculate the F test.

Under equal group size, the sample mean is

$$\bar{\bar{X}} = (6 + 7 + 7.75)/3 = 6.917$$

$$SSB = \Sigma n_j (\bar{X}_j - \bar{\bar{X}})^2 = 8(6 - 6.917)^2 + 8(7 - 6.917)^2 + 8(7.75 - 6.917)^2 = 12.333$$

$$SSW = SST - SSB = 47.833 - 12.333 = 35.5$$

Now we have all the numbers to complete the ANOVA summary table as shown in Table 12.7a.

TABLE 12.7a ● The ANOVA Summary Table for the Food Quality Ratings in Three Focus Groups

Source	SS	*df*	*MS*	*F*
Between groups	12.333	$df_B = 2$	MSB = 6.167	$F = 3.649$
Within groups	35.5	$df_W = 21$	MSW = 1.690	
Total	47.833	$df_T = 23$		

Step 4: Make the correct conclusion.

The calculated $F = 3.649$ was within the rejection zone, so we rejected H_0. The evidence was strong enough to claim that significant differences existed in the food quality ratings across three focus groups. However, a significant F test does not tell us exactly where the difference occurs. Thus, further analysis needs to be conducted to identify the sources of significant difference. Post hoc comparisons are designed to fulfill this purpose.

Post Hoc Comparisons

Post hoc means "after the fact." After the calculated F value in ANOVA turns out to be within the rejection zone, $F > F_{(k-1,\ n-k)}$, we reject H_0. We conclude that not all group means are equal. But at this point, a significant F test does not tell us exactly where the significant difference comes from. Our answer is incomplete until we can clearly identify exactly where the significant difference occurs.

We don't have to worry about post hoc comparisons in two groups because we are sure that the significant ANOVA can only mean $\mu_1 \neq \mu_2$ in two groups. In case of three or more groups, there are multiple possibilities with regard to where the significant differences may come from. Specifically, in Example 12.2, the calculated $F = 3.649$ was within the rejection zone. We concluded that not all group means were equal. To identify exactly where the source of the significant difference comes from, we have to do the comparisons on all possible pairs simultaneously: μ_1 versus μ_2, μ_1 versus μ_3, and μ_2 versus μ_3. There are three possible pairwise comparisons that need to be conducted. If $\alpha = .05$ is set for an individual pairwise hypothesis test, conducting three such pairwise comparisons inflates the FWER. Therefore, Bonferroni correction formula is used to simply adjust the α level of the individual pairwise comparison to a new level in order to control the familywise α level at .05 level.

The Bonferroni correction formula is $\alpha_B = \dfrac{\alpha}{C}$

where $C = k(k-1)/2$ = the number of all possible pairwise comparisons

In Example 12.2, we have $C = 3(3-1)/2 = 3$ possible pairwise comparisons. To keep the familywise α level = .05, the new α_B for the individual pair comparison is $0.5/3 = .017$. After conducting $k(k-1)/2$ times of independent-samples t tests, we need to use the Bonferroni adjusted p value to evaluate whether each pair comparison is significant. The results of this process are shown in Table 12.7b. We can easily conduct an independent-samples t test on any two groups as discussed in Chapter 9. However, we will need to find the p-values associated with the t values. The purpose to conduct all possible pairwise comparisons is to identify exactly where the significant differences come from. Therefore, we use the Excel function **T.DIST.2T(x,df)** to identify the p value for the calculated t, where x is the calculated t and the df is the degrees of freedom for the pairwise comparison. In a two-tailed test, Excel requires $x > 0$, so x is the absolute value of the calculated t.

TABLE 12.7b ● Three Possible Pairwise Comparisons

Pairwise Comparison	*t*	*p*
$\mu_1 - \mu_2$	−1.414	.179
$\mu_1 - \mu_3$	−2.701	.017
$\mu_2 - \mu_3$	−1.271	.224

According to Table 12.7, we can calculate $\bar{X}_1 = 6$, $s_1 = 1.512$, $\bar{X}_2 = 7$, $s_2 = 1.309$, $\bar{X}_3 = 7.75$, and $s_3 = 1.035$ with $n_1 = n_2 = n_3 = 8$ (please review Chapter 3 if you have trouble calculating mean and standard deviation from eight individual values in the table). Use these means and standard deviations number to conduct three possible pairwise comparisons. The results of the three possible pairwise comparisons are shown in

Table 12.7b. Use the Bonferroni correction α_B to evaluate the significance of each pairwise comparison. The Bonferroni correction tends to be more conservative, which means that it is stricter than the regular $\alpha = .05$. To find the *p* value associated with -1.414, type "**=T.DIST.2T(1.414,14)**" in a blank cell where we want the *p* value to appear, then hit **Enter**. The answer 0.179213 appears. Follow the principle to find the other *p* values shown in Table 12.7b.

Judging by the results of these three pairwise comparisons, there were no differences in food ratings between weight control group and all natural ingredient group ($\mu_1 - \mu_2$) or between all natural group and environmentally friendly group ($\mu_2 - \mu_3$). The differences in food quality ratings were specifically between the weight control group and the environmentally friendly group ($\mu_1 - \mu_3$).

POP QUIZ

4. When is it necessary to conduct post hoc comparisons for an ANOVA procedure?
   a. Post hoc comparisons need to be conducted in every ANOVA procedure.
   b. Post hoc comparisons need to be conducted after a significant ANOVA result.
   c. Post hoc comparisons need to be conducted after a nonsignificant ANOVA result.
   d. Post hoc comparisons are not necessary after an ANOVA procedure.

EFFECT SIZE FOR ONE-WAY ANOVA

The most common measure of effect size for the one-way ANOVA is η^2, pronounced as eta squared. Conceptually, η^2 measures the percentage of the total variance of the dependent variable explained by the independent variable or the between-group factor. Mathematically, η^2 is calculated using the formula $\eta^2 = SSB / SST$.

The hypothesis test can tell us whether the ratio of the between-group variance to the within-group variance is statistically significant. However, when a large sample size is involved, even small, trivial between-group variance can lead to a statistically significant *F* value. A statistically significant outcome does not tell us how large the difference is or whether the difference is practically meaningful. The calculation of effect size (eta squared) in one-way ANOVA tells us what percentage of the total variance of the dependent variable is explained by the independent variable.

Let's use the numbers in Example 12.2 to calculate an effect size of three advertising strategies: (1) weight control, (2) all natural ingredients, and (3) environmental friendliness on customers' perceptions of the food quality. Use the numbers presented in the ANOVA summary table in Table 12.7a to calculate the effect size.

$$\text{The effect size } \eta^2 = \frac{\text{SSB}}{\text{SST}} = \frac{12.333}{47.833} = .258$$

Although the hypothesis test generates a significant F value, the effect size for the three advertising strategies on customers' perceptions of the food quality was .258. There was 25.8% of the total variance in the food quality ratings that could be explained by the different advertising strategies. The effect size was small.

POP QUIZ

5. There were 36 participants randomly assigned to one of the three groups: (1) driving while eating, (2) driving while texting, or (3) normal driving without distraction. There were 12 participants in each group, and the ANOVA summary table is shown below.

Source	*SS*	*df*	*MS*	*F*
Between groups	504	$df_B = 2$	MSB = 252	$F = 5.500$
Within groups	1,512	$df_W = 33$	MSW = 45.82	
Total	2,016	$df_T = 35$		

What was the effect size for the distracted driving conditions on driving behavior?

a. 0.25
b. 0.33
c. 0.75
d. 5.50

STEP-BY-STEP INSTRUCTIONS FOR USING EXCEL TO RUN AN ANOVA

Let's use the data in Example 12.2 to demonstrate the process of running an ANOVA using Excel. The food quality ratings data in Table 12.7 were copied into Excel without the group means at the bottom. We can calculate SST, SSB, and SSW in one-way ANOVA using Excel.

1. Calculate group means by moving the cursor to where we want the answer to appear, **A11** and typing "**=average(A2:A9)**", then hit **Enter**. The group mean for Weight Control group shows up as 6. To calculate the other two group means, move the cursor to **A11** lower-right corner until the solid + shows up, hold down left click while dragging the cursor across **B11** and **C11** to get the means for All Natural and Environmental Friendliness groups.

2. Calculate overall sample mean, $\bar{\bar{X}}$ by moving the cursor to **D11** and typing "**=average(A11:C11)**", then hit **Enter**. The overall sample mean $\bar{\bar{X}}$ is 6.916667. When the group sizes are the same, the mean of group means is the overall sample mean as shown in Figure 12.2.

FIGURE 12.2 ● Calculating Group Means and Overall Sample Mean in Excel

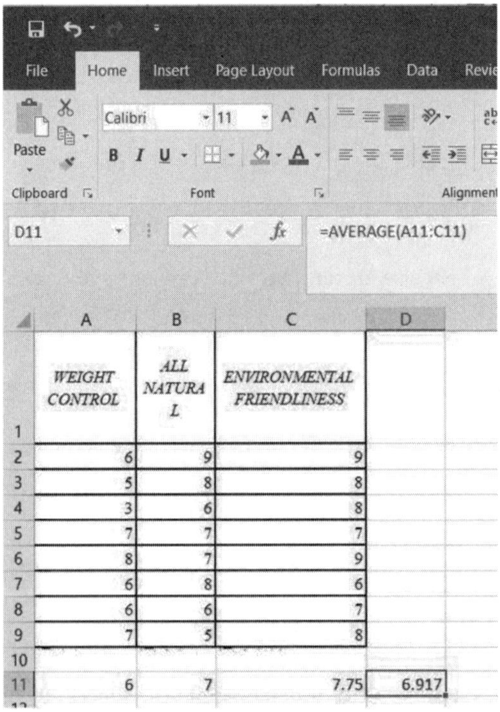

3. Calculate $SST = \Sigma\Sigma(X_{ij} - \bar{\bar{X}})^2$ by moving the cursor to **E2** to calculate the first individual value's difference from the overall sample mean then square the difference. Type "**=(A2-$D$11)^2**" in **E2**, then hit **Enter**. Every individual value across three groups need to calculate the squared difference against the overall sample mean; therefore, we need to place two **$** signs in column **D** and in row **11** to prevent the value from floating when we copy and paste the math operation across columns and rows. To calculate the rest of the squared differences from the overall sample mean, move the cursor to **E2** lower-right corner until the solid + shows up, hold down left click while dragging the cursor down to **E9**. Highlight the numbers in **E2** to **E9**, move the cursor to the lower-right corner until the solid + shows up, hold down left click while dragging columns **F** and **G**, then let go. You can see the 24 squared differences between individual values and its overall sample mean. To organize the results of the ANOVA, create an ANOVA summary table as shown in Figure 12.3. To calculate SST, simply sum up all 24 squared differences, type "**=sum(E2:G9)**" in **F14** where we want SST to appear, then hit **Enter**. We get SST = 47.833.

FIGURE 12.3 ● Calculate SST (Sum of Squared Total Deviations) in Excel

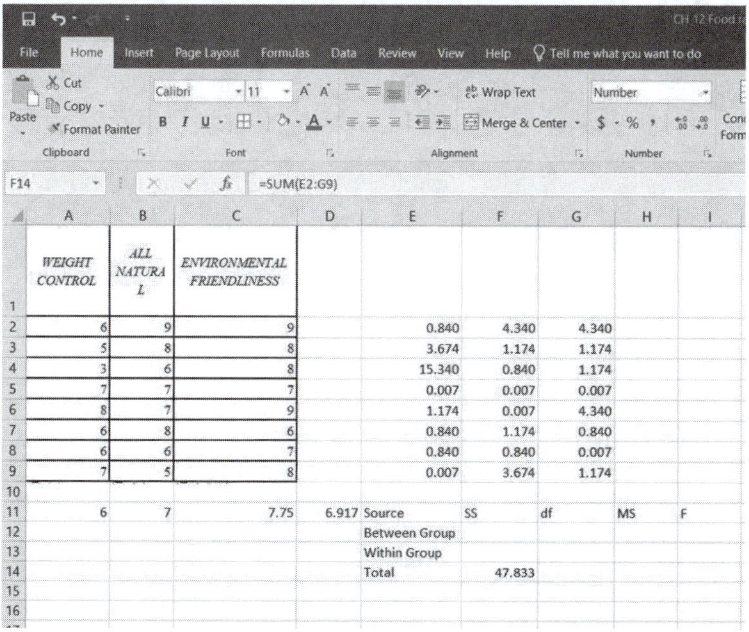

	A	B	C	D	E	F	G	H	I
1	WEIGHT CONTROL	ALL NATURAL	ENVIRONMENTAL FRIENDLINESS						
2	6	9	9		0.840	4.340	4.340		
3	5	8	8		3.674	1.174	1.174		
4	3	6	8		15.340	0.840	1.174		
5	7	7	7		0.007	0.007	0.007		
6	8	7	9		1.174	0.007	4.340		
7	6	8	6		0.840	1.174	0.840		
8	6	6	7		0.840	0.840	0.007		
9	7	5	8		0.007	3.674	1.174		
10									
11	6	7	7.75	6.917	Source	SS	df	MS	F
12					Between Group				
13					Within Group				
14					Total	47.833			
15									
16									

4. To calculate $SSB = \Sigma n_j(\bar{X}_j - \bar{\bar{X}})^2$, move cursor to **F12** and type "**=8*(6-6.917)^2+8*(7-6.917)^2+8*(7.75-6.917)^2**", then hit **Enter**. We get SSB = 12.333.

5. Based on SST = SSB + SSW, move cursor to **F13** and type "=47.833-12.333", then hit **Enter**. We get SSW = 35.500.

6. We can finish the rest of the ANOVA summary table by filling *df* for between-group 2 and *df* for within-group 21. Therefore, we get $MSB = \frac{SSB}{df_B} = 6.167$ and $MSW = \frac{SSW}{df_W} = 1.690$, then calculated $F = \frac{MSB}{MSW} = 3.649$. The completed ANOVA summary table is shown in Figure 12.4. We can use the **F.DIST.RT(x,df$_1$,df$_2$)** function in Excel to find the *p* value associated with the calculated *F*. Select a blank cell in Excel, type "**=F.DIST.RT(3.649,2,21)**", then hit **Enter.** The answer 0.04364 appears. Thus, the *p* value associated with $F = 3.649$ and $(df_B, df_W) = (2, 21)$ is $p = .04364$. The one-way ANOVA test is significant. Unfortunately, Excel does not provide post hoc comparison procedures, so we still have to go through the post hoc comparisons by hand.

FIGURE 12.4 ● Complete ANOVA Summary Table in Excel

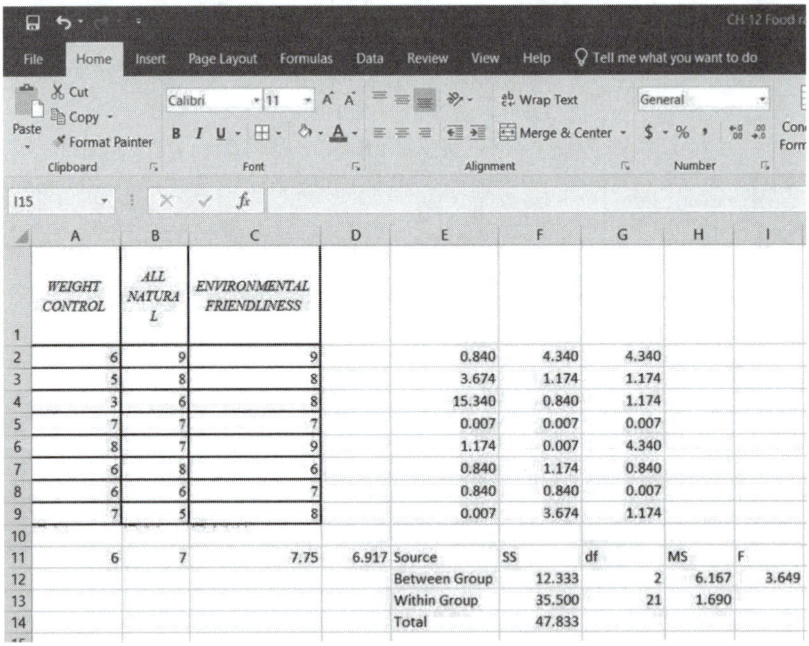

	A	B	C	D	E	F	G	H	I
1	WEIGHT CONTROL	ALL NATURAL	ENVIRONMENTAL FRIENDLINESS						
2	6	9	9		0.840	4.340	4.340		
3	5	8	8		3.674	1.174	1.174		
4	3	6	8		15.340	0.840	1.174		
5	7	7	7		0.007	0.007	0.007		
6	8	7	9		1.174	0.007	4.340		
7	6	8	6		0.840	1.174	0.840		
8	6	6	7		0.840	0.840	0.007		
9	7	5	8		0.007	3.674	1.174		
10									
11	6	7	7.75	6.917	Source	SS	df	MS	F
12					Between Group	12.333	2	6.167	3.649
13					Within Group	35.500	21	1.690	
14					Total	47.833			

7. We can also use the **Data Analysis** built-in function to conduct one-way ANOVA by clicking **Data** tab, clicking **Data Analysis**, selecting "**Anova:Single Factor**" in **Analysis Tools** as shown in Figure 12.5. Then hit **OK**.

FIGURE 12.5 ● Anova: Single Factor in Data Analysis

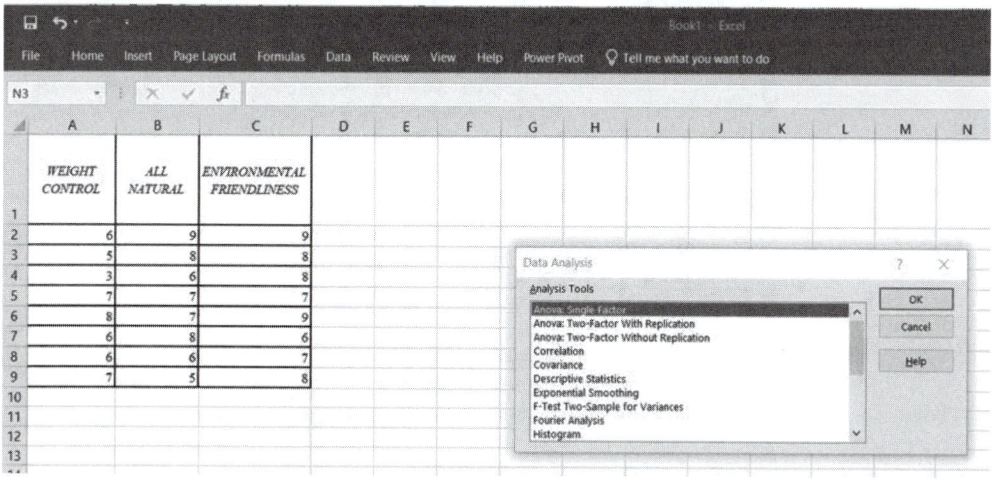

8. Then a popup window shows up asking for data input range. Click the ⇡ then highlight the entire data range **$A$1:$C$9**, select **Grouped by Column** and check the box for **Labels in First Row**. Then hit **OK** as shown in Figure 12.6.

FIGURE 12.6 ● Data Input Range for Anova: Single Factor

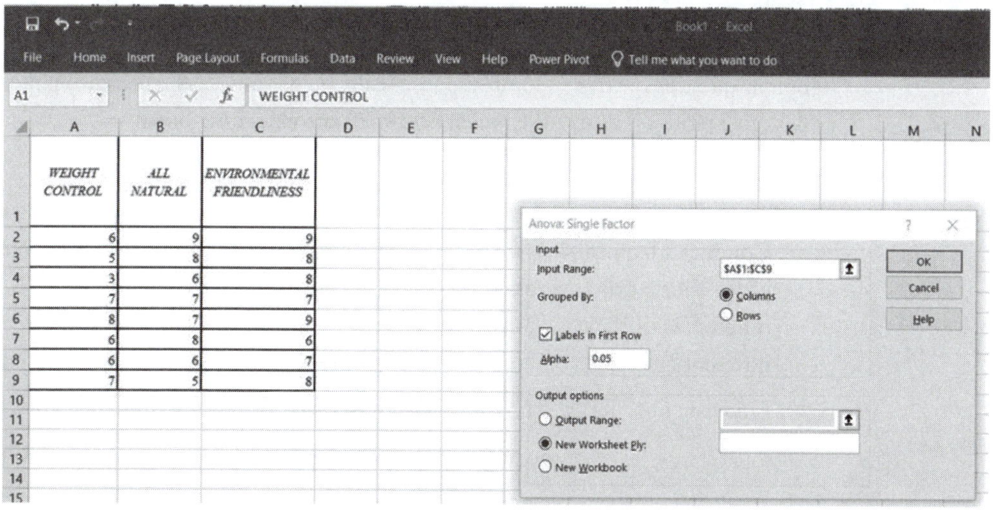

9. The results of the One-Way ANOVA shows up in Figure 12.7 which confirms the results from Figure 12.4.

FIGURE 12.7 ● Results From Anova: Single Factor

	A	B	C	D	E	F	G
1	Anova: Single Factor						
2							
3	SUMMARY						
4	*Groups*	*Count*	*Sum*	*Average*	*Variance*		
5	WEIGHT CONTROL	8	48	6	2.285714		
6	ALL NATURAL	8	56	7	1.714286		
7	ENVIRONMENTAL FRIENDLINESS	8	62	7.75	1.071429		
8							
9							
10	ANOVA						
11	*Source of Variation*	*SS*	*df*	*MS*	*F*	*P-value*	*F crit*
12	Between Groups	12.33333	2	6.166667	3.647887	0.043674	3.4668
13	Within Groups	35.5	21	1.690476			
14							
15	Total	47.83333	23				

EXERCISE PROBLEMS

1. A beverage company compared customer preferences on the color of a sports drink: red, orange, and yellow. The company randomly selected 30 participants, then randomly assigned 10 participants to each group. The customers' preferences were measured and reported in Table 12.8. The preferences were measured by the sum of four questions with a 5-point response scale, higher numbers representing more favorable ratings. Assume the preferences were normally distributed.

$$SST = 96.8$$

   a. Conduct a hypothesis test to verify the equality of the customers' preference of the sports drinks across three different colors. Are there significant differences in the customers' preferences for the sports drink across three different colors?

   b. Calculate the effect size for customers' preferences across three different colors.

TABLE 12.8 ⬢ Customers' Preference of Color of the Sports Drink		
Red	**Orange**	**Yellow**
19	14	13
18	13	14
16	13	17
17	15	15
15	15	17
14	17	18
17	15	19
18	16	18
19	16	17
17	18	16

2. Many people lost their jobs during economic slowdown caused by the COVID-19 pandemic. An industrial and organizational psychologist studied the degree of psychological stress of job seekers. A sample of 63 job seekers is randomly selected. Their ages were classified into three levels: (1) younger than 40 years, (2) 40 to 55 years, and (3) older than 55 years. The descriptive statistics of psychological stress scores for all three groups were reported in Table 12.9. Higher numbers mean higher reported psychological stress.

$$SST = 2524.4$$

a. Conduct a hypothesis test to compare the psychological stress across three different age-groups. Are there significant differences in the job seekers' psychological stress across different age-groups?

b. Calculate the effect size of job seekers' psychological stress across different age-groups.

TABLE 12.9 ⬢ Descriptive Statistics of Psychological Stress Scores for All Three Groups

Age	n	$\bar{X}$	s
Younger than 40 years	20	41.3	5.34
40 to 55 years	20	43.7	5.97
Older than 55 years	23	48.8	5.68
Total	63		

Solutions

1. a. The four-step hypothesis test for an ANOVA:

Step 1: State the pair of hypotheses.

H_0: $\mu_1 = \mu_2 = \mu_3$

H_1: Not all μ_k are equal

Step 2: Identify the rejection zone.

There are three groups, $k = 3$, 10 people in each group, $n_1 = n_2 = n_3 = 10$, and the sample size $n = 30$. The critical value of F is identified by $F_{(2,27)} = 3.35$ in the F Table for ANOVA. The rejection zone is identified as the calculated $F > 3.35$.

Step 3: Calculate the F test.

Under equal group size, the sample mean is

$$\bar{\bar{X}} = (17 + 15.2 + 16.4) / 3 = 16.2$$

$$\text{SSB} = \Sigma n_j(\bar{X}_j - \bar{\bar{X}})^2 = 10(17 - 16.2)^2 + 10(15.2 - 16.2)^2 + 10(16.4 - 16.2)^2 = 16.8$$

$$\text{SSW} = \text{SST} - \text{SSB} = 96.8 - 16.8 = 80$$

Now we have all the numbers to complete the ANOVA summary table as shown in Table 12.8a.

TABLE 12.8a ⬢ **The ANOVA Summary Table for the Customers' Preferences of Different Color Sports Drinks**

Source	*SS*	*df*	*MS*	*F*
Between groups	16.8	$df_B = 2$	MSB = 8.4	$F = 2.835$
Within groups	80	$df_W = 27$	MSW = 2.963	
Total	96.8	$df_T = 29$		

Step 4: Make the correct conclusion.

The calculated $F = 2.835$ was not within the rejection zone, so we failed to reject H_0. The evidence was not strong enough to claim that significant differences exist among sports drinks with three different colors. No further analysis is needed.

1. b. The effect size for the ANOVA is $\eta^2 = \frac{\text{SSB}}{\text{SST}} = \frac{16.8}{96.8} = .174$.

The effect size of customers' preferences of different color sports drinks was .174. Only 17.4% of the variance of customers' preferences were explained by the different colors.

2. a. The four-step hypothesis test for an ANOVA is as follows:

Step 1: State the pair of hypotheses.

H_0: $\mu_1 = \mu_2 = \mu_3$

H_1: Not all μ_k are equal

Step 2: Identify the rejection zone.

There were three groups, $k = 3$, $n_1 = 20$, $n_2 = 20$, $n_3 = 23$, and the sample size $n = 63$. The critical value of F was identified by $F_{(2,60)}$. The critical value was $F_{(2,60)} = 3.15$. The rejection zone was identified as the calculated $F > 3.15$.

Step 3: Calculate the F test.

Due to the fact that the group sizes were not the same, the sample mean needs to be calculated as a weighted mean, which was covered in Chapter 3. The weighted sample mean for three groups with different group sizes is calculated as the sum of each group mean weighted by its group size and divided by the sum of all group sizes, which is the sample size.

$$\text{The weighted sample mean } \bar{\bar{X}} = (n_1\bar{X}_1 + n_2\bar{X}_2 + n_3\bar{X}_3) / (n_1 + n_2 + n_3)$$

$$\bar{\bar{X}} = (20(41.3) + 20(43.7) + 23(48.8)) / 63 = 44.8$$

$$\text{SSB} = \Sigma n_j(\bar{X}_j - \bar{\bar{X}})^2 = 20(41.3 - 44.8)^2 + 20(43.7 - 44.8)^2 + 23(48.8 - 44.8)^2 = 637.2$$

$$\text{SSW} = \text{SST} - \text{SSB} = 2524.4 - 637.2 = 1887.2$$

Now we have all the numbers to complete the ANOVA summary table as shown in Table 12.9a.

TABLE 12.9a ● The ANOVA Summary Table for the Job Seekers' Psychological Stress Among Three Age-Groups

Source	*SS*	*df*	*MS*	*F*
Between groups	637.2	$df_B = 2$	MSB = 318.6	$F = 10.129$
Within groups	1887.2	$df_W = 60$	MSW = 31.453	
Total	2524.4	$df_T = 62$		

Step 4: Make the correct conclusion.

The calculated $F = 10.129$ was within the rejection zone, so we rejected H_0. The evidence was strong enough to claim that not all job seekers experienced the same level of psychological stress. However, a significant F test does not inform us exactly where the significant difference occurs. Thus, post hoc pairwise comparisons need to be conducted to identify the sources of difference.

Use these means and standard deviations number to conduct three possible pairwise comparisons. The results of the three possible pairwise comparisons are shown in Table 12.9b. Use the Bonferroni adjusted $\alpha_B = \frac{.05}{3} = .017$ to evaluate the significance of each pairwise comparison.

TABLE 12.9b Three Possible Pairwise Comparisons for Job Seekers' Psychological Stress

Pairwise Comparison	t	P
$\mu_1 - \mu_2$	1.340	.1882
$\mu_1 - \mu_3$	4.440	.0001
$\mu_2 - \mu_3$	2.868	.0065

Based on the results of the post hoc comparisons, the difference of psychological stress between Group 1 and Group 3 and the difference of psychological stress between Group 2 and Group 3 were statistically significant. However, the difference between Group 1 and Group 2 was not significant. Therefore, we concluded that there were significant differences in job seekers' psychological stress between the group younger than 40 years and the group older than 55 years. There were also significant differences in job seekers' psychological stress between the group between 40 and 55 years and the group older than 55 years. However, the evidence was not strong enough to claim that there is a significant difference between the two younger groups. It seems that job seekers who are older than 55 years experience significantly higher levels of psychological stress than the two younger groups.

2. b. The effect size for the ANOVA is $\eta^2 = \frac{SSB}{SST} = \frac{637.2}{2524.4} = .252$.

The effect size of job seekers' psychological stress across different age-groups was .252. There was 25.2% of the variance of the psychological stress explained by the different age-groups.

What You Learned

Conducting a hypothesis test with ANOVA is a multiple-stage process. First, we partition the variances into different sources; then, gather all the *SS*, *df*, and MS to construct an ANOVA summary table. The *F* test in the ANOVA summary table calculates the ratio of the between-group variance to the within-group variance. The between-group variance is largely attributable to the systematic group effects that are deliberately designed by the researchers, and the within-group variance is largely attributable to random individual differences or unsystematic measurement errors. The *F* test informs us the ratio of the between-group variance to the within-group variance. The critical values in the *F* Table provide the standards to judge whether the evidence is strong enough to reject H_0 in the hypothesis testing process.

The four-step hypothesis testing process for ANOVA is stated as follows.

Step 1: State the pair of hypotheses for ANOVA.

H_0: $\mu_1 = \mu_2 = \mu_3 = \ldots = \mu_K$

H_1: Not all μ_k are equal

Step 2: Identify the rejection zone.

The critical values of *F* can be identified by $F_{(k-1,\ n-k)}$ in the *F* Table for ANOVA. The rejection zone is identified as the calculated $F > F_{(k-1,\ n-k)}$.

Step 3: Calculate the *F* test for ANOVA.

The *F* test can be easily calculated by completing Table 12.5. Once the ANOVA summary table is complete, we can calculate $F = \text{MSB}/\text{MSW}$.

Step 4: Make the correct conclusion.

If the calculated *F* is not within the rejection zone, we fail to reject H_0. The evidence is not strong enough to support the claim that the group means are different. No additional analysis is needed. If the calculated *F* is within the rejection zone, we reject H_0. After rejecting H_0 in an ANOVA with three or more groups, we need to conduct post hoc pairwise comparisons to identify exactly where the significant differences come from. Bonferroni correction is a commonly used post hoc comparison method.

Key Words

Between-group variance: The between-group variance is calculated as the SSB divided by the df_B; in a special ANOVA term, the between-group variance is also called the mean squares between groups, $\text{MSB} = \text{SSB}/df_B$. 386

Between-subject design: The between-subject design refers to research in which participants are assigned to different groups based on different levels of the independent variables. Every participant is assigned to only one group. 380

Bonferroni correction: The Bonferroni correction simply adjusts the α level of the individual pairwise comparison to a new level in order to avoid inflating the familywise error rate. 390

Familywise error rate (FWER): The FWER refers to the probability of making a Type I error in a set of similar multiple comparisons. 390

One-way ANOVA: One-way ANOVA is an ANOVA that uses only one between-subject factor to classify sample data into different groups. 380

Post hoc comparisons: Post hoc comparisons are used after a significant ANOVA *F* test to identify exactly where the significant difference comes from. 389

Sum of squared total (SST): SST is the total sum of squared difference between individual scores, X_{ij}, and the sample mean, $\bar{\bar{X}}$. It is calculated as $SST = \Sigma\Sigma(X_{ij} - \bar{\bar{X}})^2$. 381

Sum of squares between groups (SSB): SSB is the sum of squares between groups and is defined by the squared differences between individual group means, $\bar{X}_j$, and the sample mean, $\bar{\bar{X}}$, multiplied by the number of participants in each group; then, sum them up. It is calculated as $SSB = \Sigma n_j(\bar{X}_j - \bar{\bar{X}})^2$. 383

Sum of squares within groups (SSW): SSW is the squared differences between individual scores, X_{ij}, within a group relative to its corresponding group mean, $\bar{X}_j$; then, sum them up. It is calculated as $SSW = \Sigma\Sigma(X_{ij} - \bar{X}_j)^2$. 385

Within-group variance: The within-group variance is calculated as the SSW divided by the df_W; in a special ANOVA term, the within-group variance is also called the mean squares within groups, $MSW = SSW / df_W$. 387

Within-subject design: The within-subject design indicates that research participants are assigned to all levels of the independent variables. Every participant experiences all levels of the independent variable and is measured repeatedly. 380

Learning Assessment

Multiple Choice: Circle the Best Answer to Every Question

1. Which one of the following statistical procedures is appropriate to compare means across three different groups?
   a. Independent-samples *t* tests
   b. Dependent-samples *t* tests
   c. ANOVA
   d. Correlation

2. The independent variable in an ANOVA needs to be
   a. an interval variable
   b. a ratio variable
   c. an interval or ratio variable
   d. a nominal or ordinal variable
3. When do you need to conduct post hoc comparisons to identify exactly where the difference comes from?
   a. After conducting an ANOVA for three or more groups
   b. After conducting an ANOVA for two groups
   c. After a significant ANOVA result for three or more groups
   d. After a significant ANOVA result for two groups
4. What are the *df* for the between-group variance, df_B, in ANOVA?
   a. $df_B = k$, where k = number of groups
   b. $df_B = k - 1$, where k = number of groups
   c. $df_B = n - k$, where n = sample size and k = number of groups
   d. $df_B = n - 1$, where n = sample size
5. What are the *df* for the within-group variance, df_W, in ANOVA?
   a. $df_W = k$, where k = number of groups
   b. $df_W = k - 1$, where k = number of groups
   c. $df_W = n - k$, where n = sample size and k = number of groups
   d. $df_W = n - 1$, where n = sample size
6. What are the *df* for the total variance, df_T, in ANOVA?
   a. $df_T = k$, where k = number of groups
   b. $df_T = k - 1$, where k = number of groups
   c. $df_T = n - k$, where n = sample size and k = number of groups
   d. $df_T = n - 1$, where n = sample size
7. Which one of the following variances is largely attributable to systematic group effects that are deliberately designed by the researchers?
   a. The total variance
   b. The between-group variance
   c. The within-group variance
   d. The subject variance
8. Which one of the following variances is largely attributable to random individual differences or unsystematic measurement errors?
   a. The total variance
   b. The between-group variance

c. The within-group variance
d. The variance due to regression

9. The dependent variable in an ANOVA needs to be
   a. a nominal, ordinal, interval, or ratio variable.
   b. an ordinal, interval, or ratio variable.
   c. an interval or ratio variable.
   d. a ratio variable.

10. When conducting multiple post hoc comparisons without keeping FWER under control,
   a. The FWER is larger than the individual pairwise comparison error rate.
   b. The FWER is smaller than the individual pairwise comparison error rate.
   c. The FWER is the same as the individual pairwise comparison error rate.
   d. There is no relationship between the individual pairwise comparison error rate and the FWER.

Free Response Questions

11. Assume that the Insurance Institute for Highway Safety conducted crash safety tests on small SUVs, midsize SUVs, and large SUVs. Five SUVs in each type were randomly selected and their crash safety ratings were reported in Table 12.10. The ratings were normally distributed and reported on a scale of 1 to 10, higher numbers representing better safety results.

TABLE 12.10 ● Crash Safety Ratings for Three Types of SUVs

Small SUV	Midsize SUV	Large SUV
6	7	9
5	8	6
7	8	7
6	6	8
7	7	7
$\bar{X}_1 = 6.2$	$\bar{X}_2 = 7.2$	$\bar{X}_3 = 7.4$

SST = 14.933

a. Conduct a hypothesis test to compare the crash safety ratings across three types of SUVs. Are there significant differences in the crash safety ratings across three different types of SUVs?

b. Calculate the effect size of the different types of SUVs on the crash safety ratings.

12. A team of psychologists compared college students' ability to concentrate under different noise conditions: constant background noise, unpredictable noise, and no noise. Psychologists randomly selected 30 participants and randomly split them into three groups. The ability to concentrate was measured by a speed test that consists of a large number of easy questions. The speed test scores are normally distributed. Higher scores mean better concentration. The speed test scores from each group were reported in Table 12.11.

TABLE 12.11 ● Speed Test Scores in Three Groups

Constant Noise	Unpredictable Noise	No Noise
21	16	15
20	15	16
19	15	19
19	17	17
17	17	19
16	14	20
19	17	21
20	14	20
21	18	19
19	19	18
$\overline{X}_1 = 19.1$	$\overline{X}_2 = 16.2$	$\overline{X}_3 = 18.4$

SST = 126.7

a. Conduct a hypothesis test to compare the speed test scores across three different types of noise conditions. Are there significant differences in the ability to concentrate under three different types of noise conditions?

b. Calculate the effect size of the different types of noise conditions on the students' ability to concentrate.

13. College Board collects information about the total annual cost of tuition, room and board from small private universities, midsize private universities, and large private universities. The tuition, room, and board from different private universities are normally distributed and reported in thousands of dollars, in Table 12.12.

$$SST = 58.5$$

a. Conduct a hypothesis test to compare the tuition, room, and board across three different sizes of private universities. Are there significant differences in the tuition, room, and board charged by the private universities with different sizes?

b. Calculate the effect size of the different sizes of private universities on their tuition, room, and board.

TABLE 12.12 ● Tuition, Room and Board for Private Universities

Small Private University	Midsize Private University	Large Private University
62	61	64
63	62	67
64	64	62
65	63	63
60	65	66
	66	64
		62
$\overline{X}_1 = 62.8$	$\overline{X}_2 = 63.5$	$\overline{X}_3 = 64$

Answers to Pop Quiz Questions

1. c
2. d
3. c
4. b
5. a

13

CHI-SQUARE TESTS FOR GOODNESS OF FIT AND INDEPENDENCE

Learning Objectives

After reading and studying this chapter, you should be able to do the following:

- Describe the difference between parametric statistics and nonparametric statistics
- Describe the formula for calculating expected frequencies for one-way frequency tables and two-way contingency tables
- Identify the degrees of freedom for chi-square tests
- Describe how to use the chi-square table to identify the critical values of chi-square
- Conduct goodness-of-fit tests using one-way frequency tables
- Conduct independence tests using two-way contingency tables
- Conduct hypothesis tests using chi-square

WHAT YOU KNOW AND WHAT IS NEW

You learned the four-step hypothesis testing procedure in Chapter 7, continued to apply it in Chapter 8, slightly modified it to a five-step hypothesis testing procedure in Chapter 9, and maintained the four-step procedure in Chapters 10 to 12. The good news is that the four-step hypothesis is still applicable in Chapter 13. The major difference is that all the previous topics from Chapters 7 to 12 are *parametric statistics* but the chi-square test is a *nonparametric statistic.*

Parametric statistics are defined as statistical procedures that make assumptions about the shape of the population distribution and about population parameters such as μ, σ, and σ^2. Such assumptions allow sample statistics to be used to estimate the population parameters. Parametric statistics require variables to be interval or ratio scales of measurement. Various *t* tests, correlation, regression, and ANOVA are all examples of parametric statistics.

In contrast, **nonparametric statistics** are defined as statistical procedures that make no assumptions about the shape of population distributions. They are also called distribution-free statistics. Nonparametric statistics can be applied to nominal or ordinal scales of measurement as well as interval or ratio variables that are extremely skewed. In this chapter, the focus is on the *chi-square test.* The chi-square (χ^2) is the first nonparametric statistic you have encountered. You will learn to like it. One of the most useful applications of the chi-square is to test the fairness of employment decisions such as hiring, promotion, and layoff. Specifically, the Civil Rights Act of 1964 prohibits discrimination on the basis of race, color, sex, religion, or national origin in employment decisions. The chi-square test is a perfect tool to verify whether female employees were promoted at the same rate as males or whether employees from different ethnic backgrounds were laid off at the same rate when the company went through a workforce reduction.

THE CHI-SQUARE

The chi-square test is my favorite statistical test because the chi-square formula is simple, logical, and easy to understand, and it generates stable results. I will use many examples and one story to demonstrate the reasons why the chi-square test is my favorite statistical procedure. But first, let's introduce the concepts and formulas of chi-square tests.

The basic concept of the **chi-square test** is to calculate the difference between the observed frequency of a particular value of a variable and a theoretically expected frequency of that value, square the difference, and divide by the expected value. This procedure is repeated for every cell of the table, then the results are summed up as follows.

$$\text{Chi-square} = \chi^2 = \sum \frac{(O-E)^2}{E}$$

where

O is the observed frequency

E is the expected frequency

This is the only formula for the chi-square analysis. All the numbers in the formula are frequencies. Frequency is a simple count of how many times each value or category occurs in a variable. These values or categories are mutually exclusive. Every individual is assigned to only one of the categories. The **observed frequency** (O) is the actual number of times each value occurs in a sample. Observed frequencies are always integers. It is not possible for a value to occur 23.45 times in a sample. The expected frequency is a new term that will be thoroughly explained in the next section.

EXPECTED FREQUENCY

The **expected frequency** (E) is defined as the theoretically predicted frequency based on the null hypothesis. Knowing how to calculate the expected frequency is a vital part of conducting a chi-square test. There are two types of chi-square tests: a chi-square goodness-of-fit test and a chi-square for independence test. A chi-square goodness-of-fit test is demonstrated by a one-way frequency table and a chi-square for independence test is demonstrated by a two-way contingency table.

Expected Frequency Under a One-Way Frequency Table

When the frequencies are counted based on the values of one variable, this process is referred to as a one-way frequency table or a goodness-of-fit test.

The expected frequency is the theoretically predicted frequency, which is calculated as $E = np$,

where

n = sample size

p = population proportion or theoretical probability

This calculation for expected frequency in a one-way frequency table for the goodness-of-fit test is simple and logical. If you think about it for a little while, you'll discover that it does not require additional explanation. Let's use an example to illustrate this calculation.

EXAMPLE 13.1

Researchers randomly selected 250 families with two children. They described the sample as 66 families with two boys, 145 families with one boy and one girl, and 39 families with two girls. Assume that the probability of having a boy is .5 and of having a girl is .5. What is the expected frequency distribution of boys and girls for families with two children?

This is the binominal probability distribution as discussed in Chapter 5.

The binomial probability formula is

$$P(r)=\frac{n!}{(n-r)!r!}p^r q^{(n-r)} \text{ for } r=0,1,2,3,\ldots,n$$

In this example, $n=2$ for two children, r is the number of girls in a family with two children; $r=0$, 1, or 2.

The probability of having zero girls and two boys is

$$P(0)=\frac{2!}{(2-0)!0!}(.5)^0(.5)^2=.25$$

The expected frequency for having zero girls and two boys in a sample of 250 families is $np=250\times .25=62.5$.

The probability of having one girl and one boy is

$$P(1)=\frac{2!}{(2-1)!1!}(.5)^1(.5)^1=.50$$

The expected frequency for having one girl and one boy in a sample of 250 families is $np=250\times .50=125$.

The probability of having two girls and zero boys is

$$P(2)=\frac{2!}{(2-2)!2!}(.5)^2(.5)^0=.25$$

The expected frequency for having two girls and zero boys in a sample of 250 families is $np=250\times .25=62.5$.

In the process of calculating the expected frequencies, you only need information about the sample size and the theoretical probability distribution. You don't need to use the observed frequencies when calculating expected frequencies. However, the comparisons between observed frequencies and expected frequencies are vital when conducting hypothesis tests, as will be shown later in the chapter.

Expected Frequency Under a Two-Way Contingency Table

When a chi-square test measures the relationship between two nominal/ordinal variables, the test is based on the differences between the expected frequencies and the observed frequencies. When there are two variables, each participant is classified based on two variables at the same time. We create a table and place one variable as the row variable and the other as the column variable. For example, a study examined the relationship between gender and smoking behavior. Smoking behavior was classified into smoker versus nonsmoker, and gender was classified into male versus female. When research participants were classified based on these two variables at the same time, it led to four ways to classify the participants: (1) male smoker, (2) male nonsmoker, (3) female smoker, and (4) female nonsmoker. The chi-square is used to test the independence of these two variables. Therefore, it is referred to as the chi-square test for independence or a two-way contingency table.

When two variables are independent of each other, it means that there is no relationship between these two variables. The frequency distribution of one variable does not depend on the values of the other variable. Let's use a true story to illustrate how to calculate the expected frequencies for the chi-square test for independence.

A company specializing in medical devices underwent a workforce reduction in its sales force. In the process of making the layoff decisions, one female employee—who was laid off despite her impressive sales records—decided to sue the company for gender discrimination. The company's attorney made the following comment: "Two thirds of the sales force were women; of course, we would lay off more women in the workforce reduction process." This comment sounded reasonable at first. However, it deserved to be carefully scrutinized.

The outcome of a workforce reduction for an individual employee was classified as either "laid off" or "not laid off." The other variable was the employee's gender, which was classified as either "male" or "female." All salespeople were classified based on layoff decision (column variable) and their gender (row variable); the observed frequencies were reported in Table 13.1, a 2×2 contingency table. There were five male salespeople who were laid off, 25 female salespeople who were laid off, 28 male salespeople who were not laid off, and 42 female salespeople who were not laid off. The row totals showed that there were 33 male salespeople and 67 female salespeople in the company before the layoff decisions. The column totals showed that there were 30 salespeople who were laid off and 70 salespeople who were not laid off.

TABLE 13.1 ● A 2×2 Contingency Table for the Observed Frequencies Based on Layoff Decision and Employee's Gender

	Laid Off	Not Laid Off	Row Total
Male	5	28	33
Female	25	42	67
Column total	30	70	100

Females constituted two thirds of the salesforce. The company trimmed 30% of the salesforce in the workforce reduction. If the layoff decision and employee's gender were independent, it implied that employee's gender was not considered when the layoff decision was made. Thus, the proportion of female employees should remain as two thirds in both the category of "laid off" employees and in the category of "not laid off" employees.

There are four cells in this 2 × 2 contingency table. The number of cells in a contingency table does not include the row totals and column totals. The expected frequency for each cell is calculated under the independence assumption,

$$E = \frac{\text{Row total} \times \text{Column total}}{n}$$

Such a calculation is mathematically designed to make the expected frequency in each cell have the same proportion as the row total and column total relative to its sample size. The expected frequency for each cell is reported in Table 13.2. The expected frequencies are not likely to be integers. The chi-square tests work best when each expected frequency is at least 5. If there are more than 20% of the cells with an expected frequency less than 5, it is not appropriate to use the chi-square. However, there is no requirement for the *observed* frequencies to be at least 5. You need to pay attention to this subtle distinction. The row totals and column totals remain the same in both the observed frequency table and the expected frequency table.

Let's scrutinize the company attorney's comment: "Two thirds of the sales force were women; of course, we would lay off more women in the workforce reduction process." This statement was partially true. The devil is in the details. The company's attorney tried to justify the fact that more women were laid off by asserting that there were more women in the sales positions to begin with. The real issue was *how many more.* If the proportion

TABLE 13.2 ● Expected Frequency Between Layoff Decision and Employee's Gender

	Laid Off	Not Laid Off	Row Total
Male	$\frac{33 \times 30}{100} = 9.9$	$\frac{33 \times 70}{100} = 23.1$	33
Female	$\frac{67 \times 30}{100} = 20.1$	$\frac{67 \times 70}{100} = 46.9$	67
Column total	30	70	100

of women who got laid off was the same as the proportion of women who did not get laid off, and both were two thirds, then I would agree that there was no relationship between layoff decision and employee's gender. The chi-square test is designed to empirically test how far apart the reality (i.e., observed frequencies) are from the independence assumption between employee's gender and layoff decisions (i.e., expected frequencies).

The expected frequency table is calculated based on the independence assumption between the two variables. The expected frequency table presents what the numbers would be were the proportion of women in the "laid off" category the same as the "not laid off" category. More generally, the expected frequencies state the theoretically predicted frequencies under the assumption that the variables are distributed exactly the same as the null hypothesis that two variables are independent. The expected frequencies show what the numbers in each cell should look like if there is no relationship between the two variables. The expected frequencies are the fair numbers free of gender discrimination in this case. The differences between the observed frequencies and the expected frequencies show the strength of the relationship between the two variables—thus, the extent of the gender discrimination in the layoff decisions.

Comparing the observed frequencies with the expected frequencies, you can see that the company should have laid off 9.9 salesmen and 20.1 saleswomen, but it laid off 5 salesmen and 25 saleswomen. The company should have kept 23.1 salesmen and 46.9 saleswomen, but it kept 28 salesmen and 42 saleswomen. Judging by the discrepancies between the observed frequencies and expected frequencies, on the face of things, it is doubtful whether the layoff decisions were carried out without considering the employees' gender. A hypothesis test using the chi-square can answer the question as to the probability of getting such observed frequencies if the layoff decision and employee's gender were independent. The outcome of such a chi-square test showed whether the company was biased against women in its workforce reduction decisions. The hypothesis test with the chi-square test for independence will be discussed in a later section.

POP QUIZ

1. The row totals and column totals from a contingency table can be calculated from
   a. only the observed frequencies of the two variables.
   b. only the expected frequencies of the two variables.
   c. either the observed frequencies or the expected frequencies; they produce the same row totals and column totals.
   d. the sum of the observed frequencies and the expected frequencies.

GOODNESS-OF-FIT TESTS

Chi-square goodness-of-fit tests refer to using one categorical variable to investigate whether the sample distribution of the values fits with a particular theoretical distribution. Once the expected frequencies are calculated, everything else simply falls into place through easy calculation and the use of chi-square tables. The chi-square formula is

$$\text{Chi - square} = \chi^2 = \Sigma \frac{(O-E)^2}{E}$$

Based on the formula, when large discrepancies exist between the observed frequencies and the expected frequencies, the calculated χ^2 is large. When small discrepancies exist between the observed frequencies and the expected frequencies, the calculated χ^2 is small. Small χ^2 indicates that the sample frequencies (i.e., the observed frequencies) fit well with the theoretical frequencies (i.e., the expected frequencies) under the null hypothesis. This is the reason why the one-way chi-square test is also called the *goodness-of-fit* test.

Every time we introduce a new test statistic, we need to know how to judge its statistical significance. When we introduced the Z test, t test, Pearson's r, regression, and ANOVA in previous chapters, we introduced the Z Table, t Table, critical values for Pearson's r Table, and F Table to identify the critical values that set off the boundaries of the rejection zones for these test statistics. It is the same way with chi-square tests. The degrees of freedom for the chi-square goodness-of-fit test are determined by (c – 1), where c is the number of columns or categories in the one-way frequency table. There is a specific chi-square curve for every degree of freedom. The critical values of chi-square that set off the boundaries of the rejection zones are presented in Appendix E. The rejection zone for the chi-square goodness-of-fit test is identified as $\chi^2 > \chi^2_{(c-1)}$. It is important to point out that the critical value that sets off the rejection zone is on the right side of the distribution.

We have already established the chi-square formula and the critical values of chi-square. Now we are ready to put everything together to conduct a hypothesis test with the chi-square goodness-of-fit test. It is important to point out that chi-square values can only be positive numbers. Large calculated χ^2 indicates a lack of fit between the observed frequencies and the expected frequencies.

Hypothesis Testing for the Chi-Square Goodness-of-Fit Test

Even though chi-square is a nonparametric statistic, the four-step hypothesis testing procedure you learned in the previous chapters on parametric statistics still applies. The four-step hypothesis testing procedure with the chi-square goodness-of-fit test includes the same four steps as in previous hypothesis tests but with slight modifications:

1. State the pair of hypotheses.
2. Identify the rejection zone.
3. Calculate the test statistic.
4. Draw the correct conclusion.

Step 1: State the pair of hypotheses.

This is the first time that you learn to state the pair of hypotheses for a nonparametric statistic. The hypotheses do not describe a population parameter as in the previous parametric statistics. The pair of hypotheses in the chi-square goodness-of-fit tests is a simple description of whether a particular sample distribution fits its theoretical distribution.

H_0: A particular (empirical) variable distribution fits its theoretical distribution (i.e., uniform distribution, binominal distribution, etc.).

H_1: A variable distribution does not fit its theoretical distribution.

Step 2: Identify the rejection zone.

Each degree of freedom has its own chi-square curve. The degrees of freedom for chi-square are calculated as $df = (c - 1)$. According to the critical values of the chi-square table, the rejection zone for a chi-square test is identified as $\chi^2 > \chi^2_{(c-1)}$.

Here are first few values in the Critical Values for the Chi-Square Table which can be found in Appendix E. When $\alpha = .05$ and $df = 1$, the rejection zone for a chi-square test is $\chi^2 > 3.841$. When $\alpha = .05$ and $df = 2$, the rejection zone for a chi-square test is $\chi^2 > 5.991$. When $\alpha = .05$ and $df = 3$, the rejection zone for a chi-square test is $\chi^2 > 7.815$.

Step 3: Calculate the test statistic.

$$\text{Chi-square} = \chi^2 = \Sigma\frac{(O - E)^2}{E}$$

Step 4: Draw the correct conclusion.

Compare the calculated χ^2 value with the rejection zone. If the calculated χ^2 is within the rejection zone, we reject H_0. If the calculated χ^2 is not within the rejection zone, we fail to reject H_0.

We use several examples to illustrate this four-step hypothesis test with chi-square tests.

EXAMPLE 13.2

Jamie threw a die 36 times and recorded the number of times each value on the die occurred in Table 13.3. Conduct a hypothesis test to see if this die is a fair die, assuming $\alpha = .05$.

TABLE 13.3 ● Observed Frequency (*O*) of the Number of Dots of a Die

X	1	2	3	4	5	6
O	9	5	7	4	3	8

The probability theory states that if this die is a fair die, each number of dots has a uniform probability, $p = 1/6$. The expected frequency for each number of dots when Jamie threw the die 36 times was $E = np = 36(1/6) = 6$. This fulfills the requirement that the expected frequency in each of the cells of the table should be at least 5. If more than 20% of the cells contain an expected frequency less than 5, it is not appropriate to use the chi-square test. Although some of the observed frequencies fell below 5, as stated before, there are no requirements for the observed frequencies to be at least 5. This did not pose any threat to the accuracy of the chi-square for goodness-of-fit test.

At first glance, none of the observed frequencies match the expected frequencies $E = 6$. You might have the impulse to jump to the conclusion that this die is not a fair die. I would strongly encourage you to temporarily control that impulse until after the hypothesis test is finished.

Let's include the expected frequencies in the table as shown in Table 13.3a.

TABLE 13.3a ● Observed Frequency (*O*) and Expected Frequency (*E*) of the Number of Dots on a Die

X	1	2	3	4	5	6
O	9	5	7	4	3	8
E	6	6	6	6	6	6

It is time to put all the numbers together to conduct a hypothesis test to see if this die is a fair die.

Step 1: State the pair of hypotheses.

H_0: The die was a fair die, which meant that it produced the same frequencies as the expected frequencies for all six values.

H_1: The die was not a fair die.

Step 2: Identify the rejection zone.

When $\alpha = .05$, with $df = (c - 1) = 6 - 1 = 5$, the rejection zone for the chi-square test is $\chi^2 > 11.07$.

Step 3: Calculate the test statistic.

$$\chi^2 = \Sigma\frac{(O-E)^2}{E} = \frac{(9-6)^2}{6} + \frac{(5-6)^2}{6} + \frac{(7-6)^2}{6} + \frac{(4-6)^2}{6} + \frac{(3-6)^2}{6} + \frac{(8-6)^2}{6} = 4.667$$

Step 4: Draw the correct conclusion.

The calculated $\chi^2 = 4.667$ was not within the rejection zone. Therefore, we failed to reject H_0. The evidence was not strong enough to claim that the die was not a fair die. Although the observed frequencies appeared to be different from the expected frequencies, the differences were, on balance of all the numbers considered together, not large enough to be statistically significant.

Another way to complete the hypothesis test procedure is to use Excel function to identify the *p* value associated with the calculated χ^2 by typing "=**CHISQ.DIST.RT(x,df)**" in a blank cell. Then hit **Enter**. The answer 0.4579 is the probability associated with the calculated χ^2. Use the *p* value compared with the α level (i.e., .05) to draw the correct conclusion. When the *p* value is less than the α level, we reject H_0. When the *p* value is greater or equals the α level, we fail to reject H_0. The $p = .4579$ is greater than .05, so we fail to reject H_0.

We have finished the chi-square test that applies to the one-way frequency table or the goodness-of-fit test. Now we are ready to move on to the chi-square test for independence between two nominal or ordinal variables. In the next section, I will demonstrate the reason why the chi-square test for independence is one of my favorite statistics.

POP QUIZ

2. Based on the formula Chi-square $= \chi^2 = \Sigma(O-E)^2 / E$, the larger the calculated χ^2 means that
   a. the greater are the discrepancies between the observed frequencies and the expected frequencies.
   b. the smaller are the discrepancies between the observed frequencies and the expected frequencies.
   c. the greater is the variance of the expected frequencies.
   d. the greater is the variance of the observed frequencies.

CHI-SQUARE FOR INDEPENDENCE TEST

The **chi-square for independence test** is a nonparametric statistical procedure to measure the relationship between two nominal or ordinal variables. The chi-square for independence test is based on a **contingency table**. A contingency table is defined as a frequency table where each individual is classified based on the column variable and the row variable at the same time. The observed frequency table presents the frequency count based on two variables. The expected frequency table presents the theoretical predicted frequency count under the H_0, which means that the frequency distribution of one variable does not depend on the values of the other variable. The expected frequency is calculated as

$$E = \frac{\text{Row total} \times \text{Column total}}{n}$$

Such a formula is mathematically designed to make sure that the expected frequencies are calculated in proportion to the row total and column total relative to the sample size.

Once the expected frequencies are figured out, the chi-square formula is calculated simply as

$$\text{Chi-square} = \chi^2 = \sum \frac{(O-E)^2}{E}$$

The degrees of freedom for the chi-square for independence test is calculated as $(r-1)(c-1)$, where r is the number of rows and c the number of columns in the contingency table. The rejection zone for a chi-square test for independence is identified as $\chi^2 > \chi^2_{(r-1)(c-1)}$ The critical values of chi-square tests are always on the right side of the distribution.

We have already established the chi-square formula and the critical values of chi-square. It seems that we are ready to put everything together to conduct a chi-square for independence test to investigate the relationship between two nominal or ordinal variables.

Hypothesis Test for the Chi-Square for Independence Test

You should feel relieved that the same four-step hypothesis testing procedure can also be applied to the chi-square for independence tests:

1. State the pair of hypotheses.
2. Identify the rejection zone.
3. Calculate the test statistic.
4. Draw the correct conclusion.

Step 1: State the pair of hypotheses.

The pair of hypotheses in the chi-square test for independence is a simple description of whether two variables are independent of each other.

H_0: Two variables are independent of each other.

H_1: Two variables are not independent of each other.

Step 2: Identify the rejection zone.

Each degree of freedom has its own chi-square curve. The degrees of freedom for the chi-square test for independence are $df = (r - 1)(c - 1)$. According to the critical values in the chi-square table, the rejection zone for a chi-square test is identified as $\chi^2 > \chi^2_{(r-1)(c-1)}$

Step 3: Calculate the test statistic.

$$\text{Chi-square} = \chi^2 = \Sigma\frac{(O-E)^2}{E}$$

Step 4: Draw the correct conclusion.

Compare the calculated χ^2 value with the rejection zone. If the calculated χ^2 is within the rejection zone, we reject H_0. If the calculated χ^2 is not within the rejection zone, we fail to reject H_0.

Examples are used to illustrate this four-step hypothesis test for the chi-square tests for independence. The time has come to finish the story about the medical device company being sued for gender discrimination in its layoff decisions by conducting a hypothesis test for the chi-square for independence test. All the relevant information is presented in Example 13.3.

EXAMPLE 13.3

A company specializing in medical devices underwent workforce reduction in its salesforce. The outcome of a workforce reduction for an individual employee was classified as either "laid off" or "not laid off." The salesperson's gender was classified as either "male" or "female." Every salesperson in the company was classified based on these two variables. The results of the workforce reduction decisions were reported in Table 13.1. Conduct a hypothesis test to see if layoff decision and employee's gender were independent, using $\alpha = .05$.

Step 1: State the pair of hypotheses.

H_0: The layoff decision and salesperson's gender were independent.

H_1: The layoff decision and salesperson's gender were not independent.

Step 2: Identify the rejection zone.

When $\alpha = .05$, with $df = (r - 1)(c - 1) = (2 - 1)(2 - 1) = 1$, the rejection zone for the chi-square test is $\chi^2 > 3.841$.

Step 3: Calculate the test statistic.

The expected frequencies were calculated by $E = \frac{\text{Row total} \times \text{Column total}}{n}$ and shown in Table 13.2.

We used each observed frequency compared with its corresponding expected frequency to complete the χ^2 calculation.

$$\chi^2 = \Sigma\frac{(O-E)^2}{E} = \frac{(5-9.9)^2}{9.9} + \frac{(28-23.1)^2}{23.1} + \frac{(25-20.1)^2}{20.1} + \frac{(42-46.9)^2}{46.9} = 5.171$$

(Continued)

(Continued)

Step 4: Draw the correct conclusion.

The calculated $\chi^2 = 5.171$ was within the rejection zone, so we rejected H_0. Or we could type "**=CHISQ.DIST.RT(5.171,1)**" in a blank cell in Excel, then hit **Enter**. The *p* value associated $\chi^2 = 5.171$ was .0230, which was less than the α level. We rejected H_0. The evidence was strong enough to support the claim that the layoff decision and salesperson's gender were not independent. The probability of obtaining such observed frequencies was .0230 under the assumption that layoff decision and salesperson's gender were independent. Therefore, the conclusion was that it was unlikely that the company did not consider the employee's gender when it went through the work-force reduction process. The evidence was strong enough to support the claim that layoff decisions and employees' gender were not independent.

It was amazing that such a simple chi-square test for independence could provide solid statistical evidence to discredit the medical device company attorney's statement that "two thirds of the sales force were women; of course, we would lay off more women in the work-force reduction process." The problem was that the company has laid off too many women and not enough men to achieve the fair proportion of two thirds of women in both the "laid off" category and the "not laid off" category.

In a very similar context, you might have heard that a big-box retailer has been accused of unfair employment practices and has many lawsuits pending against it. The next example is a demonstration of how the chi-square for independence test can be applied in such a situation.

EXAMPLE 13.4

A large big-box retailer was accused of gender discrimination in promotion decisions. The personnel records in the company's Midwest region showed that there were 3,325 male employees, and among them, 969 were promoted in the past 6 months. During the same time, there were 2,786 female employees, and among them, 701 were promoted. Do the data substantiate the claim that this large big-box retailer discriminated against female workers in promotion decisions? Conduct a four-step hypothesis testing procedure to test if promotion decisions and employee's gender were independent, using $\alpha = .05$.

Step 1: State the pair of hypotheses.

H_0: The promotion decision and employee's gender were independent.

H_1: The promotion decision and employee's gender were not independent.

Step 2: Identify the rejection zone.

When $\alpha = .05$, with $df = (r - 1)(c - 1) = (2 - 1)(2 - 1) = 1$, the rejection zone for the chi-square test is $\chi^2 > 3.841$.

Step 3: Calculate the test statistic.

It is important to be able to extract the numbers from the problem statement and accurately construct the observed frequency table with mutually exclusive categories. There were two categories in the outcomes of the promotion decision: "promoted" and "not promoted." The problem statement specified that "there were 3,325 male employees and among them 969 were promoted in the past 6 months"; so it meant that 969 males were promoted and 2,356 males were not promoted out of 3,325 male employees. The same reasoning applied to the female employees; so you know that 701 female employees were promoted and 2,085 female employees were not promoted out of 2,786 female employees. The observed frequency table is constructed in Table 13.4.

The expected frequency for each cell is calculated under the independence assumption,

$$E = \frac{\text{Row total} \times \text{Column total}}{n}$$

TABLE 13.4 ⬢ Observed Frequency for the 2 × 2 Contingency Table Between Promotion Decision and Employee's Gender

	Promoted	Not Promoted	Row Total
Male	969	2,356	3,325
Female	701	2,085	2,786
Column total	1,670	4,441	6,111

The expected frequency for the "promoted male" is $\frac{3{,}325 \times 1{,}670}{6{,}111} = 908.65$

The expected frequency for the "promoted female" is $\frac{2{,}786 \times 1{,}670}{6{,}111} = 761.35$

The expected frequency for the "not promoted male" is $\frac{3{,}325 \times 4{,}441}{6{,}111} = 2{,}416.35$

The expected frequency for the "not promoted female" is $\frac{2{,}786 \times 4{,}441}{6{,}111} = 2{,}024.65$

The expected frequency table is constructed in Table 13.4a.

(Continued)

(Continued)

TABLE 13.4a ● Expected Frequency for the 2 × 2 Contingency Table Between Promotion Decision and Employee's Gender

	Promoted	Not Promoted
Male	908.65	2,416.35
Female	761.35	2,024.65

$$\chi^2 = \sum \frac{(O-E)^2}{E} = \frac{(969-908.65)^2}{908.65} + \frac{(2356-2416.35)^2}{2416.35} + \frac{(701-761.35)^2}{761.35} + \frac{(2085-2024.65)^2}{2024.65} = 12.099$$

Step 4: Draw the correct conclusion.

The calculated $\chi^2 = 12.099$ was within the rejection zone, so we rejected H_0. Or we could type "=**CHISQ.DIST.RT(12.099,1)**" in a blank cell in Excel, then hit **Enter**. The p value associated $\chi^2 = 12.099$ was .0005 which was less than the α level. We rejected H_0. The evidence was strong enough to support the claim that the promotion decision and employee's gender were not independent. The probability of obtaining such observed frequencies was .0005 under the assumption that the promotion decision and employees' sex were independent. Therefore, the conclusion was that it was highly unlikely that the big-box retailer did not consider employees' sex when the promotion decisions were made.

The chi-square test for independence is a powerful statistic for investigating the relationship between two nominal or ordinal variables. It does not make assumptions about the distribution shape of the population or the population parameters. The calculation of chi-square only requires a frequency count based on the variables involved. As long as no more than 20% of the cells have the expected frequencies less than 5, the results of chi-square tests are accurate and reliable.

Author's Aside

The Civil Rights Act of 1964 makes it illegal for employers to discriminate against any individual based on race, sex, color, religion, or national origin. These variables are all nominal variables. Any one of these variables can be used to investigate whether there is discrimination in various employment practices such as hiring, promotion, or layoff decisions. As an industrial and organizational psychologist,

I have been involved in discrimination litigations as an expert witness who provided statistical evidence to demonstrate if a company's employment decisions were made fairly. Fair decisions are supposed to be made without considering the employee's race, sex, color, religion, or national origin. Chi-square tests use the discrepancies between observed frequencies and expected frequencies to show how far apart a company's practice is from a truly fair decision-making process. That is the reason why my favorite statistical procedure is the chi-square test.

Besides discrimination in the workplace, chi-square tests can also be applied to investigate the relationship between two nominal or ordinal variables in medical fields. The following examples illustrate such uses.

EXAMPLE 13.5

A medical insurance company evaluated the effectiveness of cardiac rehabilitation (rehab) in terms of whether heart patients returned for repeat procedures within 12 months of their original one. Cardiac rehab is a professionally supervised program to help patients recover from heart attacks, heart surgery, stenting, or angioplasty. Cardiac rehab programs usually provide education and counselling services to help heart patients increase physical fitness, reduce cardiac symptoms, improve health, and reduce the risk of future heart problems. A group of cardiac patients was randomly selected from a large hospital's cardiac rehab center. Each patient was classified based on two variables: (1) whether the patient completed the cardiac rehab and (2) whether the patient has had a repeat heart procedure within 12 months of the original procedure. The frequency counts for both variables are reported in Table 13.5. Conduct a hypothesis test to investigate the relationship between cardiac rehab and whether the patient repeated a heart procedure within 12 months, assuming $\alpha = .05$.

TABLE 13.5 ● Observed Frequency for the Contingency Table Between Cardiac Rehab and Repeat Heart Procedure Within 12 Months

	Completed Cardiac Rehab	Did not Complete Cardiac Rehab	Row Total
Repeated heart procedure	15	20	35
Did not repeat heart procedure	102	63	165
Column total	117	83	200

(Continued)

(Continued)

Step 1: State the pair of hypotheses.

H_0: The completion of cardiac rehab and a repeat heart procedure were independent.

H_1: The completion of cardiac rehab and a repeat heart procedure were not independent.

Step 2: Identify the rejection zone.

When $\alpha = .05$, with $df = (r - 1)(c - 1) = (2 - 1)(2 - 1) = 1$, the rejection zone for the chi-square test is $\chi^2 > 3.841$.

Step 3: Calculate the test statistic.

We need to construct the expected frequencies before we can calculate the chi-square. The expected frequency for each cell is calculated under the independence assumption,

$$E = \frac{\text{Row total} \times \text{Column total}}{n}$$

The expected frequency table is reported in Table 13.5a.

TABLE 13.5a ● Expected Frequency for the 2 × 2 Contingency Table Between Cardiac Rehab and Repeat Heart Procedure

	Completed Cardiac Rehab	Did not Complete Cardiac Rehab
Repeated heart procedure	20.475	14.525
Did not repeat heart procedure	96.525	68.475

$$\chi^2 = \Sigma\frac{(O-E)^2}{E} = \frac{(15-20.475)^2}{20.475} + \frac{(20-14.525)^2}{14.525} + \frac{(102-96.525)^2}{96.525} + \frac{(63-68.475)^2}{68.475} = 4.276$$

Step 4: Draw the correct conclusion.

The calculated $\chi^2 = 4.276$ was within the rejection zone, so we rejected H_0. Or we could type "**=CHISQ.DIST.RT(4.276,1)**" in a blank cell in Excel, then hit **Enter**. The p value associated $\chi^2 = 4.276$ was .0387, which was less than the α level. We rejected H_0. The evidence was strong enough to infer that completion of cardiac rehab and a repeat heart procedure were not independent. The probability of obtaining such observed frequencies was .0387 under the assumption that the two variables are independent of each other. Therefore, the conclusion was that whether the patient completed the cardiac rehab was related to whether the patient had repeat heart procedures. In other words, patients who completed the cardiac rehab were less likely to have repeat heart procedures within 12 months. Such information can be extracted from the differences between the observed frequency and the expected frequency in the cells. Patients who completed the cardiac rehab had a lower than expected number of repeated heart procedures and a higher than expected number of instances of not having repeat heart procedures. Patients who did not complete the cardiac rehab had a higher than expected number of repeat heart procedures and a lower than expected number of instances of not having repeat heart procedures. Health insurance companies should provide incentives to encourage heart patients to complete cardiac rehab by switching to an active and healthy lifestyle to reduce the risk of undergoing repeat heart procedures in the future.

Many situations in which the use of chi-square is appropriate involve dichotomous nominal variables, such as gender, layoff decisions, promotion decisions, whether patients complete cardiac rehab, and whether patients have repeated heart procedures. But chi-square can also be used with many levels or categories of each nominal or ordinal variable. The next example describes a situation where each variable can be classified into multiple levels or categories.

EXAMPLE 13.6

A large national Opinions and Lifestyle Survey was used to study the smoking habits of people of age 16 years and older. In this survey, a smoking habit was classified into three categories: (1) current smoker, (2) ex-smoker, and (3) never smoked. Economic activity was also classified into three categories: (1) employed, (2) unemployed, and (3) economically inactive (e.g., students or retired people). The frequency counts based on these two variables were reported in Table 13.6. Conduct a hypothesis test to investigate whether smoking habit and economic status were independent, assuming $\alpha = .05$.

TABLE 13.6 ● Observed Frequency for the Contingency Table Between Smoking Habit and Economic Activity

	Employed	Unemployed	Economically Inactive	Row Total
Smoker	1,390	250	913	2,553
Ex-smoker	1,192	90	1,396	2,678
Never smoked	4,038	300	3,061	7,399
Column total	6,620	640	5,370	12,630

Step 1: State the pair of hypotheses.

H_0: Smoking habit and economic activity were independent.

H_1: Smoking habit and economic activity were not independent.

Step 2: Identify the rejection zone.

When $\alpha = .05$, with $df = (r - 1)(c - 1) = (3 - 1)(3 - 1) = 4$, the rejection zone for the chi-square test is $\chi^2 > 9.488$.

Step 3: Calculate the test statistic.

We need to construct the expected frequencies before we can calculate the chi-square. The expected frequency for each cell is calculated under the independence assumption,

$$E = \frac{\text{Row total} \times \text{Column total}}{n}$$

(Continued)

(Continued)

The expected frequency table is reported in Table 13.6a.

TABLE 13.6a ● Expected Frequency for the 3 × 3 Contingency Table Between Smoking Habit and Economic Activity

	Employed	Unemployed	Economically Inactive
Smoker	1,338.15	129.37	1,085.48
Ex-smoker	1,403.67	135.70	1,138.63
Never smoked	3,878.18	374.93	3,145.89

$$\chi^2 = \sum\frac{(O-E)^2}{E} = \frac{(1390-1338.15)^2}{1338.15} + \frac{(1192-1403.67)^2}{1403.67} + \frac{(4038-3878.18)^2}{3878.18} +$$

$$\frac{(250-129.37)^2}{129.37} + \frac{(90-135.70)^2}{135.70} + \frac{(300-374.93)^2}{374.93} + \frac{(913-1085.48)^2}{1085.48} + \frac{(1396-1138.63)^2}{1138.63} +$$

$$\frac{(3061-3145.89)^2}{3145.89} = 271.24$$

Step 4: Draw a conclusion.

The calculated $\chi^2 = 271.24$ was within the rejection zone, so we rejected H_0. Or we could type "**=CHISQ.DIST.RT(271.24,4)**" in a blank cell in Excel, then hit **Enter**. The *p* value associated $\chi^2 = 271.24$ was 1.7238×10^{-57}, which was less than the α level. We rejected H_0. The evidence was strong enough to support the claim that smoking habit and economic activity were not independent. The probability of obtaining such observed frequencies was 1.7238E-57 (this means the first nonzero number showed up at the 57th place after the decimal point) under the assumption that these two variables were independent of each other. Therefore, the conclusion was that smoking habit and economic activity were not independent. For each level of economic activity, some large contributors to the overall chi-square may be seen by examining the squared discrepancy between the observed frequency and the expected frequency divided by the expected frequency in each cell. Employed people had a much lower number of ex-smokers than expected. Similarly, unemployed people have a higher number of smokers than expected. Economically inactive people have a much greater number of ex-smokers than expected. This showed that people have different smoking habits across different levels of economic activity.

The simple and straightforward nature of the chi-square test for independence allows students to handle realistic numbers from a large, national-scale survey with tens of thousands of participants without any problem.

POP QUIZ

3. A significant chi-square test for independence means that
   a. there is a causal relationship between the two variables.
   b. we can predict the dependent variable with the information on the independent variable.
   c. the two variables move in the same direction.
   d. it is highly unlikely that we would obtain the observed frequencies if the two variables were independent of each other.

EXERCISE PROBLEMS

1. A university classified students' academic performance into the following five categories: (1) academic warning (GPA ≤ 2.00), (2) academic watch (GPA 2.01–2.50), (3) continuing improvement (GPA 2.51–3.00), (4) good performance (GPA 3.01–3.50), and (5) academic excellence (GPA 3.51–4.00). It was believed that students are evenly divided into these five categories. A sample of 100 students was randomly selected from the university, and their GPAs were reported in Table 13.7. Conduct a hypothesis test to see if there was a uniform distribution among students' academic performance in this university, assuming $\alpha = .05$.

TABLE 13.7 ● Number of Students in Each Academic Performance Category

GPA≤ 2.00	GPA 2.01–2.50	GPA 2.51–3.00	GPA 3.01–3.50	GPA 3.51–4.00	Total
12	16	17	31	24	100

2. A researcher tested a randomly selected sample of 606 people for color vision and then classified them according to their gender and color vision status (normal, red–green color blindness, other color blindness). The result is reported in Table 13.8. Conduct a hypothesis test to investigate if color blindness and gender were independent, assuming $\alpha = .05$.

TABLE 13.8 ● Observed Frequency for Gender and Color Vision Category

Gender	Normal	Red-Green Color Blindness	Other Color Blindness	Row Total
Male	192	42	12	246
Female	348	6	6	360
Column total	540	48	18	606

3. A medical supply company was accused of age discrimination in its decisions on workforce reduction. The personnel files indicated that there were 51 employees aged 40 years and older, and among them, 22 were laid off in the past 2 months. During the same time, there were 37 employees younger than 40 years, and among them, 8 were laid off. Conduct a hypothesis test to investigate if the reduction in workforce decision and employees' age were independent, assuming $\alpha = .05$.

Solutions

1.

Step 1: State the pair of hypotheses.

H_0: Students' GPAs were uniformly distributed across the five categories.

H_1: Students' GPAs were not uniformly distributed across the five categories.

Step 2: Identify the rejection zone.

When $\alpha = .05$, with $df = (c - 1) = 5 - 1 = 4$, the rejection zone for the chi-square test is $\chi^2 > 9.488$.

Step 3: Calculate the test statistic.

The expected frequency for a uniform distribution in the five categories is $E = np = 100(.20) = 20$.

$$\chi^2 = \Sigma\frac{(O-E)^2}{E} = \frac{(12-20)^2}{20} + \frac{(16-20)^2}{20} + \frac{(17-20)^2}{20} + \frac{(31-20)^2}{20} + \frac{(24-20)^2}{20} = 11.30$$

Step 4: Draw the correct conclusion.

The calculated $\chi^2 = 11.30$ was within the rejection zone. Therefore, we rejected H_0. Or we could type "**=CHISQ.DIST.RT(11.30,4)**" in a blank cell in Excel, then hit **Enter**. The p value associated $\chi^2 = 11.30$ was .0234, which was less than the α level. We rejected H_0. The evidence was strong enough to claim that students' GPAs were not uniformly distributed in this university across the five academic performance categories. The frequency counts were higher than expected for the two high-GPA categories but lower than expected for the low-GPA categories.

2.

Step 1: State the pair of hypotheses.

H_0: Gender and color blindness were independent.

H_1: Gender and color blindness were not independent.

Step 2: Identify the rejection zone.

When $\alpha = .05$, with $df = (r - 1)(c - 1) = 2$, the rejection zone for the chi-square test is $\chi^2 > 5.991$.

Step 3: Calculate the test statistic.

The expected frequency is

$$E = \frac{\text{Row total} \times \text{Column total}}{n}$$

The expected frequency table is reported in Table 13.8a.

TABLE 13.8a ● Expected Frequency for Gender and Color Blindness

Gender	Normal	Red-Green Color Blindness	Other Color Blindness
Male	219.21	19.49	7.31
Female	320.79	28.51	10.69

$$\chi^2 = \Sigma\frac{(O-E)^2}{E} = \frac{(192-219.21)^2}{219.21} - \frac{(42-19.49)^2}{19.49} + \frac{(12-7.31)^2}{7.31}$$

$$+\frac{(348-320.79)^2}{320.79} + \frac{(6-28.51)^2}{28.51} + \frac{(6-10.69)^2}{10.69} = 54.523$$

Step 4: Draw the correct conclusion.

The calculated $\chi^2 = 54.523$ was within the rejection zone. Therefore, we rejected H_0. Or we could type "**=CHISQ.DIST.RT(54.523,2)**" in a blank cell in Excel, then hit **Enter**. The p value associated $\chi^2 = 54.523$ was 1.447E-12, which was less than the α level. We rejected H_0. The evidence was strong enough to claim that gender and color vision status were not independent. Males had higher numbers than expected in red–green color blindness and other color blindness. Females had lower numbers than expected in red–green color blindness and other color blindness.

3. It is important to correctly translate the problem statement into a two-way contingency table. The problem statement identified two nominal variables: (1) age and (2) layoff decisions. Although age was usually treated as an interval variable, in this problem statement, age was reported as a nominal variable. The categories of age were "age 40 and older" or "younger than 40." The categories of layoff decision were "laid off" or "not laid off." The categories in each variable were mutually exclusive.

The observed frequencies for age and layoff decision are reported in Table 13.9.

TABLE 13.9 ● Observed Frequency for Age and Layoff Decision

	Laid Off	Not Laid Off	Row Total
Age ≥ 40	22	29	51
Age < 40	8	29	37
Column total	30	58	88

Step 1: State the pair of hypotheses.

H_0: Age and layoff decision were independent.

H_1: Age and layoff decision were not independent.

Step 2: Identify the rejection zone.

When $\alpha = .05$, with $df = (r - 1)(c - 1) = 1$, the rejection zone for the chi-square test is $\chi^2 > 3.841$.

Step 3: Calculate the test statistic.

The expected frequency is

$$E = \frac{\text{Row total} \times \text{Column total}}{n}$$

The expected frequency table is reported in Table 13.9a.

TABLE 13.9a ● Expected Frequency for Age and Layoff Decision

	Laid Off	Not Laid Off
Age ≥ 40	17.39	33.61
Age < 40	12.61	24.39

$$\chi^2 = \Sigma\frac{(O-E)^2}{E} = \frac{(22-17.39)^2}{17.39} + \frac{(29-33.61)^2}{33.61} + \frac{(8-12.61)^2}{12.61} + \frac{(29-24.39)^2}{24.39} = 4.411$$

Step 4: Draw the correct conclusion.

The calculated $\chi^2 = 4.411$ was within the rejection zone. Therefore, we rejected H_0. Or we could type "**=CHISQ.DIST.RT(4.411,1)**" in a blank cell in Excel, then hit **Enter**. The p value associated $\chi^2 = 4.411$ was .0357, which was less than the α level. We rejected H_0. The evidence was strong enough to claim that age and layoff decision were not independent. Employees who were 40 years and older had higher numbers than expected in the category of being laid off. In contrast, employees who were younger than 40 years had lower numbers than expected in the category of being laid off.

What You Learned

You now have learned your first nonparametric statistic: the chi-square. Chi-square tests can be used to evaluate goodness-of-fit or independence tests. The chi-square test is a distribution-free test and only requires frequency counts of the variables involved in the tests.

The essential component of a chi-square test is to calculate the expected frequency. The expected frequency for the goodness-of-fit test is simply $E = np$, where n is the sample size and p is the theoretical probability according to the null hypothesis. The expected frequency for the independence test is

$$E = \frac{\text{Row total} \times \text{Column total}}{n}$$

Each cell in the table has an observed frequency and an expected frequency. Chi-square tests are calculated as $\chi^2 = \Sigma(O - E)^2/E$. Larger calculated chi-square values indicate larger discrepancies between the observed frequencies and the expected frequencies. Expected frequencies are based on the null hypothesis. The critical value of chi-square depends on its degrees of freedom. The degrees of freedom for goodness-of-fit tests are $(c - 1)$, where c is the number of columns or categories for the variable. The degrees of freedom for independence tests are $(c - 1)(r - 1)$, where c is the number of columns and r is the number of rows.

The four-step hypothesis testing procedure for the goodness-of-fit test is stated below.

Step 1: State the pair of hypotheses.

H_0: A variable distribution fits its theoretical distribution (i.e., uniform distribution, binominal distribution, etc.).

H_1: A variable distribution does not fit its theoretical distribution.

Step 2: Identify the rejection zone.

Each degree of freedom has its own chi-square curve. The degrees of freedom for chi-square are calculated as $df = (c - 1)$. According to the critical values found in the chi-square table, the rejection zone for a chi-square test is identified as $\chi^2 > \chi^2_{(c-1)}$.

Step 3: Calculate the test statistic.

$$\text{Chi-square} = \chi^2 = \Sigma\frac{(O - E)^2}{E}$$

Step 4: Draw the correct conclusion.

The four-step hypothesis testing procedure for the independence tests is stated below.

Step 1: State the pair of hypotheses.

H_0: The two variables are independent of each other.

H_1: The two variables are not independent of each other.

Step 2: Identify the rejection zone.

Each degree of freedom has its own chi-square curve. The degrees of freedom for the chi-square test for independence are $df = (r - 1)(c - 1)$. According to the critical values found in the chi-square table, the rejection zone for a chi-square test is identified as $\chi^2 > \chi^2_{(r-1)(c-1)}$.

Step 3: Calculate the test statistic.

$$\text{Chi-square} = \chi^2 = \Sigma\frac{(O-E)^2}{E}$$

Step 4: Draw the correct conclusion.

Key Words

Chi-square goodness-of-fit tests: Chi-square goodness-of-fit tests refer to using one categorical variable to investigate whether the sample distribution of the values in a variable fits with its theoretical distribution. 420

Chi-square for independence tests: Chi-square for independence tests are statistics applied to investigate the relationship between two nominal or ordinal variables. 423

Chi-square test: Chi-square test investigates the differences between the observed frequency and the expected frequency. When the calculated chi-square value is larger than the critical value in the chi-square table, the result is statistically significant. 414

Contingency table: A contingency table refers to a two-way frequency table where frequency counts are based on a row variable and a column variable at the same time. 423

Expected frequency: Expected frequency is the theoretically predicted frequency that assumes that the variable is distributed exactly as stated in the null hypothesis. 415

Nonparametric statistics: Nonparametric statistics are statistical procedures that make no assumptions about the shape of the population distribution. They are also called distribution-free statistics. 414

Observed frequency: Observed frequency is the frequency count of the values of a variable that we observe in a sample. 415

Parametric statistics: Parametric statistics are statistical procedures that make assumptions about the shape of the population distribution and population parameters such as μ, σ, and σ^2. 414

Learning Assessment

Multiple Choice: Circle the Best Answer to Every Question

1. The degrees of freedom for the chi-square goodness-of-fit test are
   a. $(c - 1)$, where c is the number of columns
   b. $(r - 1)(c - 1)$, where r is the number of rows and c is the number of columns

   c. $(n - 2)$, where n is the sample size
   d. $(n - 1)$, where n is the sample size

2. The degrees of freedom for the chi-square for independence test are
   a. $(c - 1)$, where c is the number of columns
   b. $(r - 1)(c - 1)$, where r is the number of rows and c is the number of columns
   c. $(n - 2)$, where n is the sample size
   d. $(n - 1)$, where n is the sample size

3. The observed frequencies for a two-way contingency table
   a. are the expected number under the independence assumption
   b. need to be greater than 5
   c. have to be integers (i.e., whole numbers)
   d. carry fractions or a decimal point

4. The expected frequencies for a two-way contingency table
   a. are the observed number from the sample
   b. need to follow the rule that no more than 20% of the cells have expected frequencies less than 5
   c. have to be integers (i.e., whole numbers)
   d. may not add up to the same row totals as the observed frequencies

5. Which one of the following statistics is appropriate to test the relationship between two nominal or ordinal variables?
   a. Regression
   b. Pearson's r
   c. ANOVA
   d. Chi-square

6. Which pair of distributions has a different curve for every degree of freedom?
   a. t distribution and χ^2 distribution
   b. Z distribution and t distribution
   c. Z distribution and F distribution
   d. Z distribution and χ^2 distribution

7. The rejection zone for a χ^2 test
   a. is located on the right side of the distribution
   b. is located on the left side of the distribution
   c. is located on both tails of the distribution
   d. depends on what hypotheses are being tested

Free Response Questions

8. The results of the male and female students' grades from multiple statistics classes are reported in Table 13.10. Conduct a hypothesis test to verify if statistics grades and students' gender are independent.

TABLE 13.10 ● Male and Female Students' Grades From Multiple Statistics Classes

Student	A	B	C	D	F
Male	25	37	44	15	7
Female	31	55	66	7	5

9. A large retail chain is under attack for racial discrimination in promotions. The personnel records in the corporate headquarters show that 3,215 white employees were hired, and among them, 966 were promoted in the past 6 months. During the same time, 288 black employees were hired, and among them, 74 were promoted; 317 Hispanic employees were hired, and among them, 81 were promoted; 39 employees whose racial background was classified as other were hired, and among them, 5 were promoted. Conduct a hypothesis test to see if employees' racial background and promotion decisions were independent in this large retail chain.

10. The Type A personality is known to be related to some health problems. Type A personality is characterized as competitive, achievement oriented, talking fast, eating fast, impatient with a painful awareness of time urgency, and free-floating hostility. Researchers studied the relationship between Type A personality and heart health. They randomly selected 200 adults. The 200 adults were classified by their Type A personality and history of heart problems, as reported in Table 13.11. Conduct a hypothesis test to see if Type A personality and heart problems are independent.

TABLE 13.11 ● Type A Personality and Heart Problems

	Heart Problems	No Heart Problems	Row Total
Type A personality	25	55	80
Not Type A personality	22	98	120
Column total	47	153	200

Answers to Pop Quiz Questions

1. c
2. a
3. d

APPENDIX A

The Standard Normal Distribution Table (*Z* Table)

The values inside the table represent the areas under the curve to the left of the Z values *or* the probability of Z values smaller than the specified z value, $P(Z < z)$. The column represents Z values to one place after the decimal point and the row represents Z values to the second place after the decimal point. The intersection of the column and the row show the probability associated with that Z value, as shown in Figure A.1.

1. $P(Z < 2) = .9772$

FIGURE A.1 ⬢ The *Z* Table Provides Probability to the Left of the Specified Value

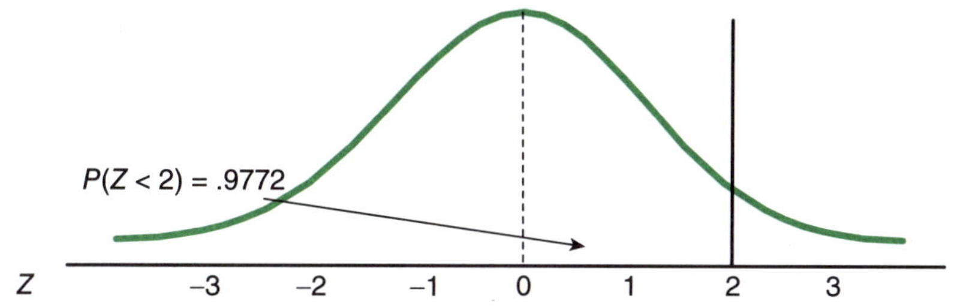

When the question asks for the areas to the right of the Z values *or* the probability of Z values larger than the specified z value, $P(Z > z)$, the probability under the curve is 1, $P(Z > z)$ can be calculated by $1 - P(Z < z)$, as shown in Figure A.2.

2. $P(Z > -1) = 1 - P(Z < -1) = 1 - .1587 = .8413$

FIGURE A.2 ⬢ The Area to the Right of the Specified *Z* Value, $P(Z > z)$ Is Calculated as $1 - P(Z < z)$

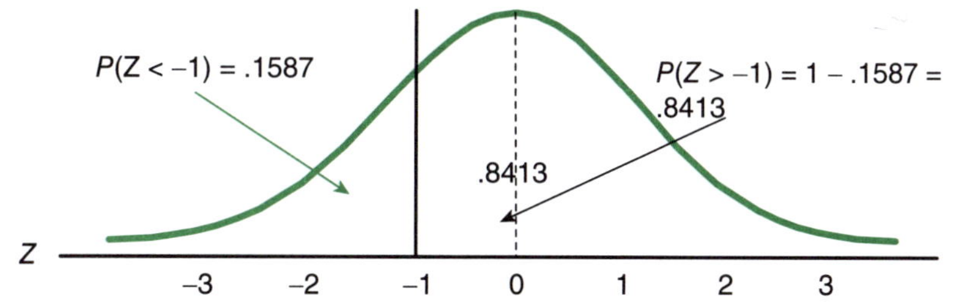

STANDARD NORMAL DISTRIBUTION TABLE: *Z* TABLE

Z	0.00	0.01	0.02	0.03	0.04	0.05	0.06	0.07	0.08	0.09
0.0	0.5000	0.5040	0.5080	0.5120	0.5160	0.5199	0.5239	0.5279	0.5319	0.5359
0.1	0.5398	0.5438	0.5478	0.5517	0.5557	0.5596	0.5636	0.5675	0.5714	0.5753
0.2	0.5793	0.5832	0.5871	0.5910	0.5948	0.5987	0.6026	0.6064	0.6103	0.6141
0.3	0.6179	0.6217	0.6255	0.6293	0.6331	0.6368	0.6406	0.6443	0.6480	0.6517
0.4	0.6554	0.6591	0.6628	0.6664	0.6700	0.6736	0.6772	0.6808	0.6844	0.6879
0.5	0.6915	0.6950	0.6985	0.7019	0.7054	0.7088	0.7123	0.7157	0.7190	0.7224
0.6	0.7257	0.7291	0.7324	0.7357	0.7389	0.7422	0.7454	0.7486	0.7517	0.7549
0.7	0.7580	0.7611	0.7642	0.7673	0.7704	0.7734	0.7764	0.7794	0.7823	0.7852
0.8	0.7881	0.7910	0.7939	0.7967	0.7995	0.8023	0.8051	0.8078	0.8106	0.8133
0.9	0.8159	0.8186	0.8212	0.8238	0.8264	0.8289	0.8315	0.8340	0.8365	0.8389
1.0	0.8413	0.8438	0.8461	0.8485	0.8508	0.8531	0.8554	0.8577	0.8599	0.8621
1.1	0.8643	0.8665	0.8686	0.8708	0.8729	0.8749	0.8770	0.8790	0.8810	0.8830
1.2	0.8849	0.8869	0.8888	0.8907	0.8925	0.8944	0.8962	0.8980	0.8997	0.9015
1.3	0.9032	0.9049	0.9066	0.9082	0.9099	0.9115	0.9131	0.9147	0.9162	0.9177
1.4	0.9192	0.9207	0.9222	0.9236	0.9251	0.9265	0.9279	0.9292	0.9306	0.9319
1.5	0.9332	0.9345	0.9357	0.9370	0.9382	0.9394	0.9406	0.9418	0.9429	0.9441
1.6	0.9452	0.9463	0.9474	0.9484	0.9495	0.9505	0.9515	0.9525	0.9535	0.9545
1.7	0.9554	0.9564	0.9573	0.9582	0.9591	0.9599	0.9608	0.9616	0.9625	0.9633
1.8	0.9641	0.9649	0.9656	0.9664	0.9671	0.9678	0.9686	0.9693	0.9699	0.9706
1.9	0.9713	0.9719	0.9726	0.9732	0.9738	0.9744	0.9750	0.9756	0.9761	0.9767
2.0	0.9772	0.9778	0.9783	0.9788	0.9793	0.9798	0.9803	0.9808	0.9812	0.9817
2.1	0.9821	0.9826	0.9830	0.9834	0.9838	0.9842	0.9846	0.9850	0.9854	0.9857
2.2	0.9861	0.9864	0.9868	0.9871	0.9875	0.9878	0.9881	0.9884	0.9887	0.9890
2.3	0.9893	0.9896	0.9898	0.9901	0.9904	0.9906	0.9909	0.9911	0.9913	0.9916
2.4	0.9918	0.9920	0.9922	0.9925	0.9927	0.9929	0.9931	0.9932	0.9934	0.9936
2.5	0.9938	0.9940	0.9941	0.9943	0.9945	0.9946	0.9948	0.9949	0.9951	0.9952
2.6	0.9953	0.9955	0.9956	0.9957	0.9959	0.9960	0.9961	0.9962	0.9963	0.9964
2.7	0.9965	0.9966	0.9967	0.9968	0.9969	0.9970	0.9971	0.9972	0.9973	0.9974
2.8	0.9974	0.9975	0.9976	0.9977	0.9977	0.9978	0.9979	0.9979	0.9980	0.9981
2.9	0.9981	0.9982	0.9982	0.9983	0.9984	0.9984	0.9985	0.9985	0.9986	0.9986
3.0	0.9987	0.9987	0.9987	0.9988	0.9988	0.9989	0.9989	0.9989	0.9990	0.9990

STANDARD NORMAL DISTRIBUTION TABLE: Z TABLE

Z	0.00	-0.01	-0.02	-0.03	-0.04	-0.05	-0.06	-0.07	-0.08	-0.09
0.0	0.5	0.496	0.492	0.488	0.484	0.4801	0.4761	0.4721	0.4681	0.4641
-0.1	0.4602	0.4562	0.4522	0.4483	0.4443	0.4404	0.4364	0.4325	0.4286	0.4247
-0.2	0.4207	0.4168	0.4129	0.409	0.4052	0.4013	0.3974	0.3936	0.3897	0.3859
-0.3	0.3821	0.3783	0.3745	0.3707	0.3669	0.3632	0.3594	0.3557	0.352	0.3483
-0.4	0.3446	0.3409	0.3372	0.3336	0.33	0.3264	0.3228	0.3192	0.3156	0.3121
-0.5	0.3085	0.305	0.3015	0.2981	0.2946	0.2912	0.2877	0.2843	0.281	0.2776
-0.6	0.2743	0.2709	0.2676	0.2643	0.2611	0.2578	0.2546	0.2514	0.2483	0.2451
-0.7	0.242	0.2389	0.2358	0.2327	0.2296	0.2266	0.2236	0.2206	0.2177	0.2148
-0.8	0.2119	0.209	0.2061	0.2033	0.2005	0.1977	0.1949	0.1922	0.1894	0.1867
-0.9	0.1841	0.1814	0.1788	0.1762	0.1736	0.1711	0.1685	0.166	0.1635	0.1611
-1.0	0.1587	0.1562	0.1539	0.1515	0.1492	0.1469	0.1446	0.1423	0.1401	0.1379
-1.1	0.1357	0.1335	0.1314	0.1292	0.1271	0.1251	0.123	0.121	0.119	0.117
-1.2	0.1151	0.1131	0.1112	0.1093	0.1075	0.1056	0.1038	0.102	0.1003	0.0985
-1.3	0.0968	0.0951	0.0934	0.0918	0.0901	0.0885	0.0869	0.0853	0.0838	0.0823
-1.4	0.0808	0.0793	0.0778	0.0764	0.0749	0.0735	0.0721	0.0708	0.0694	0.0681
-1.5	0.0668	0.0655	0.0643	0.063	0.0618	0.0606	0.0594	0.0582	0.0571	0.0559
-1.6	0.0548	0.0537	0.0526	0.0516	0.0505	0.0495	0.0485	0.0475	0.0465	0.0455
-1.7	0.0446	0.0436	0.0427	0.0418	0.0409	0.0401	0.0392	0.0384	0.0375	0.0367
-1.8	0.0359	0.0351	0.0344	0.0336	0.0329	0.0322	0.0314	0.0307	0.0301	0.0294
-1.9	0.0287	0.0281	0.0274	0.0268	0.0262	0.0256	0.025	0.0244	0.0239	0.0233
-2.0	0.0228	0.0222	0.0217	0.0212	0.0207	0.0202	0.0197	0.0192	0.0188	0.0183
-2.1	0.0179	0.0174	0.017	0.0166	0.0162	0.0158	0.0154	0.015	0.0146	0.0143
-2.2	0.0139	0.0136	0.0132	0.0129	0.0125	0.0122	0.0119	0.0116	0.0113	0.011
-2.3	0.0107	0.0104	0.0102	0.0099	0.0096	0.0094	0.0091	0.0089	0.0087	0.0084
-2.4	0.0082	0.008	0.0078	0.0075	0.0073	0.0071	0.0069	0.0068	0.0066	0.0064
-2.5	0.0062	0.006	0.0059	0.0057	0.0055	0.0054	0.0052	0.0051	0.0049	0.0048
-2.6	0.0047	0.0045	0.0044	0.0043	0.0041	0.004	0.0039	0.0038	0.0037	0.0036
-2.7	0.0035	0.0034	0.0033	0.0032	0.0031	0.003	0.0029	0.0028	0.0027	0.0026
-2.8	0.0026	0.0025	0.0024	0.0023	0.0023	0.0022	0.0021	0.0021	0.002	0.0019
-2.9	0.0019	0.0018	0.0018	0.0017	0.0016	0.0016	0.0015	0.0015	0.0014	0.0014
-3.0	0.0013	0.0013	0.0013	0.0012	0.0012	0.0011	0.0011	0.0011	0.001	0.001

APPENDIX B

The *t*-Distribution Table (*t* Table)

	One-Tailed α Level							
	.25	.20	.15	.10	.05	.025	.01	.005
	Two-Tailed α Level							
df	.50	.40	.30	.20	.10	.05	.02	.01
1	1	1.376	1.963	3.078	6.314	12.71	31.82	63.66
2	0.816	1.061	1.386	1.886	2.92	4.303	6.965	9.925
3	0.765	0.978	1.25	1.638	2.353	3.182	4.541	5.841
4	0.741	0.941	1.19	1.533	2.132	2.776	3.747	4.604
5	0.727	0.92	1.156	1.476	2.015	2.571	3.365	4.032
6	0.718	0.906	1.134	1.44	1.943	2.447	3.143	3.707
7	0.711	0.896	1.119	1.415	1.895	2.365	2.998	3.499
8	0.706	0.889	1.108	1.397	1.86	2.306	2.896	3.355
9	0.703	0.883	1.1	1.383	1.833	2.262	2.821	3.25
10	0.7	0.879	1.093	1.372	1.812	2.228	2.764	3.169
11	0.697	0.876	1.088	1.363	1.796	2.201	2.718	3.106
12	0.695	0.873	1.083	1.356	1.782	2.179	2.681	3.055
13	0.694	0.87	1.079	1.35	1.771	2.16	2.65	3.012
14	0.692	0.868	1.076	1.345	1.761	2.145	2.624	2.977
15	0.691	0.866	1.074	1.341	1.753	2.131	2.602	2.947
16	0.69	0.865	1.071	1.337	1.746	2.12	2.583	2.921
17	0.689	0.863	1.069	1.333	1.74	2.11	2.567	2.898

(Continued)

(Continued)

	One-Tailed α Level							
	.25	.20	.15	.10	.05	.025	.01	.005
	Two-Tailed α Level							
df	.50	.40	.30	.20	.10	.05	.02	.01
18	0.688	0.862	1.067	1.33	1.734	2.101	2.552	2.878
19	0.688	0.861	1.066	1.328	1.729	2.093	2.539	2.861
20	0.687	0.86	1.064	1.325	1.725	2.086	2.528	2.845
21	0.686	0.859	1.063	1.323	1.721	2.08	2.518	2.831
22	0.686	0.858	1.061	1.321	1.717	2.074	2.508	2.819
23	0.685	0.858	1.06	1.319	1.714	2.069	2.5	2.807
24	0.685	0.857	1.059	1.318	1.711	2.064	2.492	2.797
25	0.684	0.856	1.058	1.316	1.708	2.06	2.485	2.787
26	0.684	0.856	1.058	1.315	1.706	2.056	2.479	2.779
27	0.684	0.855	1.057	1.314	1.703	2.052	2.473	2.771
28	0.683	0.855	1.056	1.313	1.701	2.048	2.467	2.763
29	0.683	0.854	1.055	1.311	1.699	2.045	2.462	2.756
30	0.683	0.854	1.055	1.31	1.697	2.042	2.457	2.75
40	0.681	0.851	1.05	1.303	1.684	2.021	2.423	2.704
50	0.679	0.849	1.047	1.299	1.676	2.009	2.403	2.678
60	0.679	0.848	1.045	1.296	1.671	2	2.39	2.66
80	0.678	0.846	1.043	1.292	1.664	1.99	2.374	2.639
100	0.677	0.845	1.042	1.29	1.66	1.984	2.364	2.626
120	0.677	0.845	1.041	1.289	1.658	1.98	2.358	2.617
∞	0.674	0.842	1.036	1.282	1.645	1.96	2.326	2.576

APPENDIX C

The *F* Table

df_1/df_2	1	2	3	4	5	6	7	8	9	10	11	12	15	20	24	30	40	60	100
1	161.4	199.5	215.7	224.6	230.2	234.0	236.8	238.9	240.5	241.9	243.0	243.9	245.9	248.0	249.1	250.1	251.1	252.2	253.0
2	18.51	19.0	19.16	19.25	19.30	19.33	19.35	19.37	19.38	19.40	19.40	19.41	19.43	19.45	19.45	19.46	19.47	19.48	19.49
3	10.13	9.55	9.28	9.12	9.01	8.94	8.89	8.85	8.81	8.79	8.76	8.74	8.7	8.66	8.64	8.62	8.59	8.57	8.56
4	7.71	6.94	6.59	6.39	6.26	6.16	6.09	6.04	6.00	5.96	5.94	5.91	5.86	5.80	5.77	5.75	5.72	5.69	5.66
5	6.61	5.79	5.41	5.19	5.05	4.95	4.88	4.82	4.77	4.74	4.70	4.68	4.62	4.56	4.53	4.50	4.46	4.43	4.41
6	5.99	5.14	4.76	4.53	4.39	4.28	4.21	4.15	4.10	4.06	4.03	4.00	3.94	3.87	3.84	3.81	3.77	3.74	3.71
7	5.59	4.74	4.35	4.12	3.97	3.87	3.79	3.73	3.68	3.64	3.60	3.57	3.51	3.44	3.41	3.38	3.34	3.30	3.27
8	5.32	4.46	4.07	3.84	3.69	3.58	3.50	3.44	3.39	3.35	3.31	3.28	3.22	3.15	3.12	3.08	3.04	3.01	2.97
9	5.12	4.26	3.86	3.63	3.48	3.37	3.29	3.23	3.18	3.14	3.10	3.07	3.01	2.94	2.90	2.86	2.83	2.79	2.76
10	4.96	4.10	3.71	3.48	3.33	3.22	3.14	3.07	3.02	2.98	2.94	2.91	2.85	2.77	2.74	2.70	2.66	2.62	2.59
11	4.84	3.98	3.59	3.36	3.20	3.09	3.01	2.95	2.90	2.85	2.82	2.79	2.72	2.65	2.61	2.57	2.53	2.49	2.46
12	4.75	3.89	3.49	3.26	3.11	3.00	2.91	2.85	2.80	2.75	2.72	2.69	2.62	2.54	2.51	2.47	2.43	2.38	2.35
13	4.67	3.81	3.41	3.18	3.03	2.92	2.83	2.77	2.71	2.67	2.63	2.60	2.53	2.46	2.42	2.38	2.34	2.30	2.26
14	4.60	3.74	3.34	3.11	2.96	2.85	2.76	2.70	2.65	2.60	2.57	2.53	2.46	2.39	2.35	2.31	2.27	2.22	2.19
15	4.54	3.68	3.29	3.06	2.90	2.79	2.71	2.64	2.59	2.54	2.51	2.48	2.40	2.33	2.29	2.25	2.20	2.16	2.12

(Continued)

(Continued)

df_1/df_2	1	2	3	4	5	6	7	8	9	10	11	12	15	20	24	30	40	60	100
16	4.49	3.63	3.24	3.01	2.85	2.74	2.66	2.59	2.54	2.49	2.46	2.42	2.35	2.28	2.24	2.19	2.15	2.11	2.07
17	4.45	3.59	3.20	2.96	2.81	2.70	2.61	2.55	2.49	2.45	2.41	2.38	2.31	2.23	2.19	2.15	2.10	2.06	2.02
18	4.41	3.55	3.16	2.93	2.77	2.66	2.58	2.51	2.46	2.41	2.37	2.34	2.27	2.19	2.15	2.11	2.06	2.02	1.98
19	4.38	3.52	3.13	2.90	2.74	2.63	2.54	2.48	2.42	2.38	2.34	2.31	2.23	2.16	2.11	2.07	2.03	1.98	1.94
20	4.35	3.49	3.10	2.87	2.71	2.60	2.51	2.45	2.39	2.35	2.31	2.28	2.20	2.12	2.08	2.04	1.99	1.95	1.91
21	4.32	3.47	3.07	2.84	2.68	2.57	2.49	2.42	2.37	2.32	2.28	2.25	2.18	2.10	2.05	2.01	1.96	1.92	1.88
22	4.30	3.44	3.05	2.82	2.66	2.55	2.46	2.40	2.34	2.30	2.26	2.23	2.15	2.07	2.03	1.98	1.94	1.89	1.85
23	4.28	3.42	3.03	2.80	2.64	2.53	2.44	2.37	2.32	2.27	2.24	2.20	2.13	2.05	2.01	1.96	1.91	1.86	1.82
24	4.26	3.40	3.01	2.78	2.62	2.51	2.42	2.36	2.30	2.25	2.22	2.18	2.11	2.03	1.98	1.94	1.89	1.84	1.80
25	4.24	3.39	2.99	2.76	2.60	2.49	2.40	2.34	2.28	2.24	2.20	2.16	2.09	2.01	1.96	1.92	1.87	1.82	1.78
26	4.23	3.37	2.98	2.74	2.59	2.47	2.39	2.32	2.27	2.22	2.18	2.15	2.07	1.99	1.95	1.90	1.85	1.80	1.76
27	4.21	3.35	2.96	2.73	2.57	2.46	2.37	2.31	2.25	2.20	2.17	2.13	2.06	1.97	1.93	1.88	1.84	1.79	1.74
28	4.20	3.34	2.95	2.71	2.56	2.45	2.36	2.29	2.24	2.19	2.15	2.12	2.04	1.96	1.91	1.87	1.82	1.77	1.73
29	4.18	3.33	2.93	2.70	2.55	2.43	2.35	2.28	2.22	2.18	2.14	2.10	2.03	1.94	1.90	1.85	1.81	1.75	1.71
30	4.17	3.32	2.92	2.69	2.53	2.42	2.33	2.27	2.21	2.16	2.13	2.09	2.01	1.93	1.89	1.84	1.79	1.74	1.70
40	4.08	3.23	2.84	2.61	2.45	2.34	2.25	2.18	2.12	2.08	2.04	2.00	1.92	1.84	1.79	1.74	1.69	1.64	1.59
50	4.03	3.18	2.79	2.56	2.40	2.29	2.20	2.13	2.07	2.03	1.99	1.95	1.87	1.78	1.74	1.69	1.63	1.58	1.52
60	4.00	3.15	2.76	2.53	2.37	2.25	2.17	2.10	2.04	1.99	1.95	1.92	1.84	1.75	1.70	1.65	1.59	1.53	1.48
80	3.96	3.11	2.72	2.49	2.33	2.21	2.13	2.06	2.00	1.95	1.91	1.88	1.79	1.70	1.65	1.60	1.54	1.48	1.43
100	3.94	3.09	2.70	2.46	2.31	2.19	2.10	2.03	1.97	1.93	1.89	1.85	1.77	1.68	1.63	1.57	1.52	1.45	1.39

Note. Critical values of the *F* distribution for $\alpha = .05$; $P\{F > F\ .05\ (df_1, df_2)\} = .05$. df_1 = numerator degrees of freedom; df_2 = denominator degrees of freedom.

APPENDIX D

The Critical Values of Pearson's Correlation Table (*r* Table)

	Level of Significance for One-Tailed Test			
	.05	.025	.01	.005
	Level of Significance for Two-Tailed Test			
$df = n - 2$	.10	.05	.02	.01
1	.988	.997	.9995	.9999
2	.900	.950	.980	.990
3	.805	.878	.934	.959
4	.729	.811	.882	.917
5	.669	.754	.833	.874
6	.622	.707	.789	.834
7	.582	.666	.75	.798
8	.549	.632	.716	.765
9	.521	.602	.685	.735
10	.497	.576	.658	.708
11	.476	.553	.634	.684
12	.458	.532	.612	.661
13	.441	.514	.592	.641
14	.426	.497	.574	.628
15	.412	.482	.558	.606
16	.400	.468	.542	.590

(Continued)

(Continued)

	Level of Significance for One-Tailed Test			
	.05	.025	.01	.005
	Level of Significance for Two-Tailed Test			
df = *n* - 2	.10	.05	.02	.01
17	.389	.456	.528	.575
18	.378	.444	.516	.561
19	.369	.433	.503	.549
20	.360	.423	.492	.537
21	.352	.413	.482	.526
22	.344	.404	.472	.515
23	.337	.396	.462	.505
24	.330	.388	.453	.495
25	.323	.381	.445	.487
26	.317	.374	.437	.479
27	.311	.367	.430	.471
28	.306	.361	.423	.463
29	.301	.355	.416	.456
30	.296	.349	.409	.449
35	.275	.325	.381	.418
40	.257	.304	.358	.393
45	.243	.288	.338	.372
50	.231	.273	.322	.354
60	.211	.250	.295	.325
70	.195	.232	.274	.302
80	.183	.217	.256	.284
90	.173	.205	.242	.267
100	.164	.195	.23	.254

APPENDIX E

The Critical Values for the Chi-Square Tests (χ^2 Table)

	Right-Tailed α Level				
df	.1	.05	.025	.01	.005
1	2.706	3.841	5.024	6.635	7.879
2	4.605	5.991	7.378	9.210	10.597
3	6.251	7.815	9.348	11.345	12.838
4	7.779	9.488	11.143	13.277	14.86
5	9.236	11.07	12.833	15.086	16.75
6	10.645	12.592	14.449	16.812	18.548
7	12.017	14.067	16.013	18.475	20.278
8	13.362	15.507	17.535	20.09	21.955
9	14.684	16.919	19.023	21.666	23.589
10	15.987	18.307	20.483	23.209	25.188
11	17.275	19.675	21.920	24.725	26.757
12	18.549	21.026	23.337	26.217	28.300
13	19.812	22.362	24.736	27.688	29.819
14	21.064	23.685	26.119	29.141	31.319
15	22.307	24.996	27.488	30.578	32.801
16	23.542	26.296	28.845	32.000	34.267
17	24.769	27.587	30.191	33.409	35.718
18	25.989	28.869	31.526	34.805	37.156

(Continued)

(Continued)

	Right-Tailed α Level				
df	.1	.05	.025	.01	.005
19	27.204	30.144	32.852	36.191	38.582
20	28.412	31.410	34.170	37.566	39.997
21	29.615	32.671	35.479	38.932	41.401
22	30.813	33.924	36.781	40.289	42.796
23	32.007	35.172	38.076	41.638	44.181
24	33.196	36.415	39.364	42.980	45.559
25	34.382	37.652	40.646	44.314	46.928
26	35.563	38.885	41.923	45.642	48.290
27	36.741	40.113	43.195	46.963	49.645
28	37.916	41.337	44.461	48.278	50.993
29	39.087	42.557	45.722	49.588	52.336
30	40.256	43.773	46.979	50.892	53.672
40	51.805	55.758	59.342	63.691	66.766
50	63.167	67.505	71.420	76.154	79.490
60	74.397	79.082	83.298	88.379	91.952
70	85.527	90.531	95.023	100.425	104.215
80	96.578	101.879	106.629	112.329	116.321
90	107.565	113.145	118.136	124.116	128.299
100	118.498	124.342	129.561	135.807	140.169

ODD ANSWERS TO LEARNING ASSESSMENTS

Straightforward Statistics with Excel®, Second Edition

By Chieh-Chen Bowen

Solutions to Exercises in:
Chapter 1: Introduction to Statistics
Chapter 2: Summarizing, Organizing Data, and Measures of Central Tendency
Chapter 3: Measures of Variability
Chapter 4: Standard *Z* Scores
Chapter 5: Basic Principles of Probability
Chapter 6: The Central Limit Theorem
Chapter 7: Hypothesis Testing
Chapter 8: One-Sample *t* Test and Dependent-Sample *t* Test
Chapter 9: Independent-Samples *t* Tests
Chapter 10: Correlation
Chapter 11: Simple Regression
Chapter 12: One-Way Analysis of Variance
Chapter 13: Chi-Square Tests for Goodness of Fit and Independence

CHAPTER 1: INTRODUCTION TO STATISTICS

1. D

3. C

5. A

7. D

9. $X = 3, 4, 5, and\, 7$

   $\Sigma X = 19$

   $[\Sigma X]^2 = 19^2 = 361$

   $\Sigma X^2 = (3^2) + (4^2) + (5^2) + (7^2) = 9 + 16 + 25 + 49 = 99$

11.

X	*Y*	*XY*
3	−1	−3
4	0	0
5	1	5
7	2	14
		16

$\Sigma XY = 16$

$(\Sigma XY)^2 = 16^2 = 256$

13.

X	*Y*	*XY*
4	−1	−4
5	−1	−5
6	1	6
9	2	18
		15

$(\Sigma XY)^2 = 15^2 = 225$

X	*X* − 5	*Y*	*Y* + 1	[*X* − 5] [*Y* + 1]
4	−1	−1	0	0
5	0	−1	0	0

6	1	1	2	2
9	4	2	3	12
				14

$\Sigma[X-5][Y+1] = 14$

CHAPTER 2: SUMMARIZING, ORGANIZING DATA, AND MEASURES OF CENTRAL TENDENCY

1. A
3. B
5. C
7. A
9. C
11. C

13a.

TABLE 2.16a ● Frequency Table of Traffic Citations With Midpoint

X (Citation Amount in Dollars)	*f* (Frequency)	$X_{midpoint}$	$fX_{midpoint}$
90–109	12	99.5	1194
130–149	25	139.5	3487.5
150–169	46	159.5	7337
170–189	105	179.5	18,847.5
190–209	852	199.5	169,974
210–229	907	219.5	199,086.5
230–249	981	239.5	234,949.5
			634,876

Because the citation amount is listed in intervals, we need to find the midpoint of the interval. Then conduct $\Sigma fX_{midpoint}$ as shown in Table 2.16a.

13b.

FIGURE 2.24 ● Histogram of Traffic Citation Amount

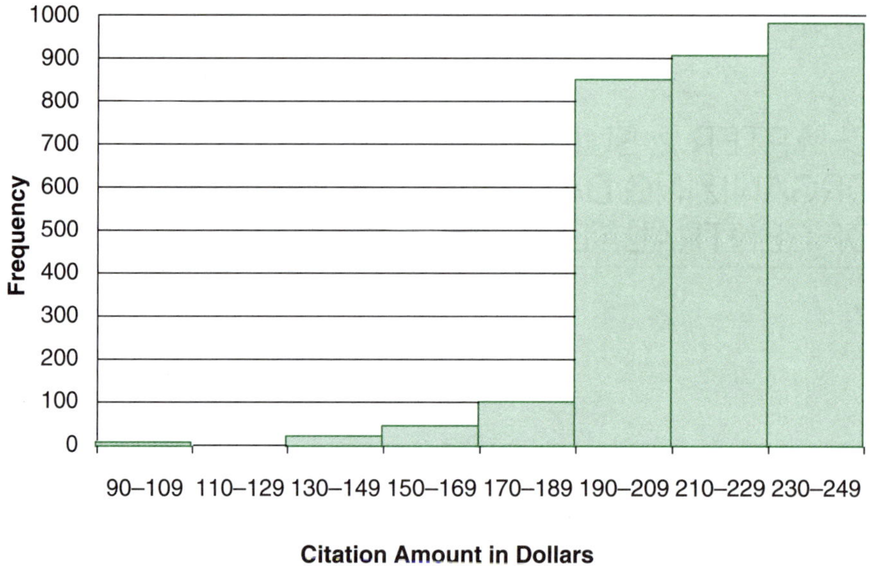

According to the histogram in Figure 2.24, most of the traffic citations are on the high end of the distribution, with a tail toward the low end, therefore it is negatively skewed.

13c. According to the table, the interval of 230–249 has the highest frequency with 981, therefore the mode is 230–249.

Because the citations are in a frequency table, we can determine the median by calculating the relative frequency.

First, we determine n by adding the frequencies to get 2928. We use this to determine the relative frequency by taking the frequency of each interval and dividing by n. This is displayed in Table 2.16b.

TABLE 2.16b ● Cumulative Relative Frequency of Traffic Citation Amounts

X	*f*	Relative Frequency	Cumulative Relative Frequency
90–109	12	.004	.004
130–149	25	.009	.015

150–169	46	.016	.029
170–189	105	.036	.065
190–209	852	.291	.356
210–229	907	.310	.666
230–249	981	.335	1.000
	2928		

The first cumulative relative frequency over .50 is .666 and its corresponding interval is 210–229. Therefore, the median is 210–229.

When estimated mean is calculated from a frequency table with equal intervals, the formula needs to be modified to $\overline{X} = \frac{\Sigma f X_{midpoint}}{n}$. We are able to plug in these numbers from what we have calculated previously in this question.

$$\overline{X} = \frac{\Sigma f X_{midpoint}}{n} = \frac{634{,}876}{2928} = 216.8$$

15.

FIGURE 2.26 ● Bar Graph of College Students' Hand Washing Behavior

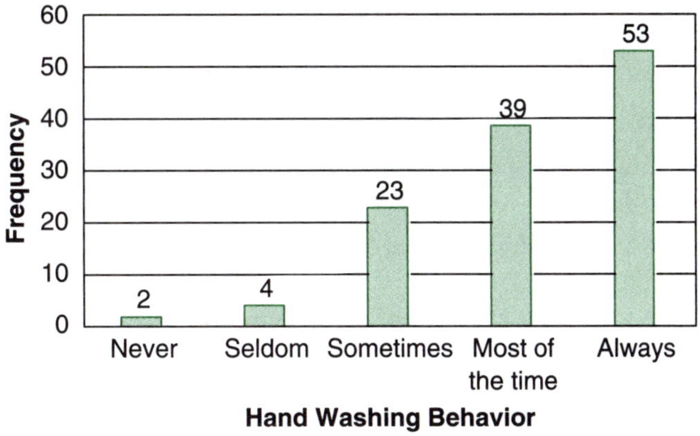

According to the bar chart in Figure 2.26, most of the hand washing behavior is on the high end of the distribution, with a tail toward the low end, therefore it is negatively skewed.

CHAPTER 3: MEASURES OF VARIABILITY

1. D
3. B
5. C
7. B
9. A
11. B
13.

TABLE 3.10a ● Women's Shoe Sizes with Columns of fX, $(X - \bar{X})$, $(X - \bar{X})^2$, and $f(X - \bar{X})^2$

X	f	fX	$X - \bar{X}$	$(X - \bar{X})^2$	$f(X - \bar{X})^2$
5	2	10	−2.8	7.84	15.68
6	5	30	−1.8	3.24	16.2
7	14	98	−0.8	0.64	8.96
8	16	128	.2	0.04	0.64
9	7	63	1.2	1.44	10.08
10	5	50	2.2	4.84	24.2
11	1	11	3.2	10.24	10.24
	50	390			86

The measure of variability are range, variance, and standard deviation.

$$\text{Range is Maximum} - \text{minimum} = 11 - 5 = 6$$

$$\bar{X} = 390 / 50 = 7.8$$

The SS formula in a frequency table needs to be modified to $SS = \Sigma f(\bar{X} - X)^2$. According to Table 3.10a, $SS = 86$.

Sample Variance $s^2 = SS/(n - 1) = 86/\ (50 - 1) = 86/49 = 1.8$

Sample standard deviation = $\sqrt{\text{variance}} = \sqrt{1.8} = 1.3$

15.

FIGURE 3.8 ● Calculating $\bar{X}$, $(X - \bar{X})$, $(X - \bar{X})^2$, and *SS* in Excel

	A	B	C
1	Students Exam Grades	X-mean	X-mean^2
2	82	8.9	79.3
3	45	-28.1	789.3
4	82	8.9	79.3
5	84	10.9	118.9
6	73	-0.1	0.0
7	74	0.9	0.8
8	93	19.9	396.2
9	88	14.9	222.2
10	83	9.9	98.1
11	54	-19.1	364.6
12	80	6.9	47.7
13	80	6.9	47.7
14	87	13.9	193.3
15	44	-29.1	846.5
16	21	-52.1	2713.9
17	50	-23.1	533.4
18	94	20.9	437.0
19	94	20.9	437.0
20	63	-10.1	101.9
21	77	3.9	15.2
22	87	13.9	193.3
23			
24	73.1	mean	7715.8

Range is Maximum – minimum = 6 – 1 = 5

To calculate the variance, we need to calculate *SS*. To calculate *SS*, we need to know $\bar{X}$ first. Start by inputting homework grades in Column A. In A24, type "=AVERAGE(A2:A22) to obtain the mean homework grade. In Column B, you can label it as (X-mean) as seen in Figure 3.8. To obtain $(X - \bar{X})$, in B2, type "=A2-A$24". Move the cursor to the lower right corner of B2 until + shows up; then left click and hold the mouse and drag it to B22 and release the mouse. In Column C, you can label it as (X-mean^2) as shown in Figure 3.8. In C2, type "=B2^2", which shows the operation to obtain $(X - \bar{X})^2$. Then move the cursor to the lower right corner of C2; left click and drag. Finally, to determine the *SS*, in C24, type "=SUM(C2:C22)".

According to Figure 3.8, $SS = 7715.8$.

Sample variance $s^2 = SS / (n-1) = 7715.8 / (21-1) = 7715.8 / 20 = 385.79$

Sample standard deviation $= \sqrt{\text{variance}} = \sqrt{385.79} = 19.6$

CHAPTER 4: STANDARD *Z* SCORES

1. D
3. A
5. C
7. C
9. A
11.

$$\text{Sample} = Z = \frac{(X - \bar{X})}{s}$$

$$\text{English} = \frac{(40-30)}{8} = \frac{10}{8} = 1.25$$

$$\text{Math} = \frac{(50-55)}{5} = \frac{(-5)}{5} = -1$$

Based on the *Z* score, Robin scored 1.25 standard deviations above the mean in English and for Math, Robin scored 1 standard deviation below the mean, therefore he did better in English.

13.

$$Z = \frac{(1781-1427)}{354} = \frac{354}{354} = 1$$

Because 1 standard deviation above the mean of a normal curve is 34% above 50%, Alex paid 84% more than other people. By subtracting 84 from 100 (the top of the curve), you can determine that 16% of people paid more than Alex on their car insurance.

CHAPTER 5: BASIC PRINCIPLES OF PROBABILITY

1. A

3. B

5. B

7. The question asks for $P(21 < X < 31)$. Because there is no direct connection between raw scores and probabilities, we can use the Z formula to transform raw scores to Z scores.

   Use the Z formula to transform raw scores to Z scores.

$$Z = \frac{(X - \mu)}{\sigma} = \frac{(31 - 27.1)}{4.5} = \frac{3.9}{4.5} = .87$$

$$Z = \frac{(X - \mu)}{\sigma} = \frac{(21 - 27.1)}{4.5} = \frac{-6.1}{4.5} = 1.36$$

   $P(21 < X < 31) = P(-1.36 < Z < .87)$

   $P(-1.36 < Z < .87)$ can be calculated as $P(Z < .87) - P(Z < -1.36)$ $= .8078 - .0869 = .7209$

   The proportion of adult females with BMI between 21 and 31 is 72.09%.

9.

$$P(X < 60) = P(Z < (60 - 68)/19) = P(Z < -0.42) = .3372$$

   The probability of randomly selecting a patient with a heart rate less than 60 is .3372.

CHAPTER 6: THE CENTRAL LIMIT THEOREM

1. B

3. B

5. D

7. C

9. D

11. D

13. The question is asking the probability of the sample mean being greater than 17, $P(\bar{X} > 17)$. Because there is no direct connection between the sample mean and probability, the sample mean needs to be converted to a Z score.

$$Z = \frac{(\bar{X} - \mu)}{\frac{\sigma}{\sqrt{n}}} = \frac{17 - 15}{\frac{5}{\sqrt{16}}} = \frac{2}{1.25} = 1.6$$

$$P(Z > 1.6) = 1 - P(Z < 1.6) = 1 - .9452 = .0548$$

The probability of the random sample of 16 customers spending longer than 17 minutes on the website to get a quote is 5.48%. There is a low probability that a random sample of 16 customers would spend longer than 17 minutes on the company's website.

CHAPTER 7: HYPOTHESIS TESTING

1. B
3. D
5. A
7. B
9. A
11. C
13.

$$\sigma_{\bar{X}} = \frac{\sigma}{\sqrt{n}} = \frac{3.9}{\sqrt{42}} = \frac{3.9}{6.5} = .601$$

15.

Step 1. State the pair of hypotheses.

Left-tailed test: $H_0: \mu = 2392$

$H_1: \mu < 2392$

Step 2. Identify the rejection zone.

A left-tailed test with $\alpha = .05$

Rejection zone: $Z < -1.65$

Step 3. Calculate the statistic.

List all the known numbers:

$\mu = 2{,}392$ square feet

$\sigma = 760$ square feet

$\bar{X} = 2{,}025$ square feet

$n = 36$

$$Z = \frac{(\bar{X} - \mu)}{\frac{\sigma}{\sqrt{n}}} = \frac{2025 - 2392}{\frac{760}{\sqrt{36}}} = \frac{367}{\frac{760}{6}} = \frac{367}{126.67} = 2.90$$

Step 4.

−2.90 is within the rejection zone of $Z < -1.65$, therefore we reject H_0. The evidence is strong enough to support the claim that the average square footage of houses in Cleveland is smaller than 2,392 square feet.

CHAPTER 8: ONE-SAMPLE *t* TEST AND DEPENDENT-SAMPLE *t* TEST

1. B
3. B
5. D
7. D
9. A
11. D

13a. The population standard deviation, σ, is unknown in this problem statement so the four-step hypothesis test to conduct a *t* test is the correct approach.

Step 1. State the pair of hypotheses.

Because the problem statement does not mention a direction (greater than, less than), we use a nondirectional hypothesis test.

$H_0 : \mu = 500$
$H_0 : \mu \neq 500$

Step 2. Identify the rejection zone.

With the nondirectional hypothesis test, we know it is a two-tailed test. The problem specifies $\alpha = .05$, and $df = n - 1 = 9 - 1 = 8$. With this information, using the t Table, the critical t value is 2.306, therefore the rejection zone is $|t| > 2.306$.

Or type "=T.INV(.025,8)" in a blank cell in Excel to find out that the critical value t is –2.306, therefore, the rejection zone is $|t| > 2.306$.

Step 3.

What we know:

$\mu = 500$

$n = 9$

Because we were not provided with $\bar{X}$ and s, therefore we need to calculate these before conducting a t test.

$$\bar{X} = \Sigma X / n$$

$$SS = \Sigma(X - \bar{X})^2$$

$$s = \sqrt{\frac{\Sigma(X - \bar{X})^2}{df}} = \sqrt{\frac{SS}{df}} = \sqrt{\frac{SS}{n-1}}$$

First we need to calculate the sample mean, then we need to calculate the sum of squares. You can see the calculations in Table 8.9.

TABLE 8.9 ● SAT Verbal Scores

X	$(X - \bar{X})$	$(X - \bar{X})^2$
600	35	1225
760	195	38025
550	–15	225
505	–60	3600
640	95	9025
540	–25	625
480	–85	7225

X	$(X-\bar{X})$	$(X-\bar{X})^2$
535	−30	900
455	−110	12100
5085		72950

$$\bar{X} = \Sigma X / n = 5085/9 = 565$$

$$SS = \Sigma(X-\bar{X})^2 = 72950$$

$$s = \sqrt{\frac{\Sigma(X-\bar{X})^2}{df}} = \sqrt{\frac{SS}{df}} = \sqrt{\frac{SS}{n-1}} = \sqrt{\frac{72950}{8}} = 95.5$$

Now we are able to enter all of these numbers into the *t*-test formula.

$$t = \frac{\bar{X}-\mu}{\frac{S}{\sqrt{n}}}$$

$$t = \frac{(565-500)}{95.5/\sqrt{9}} = \frac{65}{31.83} = 2.04$$

Step 4. Make the correct conclusion.

Compare the calculate t value of 2.04 to the rejection zone of $|t| > 2.306$. The calculated t does not fall within the rejection zone. Therefore, we fail to reject H_0. At $\alpha = .05$, the evidence is not strong enough to support the claim that the new program has an effect on SAT verbal scores.

13b. Because the population standard deviation is unknown, we use s to estimate σ.

95% CI of $\mu = \bar{X} \pm t_{\alpha/2}\left(\frac{s}{\sqrt{n}}\right) = 565 \pm 2.306\left(\frac{95.5}{\sqrt{9}}\right) = 565 \pm 73.40$

95% CI of μ is [491.60,638.40]. We are 95% confident that the average SAT Verbal score for this high school is included in the interval 491.60 to 638.40.

13c. Based on the t test calculation, we determined that the evidence was not strong enough to support the claim that the program had an effect. Similarly, when determining the confidence interval, it is determined to be 95% confident that the average will fall between 491.60 and 638.40. With the population mean, $\mu = 500$, being included within the confidence interval, the effect of the new program is not significant. Both calculations solidify the same conclusion.

15.

FIGURE 8.15 ● Blood Sugar and Seminar Results

	A	Formula Bar	C
1	t-Test: Paired Two Sample for Means		
2			
3		*After*	*Before*
4	Mean	110.333333	114.222222
5	Variance	45	49.6944444
6	Observations	9	9
7	Pearson Correlation	0.76215743	
8	Hypothesized Mean Difference	0	
9	df	8	
10	t Stat	-2.4534987	
11	P(T<=t) one-tail	0.01985971	
12	t Critical one-tail	2.89645945	
13	P(T<=t) two-tail	0.03971941	
14	t Critical two-tail	3.35538733	
15			

Step 1. The problem statement asks "if the seminar had an effect on the blood sugar levels." There are no directional words in the problem statement, therefore it is a nondirectional hypothesis. The treatment effect is defined as the measure of blood sugar after the seminar minus the measure of blood sugar before the seminar. This is an exploratory study, so a two-tailed approach is appropriate.

Difference = (blood sugar after the seminar – blood sugar before the seminar)

$$H_0 : \mu_D = 0$$
$$H_1 : \mu_D \neq 0$$

Step 2. Identify the rejection zone.

Based on the problem statement, $\alpha = .10$, and $df = n - 1 = 9 - 1 = 8$, and a two-tailed test, the critical value in the t Table is 1.86, so the rejection zone for the two-tailed test is $|t| > 1.86$.

Step 3. Calculate the test statistic.

The calculated $t = -2.453$ as shown in Figure 8.15.

Step 4. Make the correct conclusion.

The results shows that the calculated $t = -2.453$ is within the rejection zone. We reject H_0.

Based on the results in Figure 8.15, we find $t = -2.453$ and the $\bar{D} = 110.3 - 114.2 = -3.9$.

$$t = \frac{\bar{D}}{s_e} = \frac{\bar{D}}{\frac{s_D}{\sqrt{n}}}$$

$$-2.453 = \frac{-3.9}{\frac{s_D}{\sqrt{9}}}$$

$$s_D = 4.77$$

$$\text{Cohen's } d = \frac{\bar{D}}{s_D} = \frac{-3.9}{4.77} = -.817$$

The seminar had a large effect size on the measures of blood sugar.

The critical value of t for 80% CI is $t = 1.86$, $s_e = \frac{s_D}{\sqrt{9}} = \frac{4.77}{3} = 1.59$

80% CI of the $\mu_D = \bar{D} \pm t_{a/2} s_e = -3.9 \pm 1.86(1.59) = -3.9 \pm 2.957 =$

$$[-6.857, -0.943,]$$

The value of H_0 was not included in this 80% CI, therefore, we concluded that there was a significant difference of the measure of blood sugar before and after the seminar.

CHAPTER 9: INDEPENDENT-SAMPLES *t* TESTS

1. B
3. A
5. D

7a.

Step 1. Test for equality of variances

Substep 1. State the pair of hypotheses regarding variances.

$$H_0 : \sigma_1^2 = \sigma_2^2$$

$$H_1 : \sigma_1^2 \neq \sigma_2^2$$

Substep 2. Identify the rejection zone.

The larger standard deviation is s_1, 12.5, and its $df_1 = n_1 - 1 = 13 - 1 = 12$, s_2 is the smaller standard deviation, 8.2, and its $df_2 = 15 - 1 = 14$.

The critical value of the right-tailed $F_{(df_1, df_2)} = F_{(12,14)} = 2.53$.

Substep 3. Calculate the statistic.

$$F = \frac{s_1^2}{s_2^2} = \frac{12.5^2}{8.2^2} = \frac{156.25}{67.24} = 2.32$$

Substep 4. Make the correct conclusion.

The calculated $F = 2.32$ is smaller than the critical value of $F_{(df_1, df_2)} = 2.53$. Therefore, we fail to reject H_0. The evidence is not strong enough to claim that the two variances are not equal. Therefore, the correct independent samples *t*-test approach is to assume equal variances.

Step 2. State the pair of hypotheses regarding group means.

The problem statement asks, "Does giving frequent quizzes increase the retention of statistics knowledge as demonstrated by the average final exam scores?" Increase is a key word for direction. Since the group without frequent quizzes has a larger sample standard deviation, it is designated as Group 1. The group with frequent quizzes is labeled Group 2. The alternative hypothesis is to test whether μ_1 is less than μ_2. A left-tailed test is appropriate. Many students automatically associate the keyword *increase* with a right-tailed test without paying attention to which sample is labeled as Sample 1 due to the larger sample standard deviation. In a one-tailed test, it is important to state the correct direction in the alternative hypothesis, H_1.

$$\text{H}_0 : \mu_1 = \mu_2$$
$$\text{H}_1 : \mu_1 < \mu_2$$

Step 3. Identify the rejection zone.

According to the *t* Table, when $df = df_1 + df_2 = n_1 + n_2 - 2 = 26$, $\alpha = .05$; it is a left-tailed test, and the critical *t* value that sets the boundary of the rejection zone is 1.706. The rejection zone for a left-tailed test is $t < -1.706$.

Step 4. Calculate the t statistic.

When $\sigma_1^2 = \sigma_2^2$ the variances need to be pooled together.

$$SS_1 = s_1^2 df_1 = 12.5^2(13-1) = 1875$$

$$SS_2 = s_2^2 df_2 = 8.2^2(15-1) = 941.36$$

$$s_p^2 = \frac{SS_1 + SS_2}{df_1 + df_2} = \frac{1875 + 941.36}{12 + 14} = 108.32$$

$$s_{(\bar{X}_1 - \bar{X}_2)} = \sqrt{\frac{s_p^2}{n_1} + \frac{s_p^2}{n_2}} = \sqrt{\frac{108.32}{13} + \frac{108.32}{15}} = \sqrt{8.33 + 7.22} = \sqrt{15.55} = 3.94$$

$$t = \frac{(\bar{X}_1 - \bar{X}_2)}{s_{(\bar{X}_1 - \bar{X}_2)}} = \frac{(75-80)}{3.94} = \frac{-5}{3.94} = -1.27$$

Step 5. Make the correct conclusion.

The calculated $t = -1.27$ is not within the rejection zone. Therefore, we fail to reject H_0. The evidence is not strong enough to support the claim that frequent quizzes increase the retention of statistics knowledge.

7b. You need to find the 95% CI for the difference between the final exam scores of these two sections. According to the formula, $(1-\alpha)100\%\ \text{CI} = (\bar{X}_1 - \bar{X}_2) \pm t_{\alpha/2}\, s_{(\bar{X}_1 - \bar{X}_2)}$, we need to calculate the point estimate and the margin of error.

The point estimate is $(\bar{X}_1 - \bar{X}_2) = 75 - 80 = -5$

The margin of error is $t_{\alpha/2}\, s_{(\bar{X}_1 - \bar{X}_2)}$. When $df = 26$, two-tailed $\alpha = 0.5$, the critical value of $t = 2.056$ and $s_{(\bar{X}_1 - \bar{X}_2)} = 3.94$, the margin of error is 8.101.

95% CI for $(\mu_1 - \mu_2)$ was -5 ± 8.101. The low end and high end of the 95% CI were [−13.101,3.101]. Zero was included in the 95% CI; therefore, there was no significant differences between the final exam scores of these two sections.

7c. When $\sigma_1^2 = \sigma_2^2$, the effect size is calculated by Cohen's $d = (\bar{X}_1 - \bar{X}_2) / s_p$.

Based on the answer from 7a $s_p = \sqrt{108.32} = 10.41$.

$$\text{Effect size } d = \frac{(\bar{X}_1 - \bar{X}_2)}{s_p} = \frac{75-80}{10.41} = \frac{-5}{10.41} = -0.48$$

The effect average score of students without frequent quizzes is 0.48 standard deviations lower than that of students with frequent quizzes. The effect is moderate.

9a.

Step 1. Test for equality of variances.

Substep 1. State the pair of hypotheses regarding variances.

$$H_0: \sigma_1^2 = \sigma_2^2$$
$$H_1: \sigma_1^2 \neq \sigma_2^2$$

Substep 2. Identify the rejection zone.

The sample standard variance, s^2, was not provided directly in the problem statement, so you have to calculate $s^2 = SS / df$.

The variance for the group that received cell phones after age 5 is

$$s^2 = \frac{SS}{df} = \frac{1559.52}{12} = 129.96.$$

The variance for the group that received cell phones before age 5 is

$$s^2 = \frac{SS}{df} = \frac{345.96}{9} = 38.44.$$

Therefore, the group that received cell phones after age 5 is labeled as Group 1, with $df_1 = 12$. The other group is Group 2, with $df_2 = 9$. The critical value of the right-tailed $F_{(df_1, df_2)} = F_{(12,9)} = 3.07$.

Substep 3. Calculate the statistic.

$$F = \frac{s_1^2}{s_2^2} = \frac{129.96}{38.44} = 3.38$$

Substep 4. Make the correct conclusion.

The calculated $F = 3.38$ is larger than the critical value of $F_{(12,9)} = 3.07$. Therefore, we reject H_0. The evidence is strong enough to claim that the two variances are not equal. Therefore, the correct independent-samples *t*-test approach is not to assume equal variances.

Step 2. State the pair of hypotheses regarding group means.

The problem statement asks, "Do these two groups have different social skills?" There is no key word to indicate a direction in the problem statement. A two-tailed test is appropriate.

$$H_0: \mu_1 = \mu_2$$
$$H_1: \mu_1 \neq \mu_2$$

Step 3. Identify the rejection zone.

Satterthwaite's approximated $df = \dfrac{(w_1 + w_2)^2}{\dfrac{w_1^2}{n_1 - 1} + \dfrac{w_2^2}{n_2 - 1}}$ where

$$w_1 = \frac{s_1^2}{n_1} = \frac{129.96}{13} = 9.997 \text{ and}$$

$$w_2 = \frac{s_2^2}{n_2} = \frac{38.44}{10} = 3.844$$

$$df = \frac{(w_1 + w_2)^2}{\frac{w_1^2}{n_1 - 1} + \frac{w_2^2}{n_2 - 1}} = \frac{(9.997 + 3.844)^2}{\frac{9.997^2}{12} + \frac{3.844^2}{9}} = \frac{191.57}{9.97} = 19.21 \text{ rounded down to 19.}$$

According to the t Table, when $df = 19$, $\alpha = .05$; it is a two-tailed test, and the critical t value that sets the boundary of the rejection zone is 2.093. The rejection zone for a two-tailed test is $|t| > 2.093$.

If you use the rule of thumb to pick the df as the smaller of df_1 and df_2, the df = 9. A two-tailed test with $\alpha = .05$, the rejection zone is $|t| > 2.262$.

Step 4. Calculate the t statistic.

When $\sigma_1^2 \neq \sigma_2^2$, the variances need to be left alone. We do not pool the variances.

$$s_{(\bar{X}_1 - \bar{X}_2)} = \sqrt{\frac{s_1^2}{n_1} + \frac{s_2^2}{n_2}} = \sqrt{\frac{129.96}{13} + \frac{38.44}{10}} = \sqrt{9.997 + 3.844} = \sqrt{13.841} = 3.72$$

$$t = \frac{(\bar{X}_1 - \bar{X}_2)}{s_{(\bar{X}_1 - \bar{X}_2)}} = \frac{(73.8 - 60.5)}{3.72} = \frac{13.3}{3.72} = 3.58$$

Step 5. Make the correct conclusion.

The calculated $t = 3.58$ is within the rejection zone. Therefore, we reject H_0. The evidence is strong enough to support the claim that these two groups of teenagers have different levels of social skills.

9b. When $\sigma_1^2 \neq \sigma_2^2$, the effect size is calculated as

$$\text{Glass's } \Delta = \frac{(\bar{X}_1 - \bar{X}_2)}{s_{control}} \text{ or } \Delta = \frac{(\bar{X}_1 - \bar{X}_2)}{s_{larger}}$$

$$\text{Effect size } \Delta = \frac{(\bar{X}_1 - \bar{X}_2)}{s_{larger}} = \frac{73.8 - 60.5}{11.4} = \frac{13.3}{11.4} = 1.17$$

The average social skills of teenagers who received cell phones after age 5 is 1.17 standard deviations higher than the average social skills of teenagers who received cell phones before age 5. The effect is large.

9c. When equal variances are not assumed, $\sigma_1^2 \neq \sigma_2^2$, the 90% CI is

$$(\bar{X}_1 - \bar{X}_2) \pm t_{\alpha/2}\sqrt{\frac{s_1^2}{n_1} + \frac{s_2^2}{n_2}}.$$

The point estimate is $(\bar{X}_1 - \bar{X}_2) = 73.8 - 60.5 = 13.3$. The margin of error is

$$t_{\alpha/2}\sqrt{\frac{s_1^2}{n_1} + \frac{s_2^2}{n_2}} = 1.729(3.72) = 6.43$$

Therefore, the 90% CI is 13.3 ± 6.43, that is [6.87, 19.73]. Zero is not included in the CI so there is a significant difference between the group means. We are 90% confident that the difference in average social skills between these two groups is between 6.87 and 19.73, with the teenagers who were given cell phones after age 5 having better social skills.

If you use the rule of thumb to pick the df as the smaller of df_1 and df_2, the df = 9. The critical t value for a 90% CI is 1.833.

The margin of error is $t_{\alpha/2}\sqrt{\frac{s_1^2}{n_1} + \frac{s_2^2}{n_2}} = 1.833(3.72) = 6.82$

Therefore, the 90% CI is 13.3 ± 6.43, that is [6.48, 20.12]. Zero is not included in the CI so there is a significant difference between the group means. We are 90% confident that the difference in average social skills between these two groups is between 6.48 and 20.12, with the teenagers who were given cell phones after age 5 having better social skills.

CHAPTER 10: CORRELATION

1. C
3. A
5. C
7. C

9a. The scatterplot shows that there is a linear relationship between ACT and SAT scores.

FIGURE 10.15 ● Scatterplot of ACT and SAT Scores of 10 Incoming Freshmen

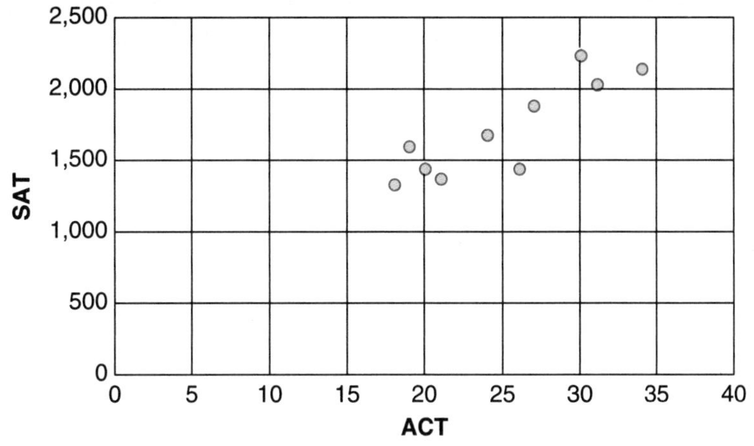

9b.

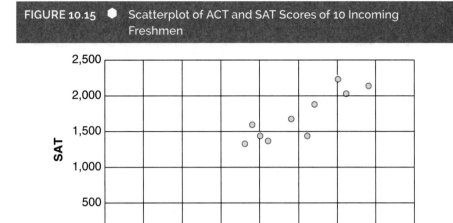

Step 1. Explicitly state the pair of hypotheses.

The problem statement asks to conduct a hypothesis test for a relationship between ACT and SAT scores. Therefore, a two-tailed hypothesis is appropriate.

$$H_0: \rho = 0$$
$$H_1: \rho \neq 0$$

Step 2. Identify the rejection zone.

According to the r Table, with a two-tailed test, $df = n - 2 = 10 - 2 = 8$, $\alpha = .05$, the critical value of $r = .632$. The rejection zone for a two-tailed test is $|r| > .632$.

Step 3. Calculate the test statistic.

According to the Pearson's r formula, we need to add the following five columns to Table 10.8a $(X - \bar{X}), (Y - \bar{Y}), (X - \bar{X})(Y - \bar{Y}), (X - \bar{X})^2, (Y - \bar{Y})^2$.

The first step is to figure out $\bar{X}$ and $\bar{Y}$ to construct the additional five columns, and the last row of every column is the total of that column.

TABLE 10.8a ● ACT and SAT Scores of 10 Incoming Freshmen With $(X-\bar{X})$, $(Y-\bar{Y})$, $(X-\bar{X})(Y-\bar{Y})$, $(X-\bar{X})^2$, and $(Y-\bar{Y})^2$

ACT	SAT	$(X-\bar{X})$	$(Y-\bar{Y})$	$(X-\bar{X})(Y-\bar{Y})$	$(X-\bar{X})^2$	$(Y-\bar{Y})^2$
27	1,890	2	165	330	4	27,225
26	1,450	1	−275	−275	1	75,625
31	2,050	6	325	1,950	36	105,625
34	2,150	9	425	3,825	81	180,625
30	2,250	5	525	2,625	25	275,625
21	1,380	−4	−345	1,380	16	119,025
20	1,450	−5	−275	1,375	25	75,625
19	1,600	−6	−125	750	36	15,625
18	1,350	−7	−375	2,625	49	140,625
24	1,680	−1	−45	45	1	2,025
250	17,250			14,630	274	1,017,650

$$\bar{X} = X / n = 250 / 10 = 25$$

$$\bar{Y} = \Sigma Y / n = 17250 / 10 = 1725$$

$$SP = \Sigma(X-\bar{X})(Y-\bar{Y}) = 14630$$

$$SS_X = \Sigma(X-\bar{X})^2 = 274$$

$$SS_Y = \Sigma(Y-\bar{Y})^2 = 1017650$$

$$r = \frac{SP}{\sqrt{SS_X SS_Y}} = \frac{14630}{\sqrt{(274)(1017650)}} = .876$$

Step 4. Make the correct conclusion.

The calculated $r = .876$ is within the rejection zone, so we reject H_0. The evidence is strong enough to support the claim that there is a significant relationship between ACT and SAT scores.

11. Assume that the amount of saturated fat in a person's diet, cholesterol level, and weight were all normally distributed. The Pearson's correlation between the amount of saturated fat in a person's diet and cholesterol level was .768, the Pearson's correlation between the amount of saturated fat in a person's diet and weight was .803, and the Pearson's correlation between cholesterol level and weight was .643. What was the correlation between the amount of saturated fat in a person's diet and cholesterol level while keeping weight constant?

The partial correlation calculates the relationship between X and Y, while keeping Z constant.

Based on the problem statement, weight is being kept constant so it will be considered Z. Diet will be X and cholesterol will be Y.

$$r_{XY} = .768$$

$$r_{XZ} = .803$$

$$r_{YZ} = .643$$

$$r_{XY.Z} = \frac{r_{XY} - (r_{XZ} r_{YZ})}{\sqrt{(1 - r_{XZ}^2)(1 - r_{YZ}^2)}} = \frac{.768 - (.803)(.643)}{\sqrt{(1 - .803^2)(1 - .643^2)}} = \frac{.768 - .516}{\sqrt{(.355)(.587)}} = \frac{.252}{.456} = .553$$

The partial correlation between diet and cholesterol level, while keeping weight constant, was .553. The relationship was strong.

CHAPTER 11: SIMPLE REGRESSION

1. A
3. A
5. B
7. D

9a. The scatterplot of car weight and fuel efficiency is shown in Figure 11.13. There is a linear relationship between car weight and fuel efficiency according to Figure 11.13.

FIGURE 11.13 ● Scatterplot of Car Weight and Fuel Efficiency

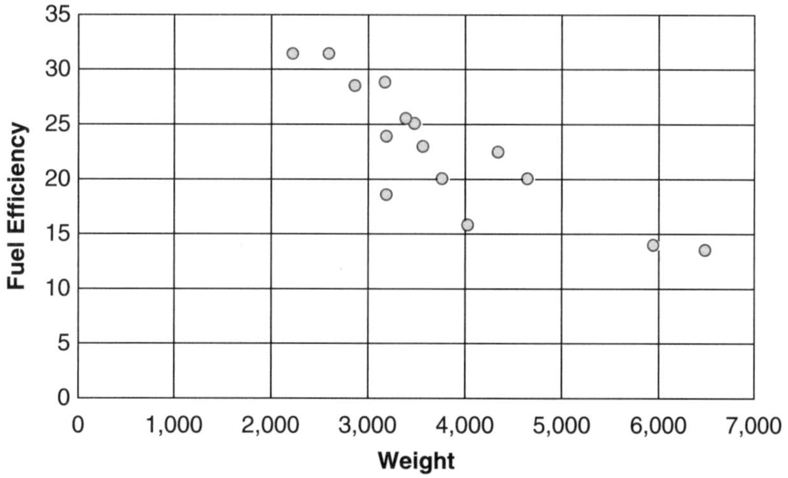

9b. We need to figure out the $\bar{X}$ and $\bar{Y}$ and then add the following five columns to Table 11.7a $(X-\bar{X}), (Y-\bar{Y}), (X-\bar{X})(Y-\bar{Y}), (X-\bar{X})^2, (Y-\bar{Y})^2$, to demonstrate the step-by-step calculation.

TABLE 11.7a ● Car Weight and Fuel Efficiency With $(X-\bar{X})$, $(Y-\bar{Y})$, $(X-\bar{X})(Y-\bar{Y})$, $(X-\bar{X})^2$, and $(Y-\bar{Y})^2$

Weight (Pounds)	Fuel Efficiency (MPG)	$(X-\bar{X})$	$(Y-\bar{Y})$	$(X-\bar{X})(Y-\bar{Y})$	$(X-\bar{X})^2$	$(Y-\bar{Y})^2$
3,190	28.8	−614.6	6	−3,687.60	377,733.16	36
3,572	23	−232.6	0.2	−46.52	54,102.76	0.04
2,888	28.5	−916.6	5.7	−5,224.62	840,155.56	32.49
3,777	20	−27.6	−2.8	77.28	761.76	7.84
3,208	18.5	−596.6	−4.3	2,565.38	355,931.56	18.49
3,393	25.6	−411.6	2.8	−1,152.48	169,414.56	7.84
4,653	20	848.4	−2.8	−2,375.52	719,782.56	7.84
3,495	25	−309.6	2.2	−681.12	95,852.16	4.84
3,208	24	−596.6	1.2	−715.92	355,931.56	1.44
4,345	22.5	540.4	−0.3	−162.12	292,032.16	0.09
2,616	31.5	−1,188.6	8.7	−10,340.82	1,412,769.96	75.69
5,950	14	2,145.4	−8.8	−18,879.52	4,602,741.16	77.44
2,235	31.5	−1,569.6	8.7	−13,655.52	2,463,644.16	75.69
4,039	15.7	234.4	−7.1	−1,664.24	54,943.36	50.41
6,500	13.4	2,695.4	−9.4	−25,336.76	7,265,181.16	88.36
57,069	**342**			**−81,280.10**	**19,060,977.60**	**484.50**

$$\bar{X} = \Sigma X / n = 57069/15 = 3804.6$$

$$\bar{Y} = \Sigma Y / n = 342/15 = 22.80$$

$$SP = \Sigma(X-\bar{X})(Y-\bar{Y}) = -81280.10$$

$$SS_X = \Sigma(X-\bar{X})^2 = 19060977.60$$

$$b = \frac{SP}{SS_X} = \frac{-81280.1}{19060977.60} = -0.00426$$

$$a = \bar{Y} - b\bar{X}$$

$$a = 22.80 - (-0.00426)(3{,}804.6) = 39.01$$

The regression line for using car weight to predict fuel efficiency is $\hat{Y} = 39.01 - 0.00426X$. For each additional pound the car weighs, the fuel efficiency decreases by 0.00426 mpg.

9c. The predicted fuel efficiency for a car that weights 4,500 pounds is
$\hat{Y} = 39.01 - 0.00426(4500) = 19.84$ mpg.

9d. Here is the four-step hypothesis test for a simple regression.

Step 1. State the pair of hypotheses.

$$H_0: \beta = 0$$
$$H_1: \beta \neq 0$$

Step 2. Identify the rejection zone.

The critical value of F in the F Table under $\alpha = .05$ is $F_{(1,13)} = 4.67$. The rejection zone
is $F > 4.67$

Step 3. Calculate the test statistic.

Conducting a hypothesis test on a simple regression requires r^2 and SS_Y.

$$r = \frac{SP}{\sqrt{SS_X SS_Y}} = \frac{-81280.10}{\sqrt{(19060977.68)(484.5)}} = -.846$$

$$r^2 = (-.846)^2 = .715$$
$$SS_Y = 484.5$$
$$SSR = r^2 SS_Y = .715(484.5) = 346.42$$
$$SSE = (1 - r^2) SS_Y = (1 - .715)(484.5) = 138.08$$

Put all the numbers into the summary hypothesis test table for car weight and fuel efficiency as shown in Table 11.7b.

TABLE 11.7b ● ANOVA Table for Car Weight and Fuel Efficiency

Source	SS	df	MS	F
Regression	346.42	1	346.42	32.620
Error	138.08	13	10.62	
Total	484.5	14		

Step 4. Draw the correct conclusion.

The calculated F is 32.620, which is within the rejection zone, so we reject H_0. Car weights help predict fuel efficiency.

CHAPTER 12: ONE-WAY ANALYSIS OF VARIANCE

1. C
3. C
5. C
7. B
9. C

11a. The four-step hypothesis test for an ANOVA:

Step 1. State the pair of hypotheses.

$H_0: \mu_1 = \mu_2 = \mu_3 = \ldots = \mu_k$

H_1 : Not all = μ_k are equal.

Step 2. Identify the rejection zone.

There are three groups, $k = 3$, five SUVs in each group, $n_1 = n_2 = n_3 = 5$, and the sample size, $n = 15$. The critical value of F is identified by $F_{(2,12)} = 3.89$ in the F Table for ANOVA. The rejection zone is identified as the calculated $F > 3.89$.

Step 3. Calculate the F test.

Under equal group size, the sample mean is

$$\bar{\bar{X}} = (6.2 + 7.2 + 7.4) / 3 = 6.93$$

$$SSB = \Sigma n_j(\bar{X}_j - \bar{\bar{X}})^2 = 5(6.2 - 6.93)^2 + 5(7.2 - 6.93)^2 + 5(7.4 - 6.93)^2$$

$$= 5(.533) + 5(.073) + 5(.221) = 2.665 + .365 + 1.105 = 4.14$$

$$SSW = SST - SSB = 14.93 - 4.14 = 10.79$$

Now we have all the numbers to complete the ANOVA summary table as shown in Table 12.10a.

TABLE 12.10a ● ANOVA Table for Crash Safety Ratings for Three Types of SUVs

Source	SS	df	MS	F
Between groups	4.14	$df_B = 2$	2.07	2.3
Within groups	10.79	$df_W = 12$	0.9	
Total	14.93	$df_r = 14$		

Step 4. Make the correct conclusion.

The calculated $F = 2.3$ is not within the rejection zone, so we fail to reject H_0. The evidence is not strong enough to claim that there are significant differences in crash safety ratings among the three different types of SUVs. No further analysis is needed.

11b.

$$\text{The effect size for the ANOVA is } \eta^2 = \frac{SSB}{SST} = \frac{4.14}{14.93} = .277$$

There are 27.7% of the variance of the crash safety ratings that can be explained by the different types of SUVs.

CHAPTER 13: CHI-SQUARE TESTS FOR GOODNESS OF FIT AND INDEPENDENCE

1. A
3. C
5. D
7. A
9. The problem stated that 3,215 white employees were hired over the past 6 months and 966 of them were promoted. This means that the rest of the white employees were not promoted. A simple subtraction shows that the number of white employees who were not promoted is 2,249. Applying the same logic, you get the number of black employees who were not promoted as 214, the number of Hispanic employees who were not promoted as 236, and the number of employees whose racial background was classified as other and who were not promoted as 34.

TABLE 13.12 ● Race and Promotions

	Promoted	Not Promoted	Row Total
White	966	2,249	3,215
Black	74	214	288
Hispanic	81	236	317
Other	5	34	39
Column Total	1,126	2,733	3,859

Once the observed frequency table is set, the four-step hypothesis-testing procedure can be carried out.

Step 1. State the pair of hypotheses.

H_0: Promotion and employee's race are independent.

H_1: Promotion and employee's race are not independent.

Step 2. Identify the rejection zone.

When $\alpha = .05$, with $df = (r-1)(c-1) = 3$, the rejection zone for the chi-square test is $\chi^2 > 7.815$.

Step 3. Calculate the test statistic.

$$E = \frac{\text{Row total} \times \text{Column total}}{n}$$

The expected frequency table is reported in Table 13.12a.

TABLE 13.12a ● Expected Frequency Table for Race and Promotions

	Promoted	Not Promoted
White	938.09	2,276.91
Black	84.03	203.97
Hispanic	92.50	224.50
Other	11.38	27.62

$$\chi^2 = \frac{(O-E)^2}{E}$$

$$= \frac{(966-938.09)^2}{938.09} + \frac{(2249-2276.91)^2}{2276.91} + \frac{(74-84.03)^2}{84.03} + \frac{(214-203.97)^2}{203.97} + \frac{(81-92.50)^2}{92.50} + \frac{(236-224.50)^2}{224.50} + \frac{(5-11.38)^2}{11.38} + \frac{(34-27.62)^2}{27.62} = 9.932$$

Step 4. Draw the correct conclusion.

The calculated $\chi^2 = 9.932$ is within the rejection zone. Therefore, we reject H_0. The evidence is strong enough to suggest that promotion decisions and employee's race are not independent. It is highly unlikely that we would obtain these observed frequencies in Table 13.12a if the promotion decision were made without considering employee's race.

GLOSSARY

Absolute zero: Absolute zero means a complete absence of the attribute that you are measuring. It is not an arbitrarily assigned number.

Addition Rule 1 of probability: When Event A and Event B are mutually exclusive, the probability of either Event A or Event B happening is the sum of the probabilities of each event: $P(\text{A or B}) = P(\text{A} \cup \text{B}) = P(\text{A}) + P(\text{A})$.

Addition Rule 2 of probability: When Event A and Event B are not mutually exclusive, the probability of either Event A or Event B, or both, happening is the sum of the probabilities of each event minus the overlapping part of the two events: $P(\text{A or B}) = P(\text{A} \cup \text{B}) = P(\text{A}) + P(\text{B}) - P(\text{A} \cap \text{B})$.

Alternative hypothesis, H_1: The hypothesis that researchers turn to when the null hypothesis is rejected. The alternative hypothesis is also called the research hypothesis.

Analysis of variance (ANOVA): ANOVA is a statistical procedure that partitions variances due to different sources.

Bar graph: A bar graph uses bars of equal width to show frequency or relative frequency of discrete categorical data (i.e., nominal or ordinal data). Adjacent bars are not touching each other.

Between-group variance: The between-group variance is calculated as the SSB divided by the df_B; in a special ANOVA term, the between-group variance is also called the mean squares between groups, $\text{MSB} = \text{SSB}/df_B$.

Between-subject design: The between-subject design refers to research in which participants are assigned to different groups based on different levels of the independent variables. Every participant is assigned to only one group.

Binomial probability distribution: A binomial probability distribution is a probability distribution that applies to variables with only two possible outcomes (i.e., success vs. failure) in each trial. The trials are independent of one another. The probability of a success and the probability of a failure remain the same in all trials.

Bonferroni correction: The Bonferroni correction simply adjusts the α level of the individual pairwise comparison to a new level in order to avoid inflating the familywise error rate.

Central limit theorem: The central limit theorem states that when the same random sampling procedure is repeated to produce many sample means, the sample means form a normal distribution with a mean of μ and a standard deviation of $\sigma_{\bar{X}} = \sigma/\sqrt{n}$ as long as the samples are selected from a normally distributed population or the sample size is large (i.e., $n > 30$).

Central tendency: Central tendency is defined as utilizing a single value to represent the center of a distribution. There are three commonly used measures for central tendency: mean, median, and mode.

Chi-square for goodness-of-fit tests: Chi-square goodness-of-fit tests refer to using one categorical variable to investigate whether the sample distribution of the values in a variable fits with its theoretical distribution.

Chi-square for independence tests: Chi-square for independence tests are statistics applied to investigate the relationship between two nominal or ordinal variables.

Chi-square test: Chi-square test investigates the differences between the observed frequency and the expected frequency. When the calculated chi-square value is larger than the critical value in the chi-square table, the result is statistically significant.

Cluster sampling: Cluster sampling works best when "natural" grouping (clustering) occurs in the population. Random sampling is conducted to select which clusters are included in the sample. Once a cluster is selected in the sample, all individuals in the cluster are included in the sample.

Coefficient of determination: The coefficient of determination is defined as r^2, which is the percentage of variance overlap between *X* and *Y* or the percentage of variance of *Y* explained by *X*.

Cohen's *d*: Cohen's $d = \frac{\bar{X}_1 - \bar{X}_2}{s_{\text{pooled}}}$ is the formula to calculate effect sizes for independent-samples *t* tests when equal variances are assumed.

Confidence interval: A confidence interval is defined as an interval of values calculated from sample data to estimate the value of a population parameter.

Confidence level: Confidence level is expressed as $(1 - \alpha)$, which is the complement to α, the significance level.

Confounding variable: A confounding variable is an extraneous variable in a study that correlates with both the independent variable and the dependent variable.

Contingency table: A contingency table refers to a two-way frequency table where frequency counts are based on a row variable and a column variable at the same time.

Continuous variables: Continuous variables are values that do not have separation from one integer to the next. Continuous variables usually are expressed with decimals or fractions.

Convenience sample: A convenience sample is one in which researchers use anyone who is willing to participate in the study. A convenience sample is created based on easy accessibility.

Counterbalance: Counterbalance is a cautionary step in conducting repeated measures of the same sample to make sure that the order of presentation of conditions is balanced out so the order does not confound the research results.

Cumulative relative frequency: Cumulative relative frequency is defined as the accumulation of the relative frequency for a particular value and all the relative frequencies of lower values.

Data: Data are defined as factual information used as a basis for reasoning, discussion, or calculation, so that meaningful conclusions can be drawn.

Data visualization: Data visualization is the graphic representation of data.

Degrees of freedom: The degrees of freedom in a statistical procedure is the number of values involved in the calculation that can vary freely in the sample.

Degrees of freedom due to error: The degrees of freedom due to error in a simple regression are calculated as $(n - 2)$.

Degrees of freedom due to regression: The degrees of freedom due to regression are the number of predictors in the regression equation. In simple regression, the degree of freedom due to regression is 1.

Dependent variable: A dependent variable is the variable that is the focus of researchers' interests and is affected by the different levels of an independent variable.

Dependent-sample *t* tests: Dependent-sample *t* tests are used to test the differences between two measures of one variable from the same sample before and after a treatment or the difference between a variable from two samples with an explicit one-on-one paired relationship.

Descriptive statistics: Descriptive statistics are statistical procedures used to describe, summarize, organize, and simplify relevant characteristics of sample data.

Discrete variables: Discrete variables are values that have clear separation from one integer to the next. The answers for discrete variables can only be integers (i.e., whole numbers).

Disjoint events: Disjoint events are mutually exclusive events. There are no overlapping parts between the events.

Distribution: Distribution is the arrangement of values of a variable as it occurs in a sample or a population.

Effect size: The effect size is a standardized measure of the difference between the sample statistic and the hypothesized population parameter in units of standard deviation.

Empirical rule: The empirical rule describes the following attributes for variables with a normal distribution: About 68% of the values fall within 1 standard deviation of the mean, about 95% of the values fall within 2 standard deviations of the mean, and about 99.7% of the values fall within 3 standard deviations of the mean.

Equal intervals: Equal intervals are created in a frequency table by including the same number of values in each interval.

Error: Error in regression is the difference between the actual *Y* and the predicted $\hat{Y}$, Error $= (Y - \hat{Y})$.

Event: An event is defined as a set of outcomes from an experiment or a procedure.

Expected frequency: Expected frequency is the theoretically predicted frequency that assumes that the variable is distributed exactly as stated in the null hypothesis.

Experimental research: Experimental research is usually conducted in a tightly controlled environment (i.e., research laboratories). The three important features in experimental research are (1) control, (2) manipulation, and (3) random assignment.

Extraneous variable: The extraneous variables are variables that are not included in the study but might have an impact on the relationship between variables included in the study.

Familywise error rate (FWER): The FWER refers to the probability of making a Type I error in a set of similar multiple comparisons.

Folded *F* test: The folded *F* test means that in calculating the *F* test, the larger variance is designated as the numerator and the smaller variance as the denominator; the calculated *F* value will always be larger than 1 to avoid the left-tailed *F* values.

Frequency: Frequency is a simple count of a particular value or category occurring in a sample or a population.

Frequency distribution table: A frequency distribution table lists all distinct values or categories arranged in an orderly fashion in a table, along with tally counts for each value or category in a data set.

Glass's delta (Δ): Glass's Δ is the formula to calculate effect sizes for independent-samples *t* tests when equal variances are not assumed.

Heteroscedasticity: Heteroscedasticity refers to the condition when the variances of the errors are not the same across all values of *X* or across all $\hat{Y}$. This is commonly referred to as violations of the homogeneity of variance of *Y* across all values of *X*.

Histogram: A histogram is a graphical presentation of a continuous variable. The bars of a histogram are touching each other to illustrate that the values are continuous.

Homoscedasticity: Homoscedasticity is the condition under which the variance of *Y* is the same across all possible values of *X*. In other words, there is equal spread of *Y* for every value of *X*.

Hypothesis testing: Hypothesis testing is the standardized process to test the strength of the evidence for a claim about a population.

Independent events: Two events A and B are independent when the occurrence of one event does not affect the probability of the occurrence of the other event.

Independent samples: Independent samples refer to samples selected from different populations where the values selected from one population are not related to those from the other population. Independent samples contain different individuals in each sample.

Independent variable: An independent variable is the variable that is deliberately manipulated by the researchers in a research study.

Inferential statistics: Inferential statistics are statistical procedures that use sample statistics to generalize to or make inferences about a population.

Interval estimate: An interval estimate is also called the margin of error and is calculated as the multiplication between a two-tailed critical *t* value, $t_{\alpha/2}$ and the standard error of the statistics.

Interval scale: An interval scale not only arranges observations according to their magnitudes but also distinguishes the ordered arrangement in equal units.

Law of large numbers: The law of large numbers is another probability theorem describing the relationship between sample means and population means. When the number of trials approaches infinity, the observed sample mean approaches the population mean.

Likert scales: Likert scales are often used to measure people's opinions, attitudes, or preferences. Likert scales measure attributes along a continuum of choices such as 1 = *strongly disagree*, 2 = *somewhat disagree*, 3 = *neutral*, 4 = *somewhat agree*, or 5 = *strongly agree* with each individual statement.

Line graph: A line graph is a graphical display of quantitative information with a line or curve that connects a series of adjacent data points.

Mean: the mean is defined as the arithmetic average of all values in a data distribution.

Mean squared deviation: The mean squared deviation is another term for the variance.

Median: the median is defined as the value in the middle position of a sorted variable arranged from the lowest value to the highest value.

Mode: The mode is defined as the value or category with the highest frequency in a data distribution.

Multiplication rule for independent events: When Event A and Event B are independent, the probability of both events happening is $P(AB) = P(A) \times P(B)$.

Mutually exclusive and collectively exhaustive: Mutually exclusive and collectively exhaustive means that events have no overlap—if one is true, the others cannot be true—and they cover all possible outcomes.

Mutually exclusive events: When Event A and Event B are mutually exclusive, it means that only one of the events can happen: If one happens, the other does not happen.

***n* factorial:** *n* factorial is calculated as the multiplicative product of all positive integers less than or equal to n, $n! = n \times (n-1) \times (n-2) \times (n-3) \cdots 3 \times 2 \times 1$. 0! is defined as 1.

Negatively skewed distribution: When the data points mostly concentrate on the high-end values with a long tail to the low-end values, the distribution is negatively skewed.

Nominal scale: In a nominal scale, measurements are used as identifiers, such as your student identification number, phone number, or social security number.

Nonexperimental research: Nonexperimental research is conducted to observe and study research participants in their natural settings without deliberately controlling the environment or manipulating their behaviors or preferences.

Nonparametric statistics: Nonparametric statistics are statistical procedures that make no assumptions about the shape of the population distribution. They are also called distribution-free statistics.

Normal distribution: A normal distribution curve peaks at the mean of the values and symmetrically tapers off on both sides with 50% of the data points above the mean and 50% of the data points below the mean. Draw a line straight through the mean, and the left side and the right side are mirror images of each other.

Null hypothesis, H_0: The null hypothesis is the no effect hypothesis. It states the effect that the researchers are trying to establish does not exist.

Observed frequency: Observed frequency is the frequency count of the values of a variable that we observe in a sample.

One-sample *t* test: A one-sample *t* test is designed to conduct a test between a sample mean and a hypothesized population mean when σ is unknown, $t = \frac{\bar{X} - \mu}{s_{\bar{X}}} = \frac{\bar{X} - \mu}{\frac{s}{\sqrt{n}}}$, with $df = n - 1$.

One-tailed test (directional test): A one-tailed test is a directional hypothesis test, which is usually conducted when the research topic suggests a consistent directional relationship with the backing of either theoretical reasoning or repeated empirical evidence.

One-way ANOVA: One-way ANOVA is an ANOVA that uses only one between-subject factor to classify sample data into different groups.

Ordinal scale: In an ordinal scale, measurements are used not only as identifiers but also to carry orders in a particular sequence.

Ordinary least squares (OLS): OLS depicts that the regression line is the best fitting line that goes through $(\bar{X}, \bar{Y})$ and minimizes the vertical distances between the actual Y values and the predicted $\hat{Y}$.

Outliers: Outliers refer to extreme values in a distribution. Outliers usually stand far away from the rest of the data points.

Parameters: Parameters are defined as numerical characteristics of a population.

Parametric statistics: Parametric statistics are statistical procedures that make assumptions about the shape of the population distribution and population parameters such as μ, σ, and σ^2.

Partial correlation: A partial correlation is the purified correlation between two variables while holding a third variable constant.

Pearson's correlation, *r*: Pearson's r is a statistical procedure that quantifies the extent that two variables move in the same direction or opposite directions. It provides direction and strength of the relationship.

Pie chart: It is a circular graph that uses slices to show different categories. The size of each slice is proportional to the frequency or relative frequency of each category.

Point estimate: A point estimate is to use a single value from the sample statistics to estimate a population parameter.

Population: A population is defined as an entire collection of everything or everyone that researchers are interested in studying or measuring.

Positively skewed distribution: When the data points mostly concentrate on the low-end values with a long tail to the high-end values, the distribution is positively skewed.

Post hoc comparisons: Post hoc comparisons are used after a significant ANOVA F test to identify exactly where the significant difference comes from.

Power: Power is defined as the sensitivity to detect an effect when an effect actually exists. The power of a statistical test is the probability of correctly rejecting H_0 when H_0 is false. Power is positively correlated with all of the three factors: (1) sample size, (2) σ level, and (3) the effect size.

Probability density function: A probability density function describes the probability for a continuous, random variable to be in a given range of values.

Probability distribution: There are three requirements for a probability distribution:

1. The variable X is a discrete, random, and numerical variable. The number of possible values of X is finite. Each value is associated with a probability.
2. The probability of each value needs to be within 0 and 1, inclusive. This is expressed as $0 \leq P(X) \leq 1$.
3. The sum of the probabilities of all the values equals 1. It is expressed as $\Sigma P(X) = 1$.

***p* value associated with a test statistic:** The p value associated with a test statistic is defined as the probability of obtaining the magnitude of the calculated test statistic assuming that H_0 is true.

Quasi-experimental research: Quasi-experimental research has some but not all of the features of experimental research. More specifically, if one or more of the control, manipulation, and random assignment features are not feasible but others remain intact, the research becomes quasi-experimental.

Random sample: A random sample is an ideal way to select participants for scientific research. A random sample occurs when every member in the population has an equal chance of being selected.

Range: The range is defined as the maximum minus the minimum in a variable.

Ratio scale: A ratio scale contains every characteristic that lower-level scales of measurement have, such as identifiers, direction of ranking, equal units, and something extra: an absolute zero.

Real limits: Real limits cover a range of possible values that may be reflected by a single continuous measure. The lower limit is the value minus ½ of the unit and the upper limit is the value plus ½ of the unit.

Regression equation: The regression equation is the equation for the regression line, which is expressed as $\hat{Y} = a + bX$, where a is the Y-intercept and b is the slope.

Rejection zone: The rejection zone is bounded by the critical value of a statistic. It is the standard used to judge the strength of the empirical evidence. When the calculated test statistics fall in the rejection zone, the correct decision is to reject H_0.

Relative frequency: Relative frequency is defined as the frequency of a particular value or category divided by the total frequency (or sample size). Relative frequency is also called proportion.

Sample: A sample is defined as a subset of the population from which measures are actually obtained.

Sample space: A sample space is a complete list of all possible outcomes.

Sample statistics: Sample statistics are defined as numerical attributes of a sample.

Sampling distribution of the sample means: The sampling distribution of the sample means is defined as the distribution of all means from samples of the same sample size, n, randomly selected from the population (i.e., $\bar{X}_1, \bar{X}_2, \bar{X}_3, \ldots$).

Sampling error: A sampling error is defined as when a sample is randomly selected, a natural divergence, difference, distance, or error occurs between sample statistics and population parameters.

Scales of measurement: Scales of measurement illustrate different ways that variables are defined and measured. Each scale of measurement has certain mathematical properties that determine the appropriate application of statistical procedures.

Scatterplot: A scatterplot between two variables is a graph of dots plotted on a two-dimensional chart. Each dot is a data point that contains a pair of values on the X and Y axes.

Simple event: A simple event is defined as an elementary event that cannot be broken into simpler parts.

Simple random sample. A simple random sample is a subset of individuals (a sample) chosen from a larger set (a population). Each individual is chosen randomly and entirely by chance, and each subset of k individuals has the same probability of being chosen for the sample as any other subset of k individuals.

Simple regression: The simple regression is the most commonly used linear predictive analysis when making quantitative predictions of the dependent variable, Y, based on the values of the independent variable, X.

Skewed distribution: A skewed distribution happens when data points are not symmetrical, and they concentrate more on one side of the mean than on the other.

Slope: The regression slope is the value change in Y when X increases by one unit.

Standard deviation: Standard deviation is a way to quantify individual differences, and it is mathematically defined as $\sigma = \sqrt{\text{Variance}}$ for a population or $S = \sqrt{\text{Variance}}$ for a sample.

Standard error of the estimate: The standard error of the estimate is the standard deviation of the prediction errors. It is to measure the standard distance between the regression line and the actual Y values.

Standard error of the mean: The standard error of the mean is defined as the standard deviation of all sample means. Its calculation is the standard deviation of the population divided by the square root of the sample size, $\sigma_{\bar{X}} = \sigma/\sqrt{n}$.

Statistics: Statistics is a science that deals with the collection, organization, analysis, and interpretation of numerical data.

Stratified sampling: Stratified sampling is the process of grouping members of the population into relatively homogeneous subgroups before sampling. A random sample from each stratum is independently taken in the same proportion as the stratum's size to the population. These subsets of the strata are then pooled to form a random sample.

Sum of squared deviations (*SS*): *SS* is the acronym for sum of squared deviations, population $SS = \Sigma(X - \mu)^2$, and sample $SS = \Sigma(X - \bar{X})^2$. *SS* is an important step in calculating variance and standard deviation. It is also called sum of squares.

Sum of squared total deviations (SST): SST is the total sum of squared difference between individual scores, X_{ij}, and the sample mean, $\bar{\bar{X}}$. It is calculated as $SST = \Sigma\Sigma(X_{ij} - \bar{\bar{X}})^2$.

Sum of squares between groups (SSB): SSB is the sum of squares between groups and is defined by the squared differences between individual group means, $\bar{X}_j$, and the sample mean, $\bar{\bar{X}}$, multiplied by the number of participants in each group, then, add them up. It is calculated as $SSB = \Sigma n_j(\bar{X}_j - \bar{\bar{X}})^2$.

Sum of squares within groups (SSW): SSW is the sum of squares within groups and is defined by the squared differences between individual scores, X_{ij}, within a group relative to its corresponding group mean, $\bar{X}_j$, then, add them up. It is calculated as $SSW = \Sigma\Sigma(X_{ij} - \bar{X}_j)^2$.

Systematic sample: A systematic sample is achieved by selecting a sample from a population using a random starting point and a fixed interval. Typically, every "*k*th" member is selected from the total population for inclusion in the sample.

***t* Table:** The *t* Distribution Table provides critical values of *t* that set the boundaries of rejection zones for *t* tests given three pieces of information: (1) one-tailed test or two-tailed test, (2) the *df* of the test, and (3) the α level.

Two-tailed test (nondirectional test): A two-tailed test is a nondirectional hypothesis test, which is usually done when the research topic is exploratory or there are no reasons to expect consistent directional relationships from either theoretical reasoning or previous studies.

Type I error: A Type I error is a mistake of rejecting H_0 when H_0 is true. In other words, a Type I error is to falsely claim an effect when the effect actually does not exist. The symbol α represents the probability of committing a Type I error.

Type II error: A Type II error is a mistake of failing to reject H_0 when H_0 is false. In other words, a Type II error is failing to claim an effect when the effect actually exists. The symbol β represents the probability of committing a Type II error.

Uniform distribution: In a uniform distribution, every value appears with the same frequency, proportion, or probability.

Variability: Variability describes the extent to which observed values of a variable are dispersed, spread out, or scattered.

Variability measures: There are three commonly used measures for variability: (1) range, (2) variance, and (3) standard deviation.

Variable: A variable refers to a measurable attribute. These measures have different values from one person to another, or the values change over time.

Variance: Variance is a way to quantify individual differences, and it is mathematically defined as the mean squared deviation, population variance, $\sigma^2 = SS/N$, and sample variance $s^2 = SS/(n-1)$.

Weighted mean: The weighted mean is defined as calculating the mean when data values are assigned different weights, *w*. The formula for weighted mean is $\bar{X} = \frac{\Sigma wX}{\Sigma w}$.

Within-group variance: The within-group variance is calculated as the SSW divided by the df_W; in a special ANOVA term, the within-group variance is also called the mean squares within groups, $MSW = SSW / df_W$

Within-subject design: The within-subject design indicates that research participants are assigned to all levels of the independent variables. Every participant experiences all levels of the independent variable and is measured repeatedly.

***Y*-intercept:** The *Y*-intercept is the value of *Y* when *X* equals 0.

***Z* scores:** *Z* scores are standard scores that describe the differences of individual raw scores from the mean in units of standard deviation.

***Z* test:** A *Z* test measures the difference between a sample mean and its population mean in units of standard error of the mean, $Z = \frac{\bar{X} - \mu}{\sigma/\sqrt{n}}$.

REFERENCES

American Psychological Association. (2010). *Publication manual of the American Psychological Association* (6th ed.).

Berry, W. D. (1993). *Understanding regression assumptions* (Sage University Paper Series on Quantitative Applications in the Social Sciences, 07-092). Sage.

Chen, X., Ender, P., Mitchell, M., & Wells, C. (2003). *Regression with Stata*. Institute for Digital Research and Education, University of California at Los Angeles. http://www.ats.ucla.edu/stat/stata/webbooks/reg/chapter2/statareg2.htm

Cohen, J. (1988). *Statistical power analysis for the behavioral sciences* (2nd ed.). Erlbaum.

College Board. (2019). *SAT suite of assessments annual report*. https://reports.collegeboard.org/pdf/2019-total-group-sat-suite-assessments-annual-report.pdf

Glass, G. V., McGaw, B., & Smith, M. L. (1981). *Meta-analysis in social research*. Sage.

Hedges, L. V. (1981). Distribution theory for Glass' estimator of effect size and related estimators. *Journal of Educational Statistics*, *6*(2), 107–128. https://doi.org/10.3102/10769986006002107

IkamusumeFan. (2013). File:T distribution 5df enhanced.*svg*. http://en.wikipedia.org/wiki/File:T_distribution_5df_enhanced.svg#filelinks

Miller, G. A. (1956). The magical number seven, plus or minus two: Some limits on our capacity for processing information. *Psychological Review*, *63*(2), 81–97. https://doi.org/10.1037/h0043158

Park, H. M. (2010). *Hypothesis testing and statistical power of a test*. University Information Technology Services, Indiana University. http://iu.edu/~statmath/stat/all/power/power.pdf

Satterthwaite, F. W. (1946). An approximate distribution of estimates of variance components. *Biometrics Bulletin*, *2*(6), 110–114. https://doi.org/10.2307/3002019

T. Rowe Price. (2019). *Detailed results: 11th Annual Parents, Kids and Money Survey—College Savings Results*. https://www.slideshare.net/TRowePrice/t-rowe-prices-11th-annual-parents-kids-money-surveycollege-savings-results

INDEX